BEWICK'S SWAN

For Val and Dewi,

and for

Peter and Phil

BEWICK'S SWAN

EILEEN REES

Illustrations by
DAFILA SCOTT

T & A D POYSER
London

Published 2006 by T & A D Poyser, an imprint of
A&C Black Publishers Ltd, 38 Soho Square, London W1D 3HB

ISBN-10: 0-7136-6559-9
ISBN-13: 978-0-7136-6559-8

A CIP catalogue record for this book is available from the British Library.

www.acblack.com

Typeset and designed by Alliance Interactive Technology, Pondicherry, India

Printed and bound in China through Phoenix Offset Ltd

10 9 8 7 6 5 4 3 2 1

Contents

Acknowledgements

In writing this book, I greatly underestimated the time that it takes to transfer thoughts to paper. I am therefore most grateful to the Wildfowl & Wetlands Trust, not only for the unique opportunity of studying Bewick's Swans in different parts of the world, but for tolerating the working hours spent on completing the manuscript during the final push to meet the final deadline. Meeting this deadline was greatly facilitated by encouragement from Valerie Rees, John Bowler and Chris Spray, who all took the trouble to read a full draft, and made a number of invaluable suggestions. John also kindly permitted use of data from his study of the swans' feeding ecology, which appear here in Chapter 4.

In addition to my mother, John and Chris, several other people made major contributions to one or more chapters. Wim Tijsen and Henk Schobben provided valuable information on individual swans seen in the Netherlands, including the story of Sligo and the Bean Goose in Chapter 7. Bart Nolet went through Chapter 4 in some detail; this section is greatly improved by his comments and published research. I am similarly indebted to Anna Belousova for her comments on the swans' breeding biology, to Kees Koffijberg for his updates on the numbers and distribution of Bewick's Swans in the Netherlands, to Axel Degen for information on the swans' use of key sites in Germany, to Simon Delany for his comments on population trends, and to Jan van Gils and Baz Hughes for their advice on Chapter 8. Konstantin Litvin kindly reviewed large parts of the text relating to Russia, and filled some of the many gaps in my knowledge of the Russian swan literature. I also thank Jan Beekman, Mennobart van Eerden, Trinus Haitjema, Marcel Klaassen and Richard Ubels for many enjoyable and fruitful discussions about Bewick's Swans over the years.

Stepping back to the start of the Bewick's Swan study at WWT Slimbridge, much of the credit for our current understanding of the Bewick's Swans can be attributed to the late Sir Peter Scott and his family. They initiated the detailed long-term study of individual birds at Slimbridge, which has since developed into a population study that extends along the swans' European flyway. In the early years, the individuals were recorded by Peter, his wife Philippa (who still takes a keen interest in the swans) and daughter Dafila. Dafila not only completed her doctoral thesis on the Bewick's Swans at WWT Welney (cited in Chapter 6), but has provided the line drawings which so greatly enhance the look of this book.

Several members of staff at Slimbridge have also been responsible for identifying the swans by their bill markings and, through their diligence and enthusiasm, have maintained the study for over 40 years. In chronological order, these are: Pat Pollard, Maya Scull, Tom Pitcairn, Mary Evans, John Bowler, Sue Carman and Julia Newth, with volunteer Steve Heaven also proving expert at swan identification. Each observer has been generous in training the next, thus ensuring the continuity of the study. I am personally indebted to Mary Evans for her training. Additionally, Jenny Earle maintains the Bewick's Swan database, and coordinates the network of people reading swan rings across Europe, assisted by volunteer Alison Bloor. Without the efforts of the ring-reading network and the swan team at Slimbridge, there would be far less information on the Bewick's Swans, and this would be a much slimmer volume.

Other colleagues at WWT who have provided invaluable help, either with the Bewick's Swan study or with this book, include Linda Butler, Colin Butters, David Chamberlain, Sharmilla Choudhury, Ailsa Hurst, Nigel Jarrett, Robin Jones, Charlie Liggett, Geoffrey Matthews, Carl Mitchell, Malcolm Ogilvie, Kevin Peberdy, Dave Paynter, Mark Roberts, David Salmon, Darrell Stevens, Chris Tomlinson, the staff members that have fed the swans and all those who have helped at swan catches. Ruth Cromie and Martin Brown corrected sections on animal health in Chapter 8. Janet Kear not only participated in the Bewick's Swan study in its early years, but subsequently gave regular encouragement to the author, and I remain indebted to her.

For their comments and advice on various topics I am grateful to Linda Birch, Helmut Eggers, David Gardner-Medwin, Marcel Klaassen, Colin Pennycuick, Christopher Perrins, Dewi Rees, Jevgeni Shergalin, Mike Smart, Pat Smiddy and Mike Wilson. David Parkin and Nick Harvey kindly took the trouble to explain results of their studies on swan population genetics. I also thank Jeroen Nienhuis for further information from the Netherlands; Bjarke Laubek and Einar Flensted-Jensen for information from Denmark; Koen Devos, Ekhart Kuijken and Christine Verscheure for swan counts from Belgium; Fabrice Croset for data from France; Graham McElwaine for updates from Ireland; Saulius Švažas and Maria Wieloch for information from Lithuania and Poland respectively; and Leho Luigujõe and Andres Kuresoo who, with Taivo Kastepold and the late Valdur Paakspuu, have ensured that we are aware of the swans' key staging sites in Estonia.

One of the joys of the Bewick's Swan study in recent years has been the excellent communication with Russian scientists, both in exchanging information and in undertaking joint expeditions to the swans' breeding grounds in arctic Russia. Initial information became available through the pioneering work of Yuri Mineyev, who has been studying Bewick's and Whooper Swans in European Russia since the 1970s. Evgeny Syroechkovski Jnr. has not only recorded recent breeding distribution in Chukotka but, since 1995, has ensured that the work of his compatriots is published regularly in the Bulletin of the Goose Study Group of Eastern Europe and Northern Asia (*Casarca*). I thank him for his communication and cooperation. I am especially grateful to friends and co-workers from the All-Russian Research Institute and from the Nenetskiy State Nature Reserve for ensuring that the joint studies undertaken within the reserve are not only fruitful but a delight. In particular I thank Anna Belousova, Yuri Morozov and Andrei Glotov for their valued friendship. I also thank Sergei Mukhortov and the staff at the Nenetskiy State Nature Reserve, Nicolai Kotkin, and the late Sergei Petrushenko for their help during expeditions to the Russian arctic. Of key importance was Yuri Shchadilov who, although sadly no longer with us, was instrumental in developing the long-term study of Bewick's Swan breeding biology undertaken by WWT and the All-Russian Research Institute. Several other people have provided useful information from other parts of Russia, including Konstantin Litvin (Vaygach Island), Tatjana Hochlova (Karelia), Vladimir Borisov (Pskov region), and Vladimir Pozdnyakov (Lena Delta), and I am most grateful to them.

This book is biased towards the northwest European Bewick's Swan population, although the eastern population has received increasing attention in recent years. Sections on the latter would have been even lighter without input from Mark Barter and Ma Ming (for China), Nial Moores (for Korea), and John O. Albertsen, Mark Brazil and Yoshi Miyabayashi (for Japan); I am most grateful to them for letting me know about the swans' eastern flyway. Even less is known about the swans wintering in the Caspian region, but Eldar Rustamov has provided counts from Turkmenistan, which have added to the pool of knowledge about the swans wintering in this area.

The Bewick's Swan research and conservation programme at Slimbridge is supported by the Wildfowl & Wetlands Trust, and I thank WWT, its Members and those that have joined the WWT Swan Adoption Programme for their contribution over the years. Fieldwork in Russia has been supported by, in chronological order: the Winston Churchill Memorial Trust, the Royal Society, British Airways Communities and Conservation, the Peter Scott Trust for Education and Research in Conservation, and the BBC Natural History Unit (Radio). The BBC (notably Julian Hector) not only supported the satellite tracking of Bewick's Swans from Russia in 2003, but Brett Westwood and Simon Roberts (BBC) and Colin Pennycuick (Bristol University) proved entertaining members of the expedition to Russia that year.

I am extremely grateful to Andy Richford for his original interest in a book on the Bewick's Swan, to Jim Martin at T & A D Poyser for seeing it through to publication, and to editors Tim Harris and Wendy Smith. It is also a pleasure to thank James McCallum for his evocative portrayal of the swans in the cover illustrations. The American Ornithologists Union, the Association of Animal Behaviour, *Ardea*, *The Auk*, the British Ecological Society, British Ornithologists' Union, the BTO, the Waterbird Society, Wetlands International, and the Wildfowl & Wetlands Trust all kindly granted permission to reproduce figures first published in their journals. Additionally, the photographs taken by Mark Barter, Mariusz Blank, Joe Blossom, John Bowler, Mark Brazil, Andrew Cooper, Mark Hulme, Paul Marshall, Dave Paynter, Philippa Scott and Wim Tijsen portray the beauty of the swans better than any words.

Finally, I thank Ronald Graham, not only for useful comments on the text, but particularly for his tolerance of the many weekends spent writing this book, and for his constant support and encouragement during a lifetime's study.

Preface

Each autumn, as the snow returns to the High Arctic, thousands of Bewick's Swans fly south to their wintering haunts in northwest Europe and east Asia. In Russia there is an old saying, *Lebed' nesyot sneg na nosu*, which means 'the swan brings snow on its bill'. Yet despite the colder weather their arrival is eagerly anticipated by birdwatchers and the wider public in countries ranging from Ireland to Japan. The first few birds may appear singly or in family groups but, as winter sets in, "swanfalls" of tens or even hundreds of Bewick's Swans arrive in waves to delight those who watch them with their grace, beauty and melodious calls.

At the Wildfowl & Wetlands Trust at Slimbridge in southwest England, there is additional interest in the swans' return. Here the life-histories of individual Bewick's Swans have been studied since the early 1960s, and the latest news of old friends – whether they have survived the migration, found a mate or raised young – is eagerly anticipated. The long-term study of the Slimbridge-wintering flock has provided substantial insight into the behaviour, ecology and dynamics of the species. More recent studies in other parts of the swans' migratory flyway have emphasised the range of environmental conditions that they encounter at different times of the year. This book aims to draw together the results of these studies to describe the many and varied aspects of the lives of these elegant birds, including threats to their welfare and the conservation measures introduced over the years.

The swans are the stars of this book, but there are also many untold human stories behind our current knowledge of the swans' lives. The acknowledgements go only a small way to show the dedication of people involved in the research, management and conservation of Bewick's Swans in very different parts of the world. As long ago as 1875, the Victorian naturalist and explorer Henry Seebohm made a remarkable journey by train, sledge and boat to the delta of the Pechora River in the Russian arctic. There he found Bewick's Swans nesting in what remains an important breeding area for the species to this day. Another eminent naturalist, Sir Peter Scott, claimed the Bewick's Swan as his favourite bird when, in February 1964, wild swans wintering on the saltmarshes at Slimbridge started visiting the lake in front of his house. Having previously favoured Pink-footed Geese, then Red-breasted Geese and Lesser White-fronted Geese, Sir Peter became fascinated by the migratory swans which he and his family found that they could recognise as individuals by the unique variation in their black-and-yellow bill markings.

My own involvement came much later – WWT's long-term study was in its 15th winter by the time I joined in 1977. Despite having to check prior to my interview in Myrfyn Owen's book *Wildfowl of Europe* to determine what a Bewick's Swan was, in the first of many winters studying the birds I fell immediately for their fidelity (both to their families and to particular wintering sites) and delicate beauty. Although now based at WWT Martin Mere – a stronghold of the larger Whooper Swans – watching the Bewick's Swans sweep in at dusk to feed, preen and roost on Swan Lake at Slimbridge inspires great joy and contentment during my weekly visits to WWT's headquarters. Ornithologists elsewhere have similarly fallen under their spell, with those involved in ring-reading – particularly the Dutch ring-readers – travelling widely to track the movements of colour-ringed swans during winter and the early stages of spring migration. In Japan, where the swans are much admired, supplementary food has been provided at an increasing number of sites over the last 50 years, and visiting these sites continues to be very popular with the general public.

Expeditions to the swans' breeding grounds on the Russian tundras are now a little easier than in Seebohm's time. Nevertheless they pose some new challenges and have only been open to western scientists since the end of the Cold War. It is now possible to fly from Moscow via Arkhangelsk to Nar'Yan Mar on the Pechora Delta, and on clear days the views of the vast tundra landscape along the arctic coastline is truly a wondrous sight. The helicopter from Nar'Yan Mar to the swans' breeding areas passes over a myriad of lakes, channels and small pools, which not only provide breeding territories for the swans but are used by the many other wildfowl and waders that live on the tundra during the summer months. Clouds of mosquitoes, the bane of researchers, provide food for many bird species in both larval and adult forms. Fish, the main food of researchers, are netted in adjacent waters.

In 2003 we gained permission from the Radiofrequency Centre in Moscow to deploy satellite-transmitters on five Bewick's and one Whooper Swan to monitor their autumn migration, the first time that the use of such transmitters has been officially permitted in Russia. The satellite-tracking programme was a major feature in BBC Radio 4's migration week, and generated widespread interest in the swans and their welfare. Education programmes and media coverage within Russia similarly aim to inspire the next generation with an interest in the birds and their habitats, which is essential for safe-guarding their future as man encroaches into their remote breeding grounds.

Since migrant birds are no respecters of political boundaries, effective conservation requires not only sound knowledge of their ecological requirements throughout the year, but also the political will in countries throughout their range to introduce appropriate conservation measures where necessary. Therefore, the designation of the swans' breeding grounds north of the Pechora Delta as a National Nature Reserve (the *Nenetskiy zapovednik*) by the Russian Government in 1998 was particularly satisfying to all concerned. The expansion of the Bewick's Swan study, from focusing on the swans wintering at Slimbridge to studies at their migration sites and breeding grounds, and the development of research programmes by scientists in other countries, has greatly improved our understanding of their lives.

Although our knowledge of the swans has increased substantially since the early 1960s, gaps do still remain. Thus, both for myself and for others delighted by these beautiful birds, what commenced as a study of the swans' lives may well become a lifetime's study.

Eileen Rees
WWT Martin Mere, November 2005

CHAPTER 1

Swans and the Bewick's Swan

'The name of those fabulous animals (pagan, I regret to say) who used to sing in the water, has quite escaped me.' Mr George Chuzzlewit suggested 'Swans'. 'No,' said Mr Pecksniff. 'Not swans. Very like swans, too. Thank you. … Wait. Sirens! Dear me! Sirens, of course.'

Charles Dickens (*Martin Chuzzlewit*)

Of all the species of bird on this planet, the Bewick's Swan *Cygnus columbianus bewickii* must be one of the most appealing. It starts with an unfair advantage over many other species simply by being one of the swans, birds whose grace and pure beauty has been the subject of legends and fairy tales since the dawn of time. It then steals a march over the magnificent but more sedentary Mute Swan *Cygnus olor* by being migratory, which adds greatly to its mystique. The Bewick's Swans' summer haunts on the High Arctic tundra are still largely inaccessible, so few people have the good fortune to follow them on their long-distance journeys to their nesting areas. Next, it is the smallest and most delicate of the northern migratory swans, even more delicate than its close relation the Whistling Swan *Cygnus columbianus columbianus*, which occupies the same ecological niches in North America as those used by the Bewick's Swan across Eurasia. Not only are Bewick's Swans beautiful – they are also musical. The concept of the 'silent swan', which sings only just before death, is probably based on the Mute Swan; in fact the Mute does vocalise but through inelegant snorts and grunts. It clearly does not refer to the Bewick's Swans, which bugle loudly during territorial disputes and utter gentler crooning or contact calls to their mates and offspring. Then there is the birds' life-style – their faithfulness to their mates and care of their young are reminiscent of another species that is long-lived with slow-growing offspring – mankind. Yet although the swans have generally delighted humans over many years, we will see here that human activity has had major costs as well as some benefits to the species.

Studies of individual animals are immensely rewarding, not only for the quality of the scientific findings, but also on a more personal level for the researchers involved. The value to science and conservation lies in the fact that differences in behavioural patterns are often associated with characteristics (such as age, sex, dominance and breeding success) of the individuals concerned; these in turn help to explain how or why animals behave as they do. Individual-based studies are being used increasingly to explain and predict changes within populations, and particularly the reasons and processes underlying changes in population size and shifts in distribution. Knowledge of the individuals' life histories is especially valuable in this detective work, since in many cases the behaviour is related to their past experiences. Thus, breeding success may be influenced by the length of time that members of a pair have been together, or by their familiarity with the breeding territories, or by the number of offspring raised the previous year. The subtleties of the ways in which animals relate to each other and to their environment are less easily understood by monitoring 'unknown' individuals, even when these are part of a group, because relevant information may be lacking. Nevertheless, zoologists thinking of embarking on a study of individual animals should maybe think twice about doing so. Not only will many years of work be required to obtain useful information about a species if it turns out to have a long lifespan, but detailed studies often lead to new ideas that stimulate further research. Moreover, the level of involvement can result in an increasing concern for the welfare of the individuals being monitored, who can become old friends over the years. This has certainly been the case with the Bewick's Swan. Almost all those who have started studying the species have ended up doing so for five years or more. Although the author joined the long-term study of Bewick's Swans' life-history strategies on a temporary contract in 1977, it now seems to be becoming a lifetime's undertaking.

It all started in a very small way, when the eminent naturalist and conservationist Sir Peter Scott had the idea of attracting to the lake in front of his house the small flock of wild Bewick's Swans which wintered on the River Severn at Slimbridge in Gloucestershire. Swans are gregarious except when breeding so, during the 1963–64 winter, seven conspecifics (three Bewick's and four Whistling Swans) from the collection of captive birds at the Wildfowl Trust (now the Wildfowl & Wetlands Trust, WWT) were put on the lake, in the hope that wild swans would be decoyed from the river by their calls. The ploy was entirely successful. By the end of the winter the 24 wild swans present on the estuary were visiting the lake regularly, where they could be observed closely from buildings and hides. The Scott family immediately took a keen interest in the new arrivals in front of their studio window, and quickly realised that the birds could be told apart by differences in their black-and-yellow bill patterns. Moreover, when the swans arrived at Slimbridge the following winter, the Scotts found that they could recognise some individuals that had been present the previous season, in some cases accompanied by their new mates and offspring. This ability to identify individuals by their natural markings, together with the swans' tendency to return to the same sites over several winters, formed the basis for a long and detailed study of the species. Initially, observations centred on the swans wintering at Slimbridge, but the study expanded with a view to monitoring the same birds at other wintering sites and along their migratory route to the northern wintering grounds, in collaboration with scientists from other parts of northwest Europe.

Even Sir Peter may not have realised the extent to which the monitoring of individual swans would develop into a long-term study, due at least in part to the swans' long lifespans. Although the Slimbridge study has been underway since February 1964, the longevity of the birds means that even after 30 years we still do not know some elementary facts, such as their potential life expectancy. A swan named 'Casino' returned to Slimbridge for her 27th winter at the site in 1997–98, looking fit and well and accompanied by her six-year-old offspring 'Croupier'. Since she was first recorded as a cygnet in 1971, Casino was 26 years old at the time. Eleven more birds have reached at least 25 years of age in the wild and, at the time of writing (autumn 2005), 68 have been recorded at 20 years or more. One captive swan, named 'Mrs Noah', was at least 33 years old when she died at Slimbridge, apparently of old age. New benefits of long-term data from both individual and population studies are coming to light, however, which may be used to explain and predict population changes associated with changes in habitat and climate over several years or decades. Thus, information on the feeding habits of the swans during the 1960s and 1970s, compared with more recent observations, show that the swans' diet has also changed over time in response to habitat loss and changes in farming practice. Such evidence, which is invaluable for monitoring the ability of populations to adapt to environmental changes, can only be obtained by monitoring a species systematically over long periods of time.

Although the Bewick's Swan study at Slimbridge was the first to study the species in any depth, scientists at other sites along the swans' migratory route from northwest Europe to Arctic Russia have embarked on separate and collaborative projects, which have added substantially to our knowledge and understanding of the swans' life-cycle. Amateur and professional ornithologists in the Netherlands originally became involved in Bewick's Swan research due to their concern following the disappearance of aquatic vegetation at the swans' feeding sites in the IJsselmeer area during the 1960s, which resulted in the birds changing to feed on farmland (Poorter 1991). The marking of Bewick's Swans with leg-rings since the late 1960s also engaged Dutch ring-reading enthusiasts, who continue to spend much of their valuable leisure time during the short winter days in reading the swans' ring codes and reporting their sightings. The initiation of the Dutch Bewick's Swan study in 1982 resulted in the development of a new long-term study of the species, which is still underway and expanding. The Netherlands is the most important wintering area for the species in Europe, and Dutch colleagues have been responsible for coordinating the international Bewick's Swan censuses across Europe, made every five years since the first 'complete' census of 1986. This is particularly valuable for monitoring total numbers within the population and for verifying the population trends derived from annual counts at selected sites, including those sent to Wetlands International (previously the International Waterfowl and Wetlands Research Bureau) as part of their international waterfowl counts programme.

The intensive studies of Bewick's Swan, not only by the Wildfowl & Wetlands Trust but increasingly by scientists at other sites along its European flyway, means that our insight into the life of this beautiful bird is very comprehensive. Comparisons of the results obtained in Britain, the Netherlands and at migratory sites in Estonia emphasise the importance of not restricting research to a single area or to just one season. Differences in the food selected, for instance, indicate either that the swans' ecological requirements change throughout the year, or that they are adapted to the conditions in which they find themselves, as indicated in Chapter 4. The opportunity for Western Europeans, used to seeing Bewick's Swans only in winter, to join Russian scientists in studying the birds at their nest sites on the tundras of Arctic Russia revolutionised our view of the birds, because the swans' habitat and behaviour is so totally different in the breeding season. Less is known about the eastern population, which migrates from the Arctic tundras of far eastern Russia to winter in Japan, China and Korea, but greater international communication and collaboration is also helping to fill in these gaps in our knowledge. This book therefore aims to draw together the substantial volume of information gained about Bewick's Swans over the years, to describe and explain the varied aspects of the swans' migrations and life histories and, it is hoped, to provide the reader with greater insight into the remarkable life of the Bewick's Swan.

1.1 SWAN TAXONOMY

1.1.1 Comparison of Bewick's Swans with other swan species

Swans need very little introduction. Their large size, long necks, short legs and large webbed feet make them amongst the easiest of all birds to recognise, familiar to most people except perhaps in Africa where only a few hundred feral Mute Swans are to be found in the southern part of the continent. Taxonomists appear to be united in considering that waterfowl, consisting of the swans, geese and ducks, belong to one family, the Anatidae. Classification of members of the Anatidae family into genera, species and subspecies is still being debated, however, including the number of swan genera and species thought to exist. The most widely held view is still that of Delacour & Mayr (1945), who considered that there are eight species or subspecies of swans in the world, five in the northern hemisphere and three in the southern hemisphere. They are also classified as falling within two genera since one species, the Coscoroba Swan *Coscoroba coscoroba* from South America, is so distinctive that it has been allocated its own genus. Indeed, the question of whether it is truly a swan is still under investigation. The seven other species or subspecies are currently grouped together in the *Cygnus* genus, although the Eurasian Mute Swan, the Black-necked Swan *Cygnus melanocoryphus* from South America and the Black Swan *Cygnus atratus* from Australia (introduced to New Zealand in the 19th century, Bowler 2005a) clearly differ from the northern migratory swans, both in physical appearance and in their behaviour patterns (Kear 2005). The Mute Swan, like the other northern hemisphere swans, has white plumage in adulthood but can be readily distinguished by its reddish-orange bill with a black knob at the forehead, the graceful curve to its long neck and the longer upturned tail. The Black-necked Swan, as its name suggests, has a striking black head and neck on a white body, with a red caruncle at the base of the bill. The Black Swan has predominantly black or near-black plumage with a bright red bill and eyes; white primary feathers and distal secondary feathers are visible in flight.

Trumpeter Swans *Cygnus buccinator*, Whistling Swans, Whooper Swans *Cygnus cygnus* and Bewick's Swans, known collectively as the northern migratory swans, are thought to be more closely related to each other than to the other swan species. They are similar in physical appearance, with the adults having all-white plumage, an upright stance with necks held rather straight, and varying amounts of black and yellow on the bill. Unlike the Mute and Black-necked Swans, they do not have a frontal knob or caruncle at the base of the bill. Also, unlike the Mute Swan and Black Swan, they do not arch their wings in aggression, but perform a series of ritualised displays, raising and flapping their wings and calling loudly. Trumpeter and Whistling Swans breed across the sub-Arctic and Arctic regions of North America respectively, in areas analogous to the breeding distribution of Whooper Swans and Bewick's Swans in Eurasia. All four species migrate south in autumn/winter to more temperate latitudes. The ranges of Bewick's and Whooper Swans overlap, particularly in winter, but Bewick's x Whooper hybridisation has never been recorded in the wild. Trumpeter and Whistling Swans are less likely to be found at the same sites, probably due to the somewhat limited distribution of the Trumpeter Swan, following a drastic reduction in numbers due to excessive hunting in the 19th century.

The Bewick's Swan is the smallest of the northern swans, and arguably the smallest of all swans depending on how the Coscoroba Swan is classified. Bewick's Swan's relationship with the nominate *Cygnus columbianus columbianus* is still unclear, with some authorities considering Bewick's and Whistling Swans to be two separate species (for example, Stejneger 1882, Wetmore 1951, Vaurie 1965), whereas others believe them to be conspecific but different subspecies (Hartert 1920; Delacour & Mayr 1945; Parkes 1958; Mayr & Short 1970). The inclusive name Tundra Swan was suggested by Palmer (1976) for both Bewick's and Whistling Swans, and this term has been commonly used for the Whistling Swan in North America from the 1980s onwards. There have been increasing moves towards referring to Bewick's Swans as Tundra Swans in Europe in recent years, and in 2004 the British Ornithologists' Union recommended that Bewick's and Whistling Swans should be treated as conspecifics (Sangster *et al.* 2004). Biometric data recorded for Bewick's and Whistling Swans indicate that the latter is generally larger, but that the body weights and measurements do overlap considerably (Appendix 1). Moreover, there are few measurements of Bewick's Swans from the easternmost parts of their range, where hybridisation with Whistling Swans occurs. The overlap of the Bewick's and Whistling Swans' breeding ranges in far east Russia, and sightings of mixed pairs and hybrids since the 1970s, confirms that interbreeding does occur in the wild, although its frequency is still not known. This is a major reason for grouping Bewick's and Whistling Swans as a single species, despite the obvious visual differences in their bill markings.

It has been suggested that birds from the eastern Bewick's Swan population have larger bills, broader near the tip and higher near the base than those further west, and with slightly more yellow on the bill. Whether they should be described as a separate race (known as Jankowski's Swan *Cygnus columbianus jankowskii*, Alphéraky 1904; or *Cygnus c. bewickii jankowskii*, Delacour 1954) therefore has been debated, but Bewick's Swans have long been classified as a single group, with populations that follow different migration routes not divided taxonomically. There is substantial variation in the skull and bill measurements of swans from the western population, and the few records obtained for the eastern population (notably the bill lengths of four adults and two cygnets received by the Wildfowl Trust from Hong Kong in 1973) fall within this range. The bill patterns of six '*jankowskii*' swans at the Wildfowl & Wetlands Trust also appeared typical for *bewickii* (Evans & Sladen 1980). Given that differences in body size have been recorded for other bird populations that have not then been described as different races (for example, Whooper Swans and some of the goose populations), it remains difficult to justify separating Jankowski's Swan on the basis of its size alone.

1.1.2 Separation of Bewick's Swans from Whooper Swans

Bewick's Swan is named after the famous 18th-century engraver and ornithologist Thomas Bewick of Newcastle, who died in 1828 two years before the bird was described in print. It is one of the last large birds occurring in Britain to have been recognised and named by the scientific community, having previously been classed together with Whooper Swans as 'Wild Swans'. That this group included some smaller birds (which he recognised as a 'variety' rather than a separate species) was first noted by Pallas in 1811. In 1824, John Latham again mentioned a 'Lesser Swan ... not so large as the Hooping Swan ... in fact, it imitates the Wild Swan in miniature...'. Latham had only a single skin, however, so he was unwilling to propose it as a distinct species. The breakthrough came in October 1829, when Richard Wingate (a taxidermist from Newcastle) presented a paper to the Natural History Society of Northumberland, Durham and Newcastle upon Tyne (now The Natural History Society of Northumbria) which described Bewick's Swan in detail but failed to give it a name. Between June and October 1830, his friend P.J. Selby published a paper in the Society's *Transactions*, naming the swan as 'Cygnus Bewickii of Wingate' and acknowledging the help of Sir William Jardine and William Yarrell. Meanwhile, Yarrell wrote another account of the new species, which was published in the *Transactions of the Linnean Society* in the same year (Yarrell 1830), in which he 'proposed to call it Bewick's Swan, thus devoting it to the memory of one whose beautiful and animated delineations of subjects in natural history entitle him to this tribute'. Because Yarrell's paper was published a couple of months before Selby's, Yarrell is usually credited with describing Bewick's Swan. He did not claim to be aware of the Bewick's Swan first, but he did say in January 1830 (and wrote in his paper) that he had suspected the new species on the basis of the characteristic anatomy of the trachea, bronchi and sternum in a specimen prepared 'six years ago' (i.e. about 1824); his first spoken declaration of this was in November 1829, a month after Wingate (D. Gardner-Medwin pers. comm.). This may have caused some bad feeling at the time, but all parties clearly were happy with the choice of name. It is not known whether Bewick ever saw the bird to which his name was given, but his telescope (dated 1794) has been used for watching Bewick's Swans at Slimbridge. His son, Robert Elliot Bewick, engraved an image of a Bewick's Swan for the 1847 posthumous edition of his father's *The History of British Birds* (D. Gardner-Medwin pers. comm.).

In the field, Bewick's Swans are more likely to be confused with Whooper Swans than with any of the other swans because they are superficially similar in appearance and their ranges overlap. The Bewick's Swan is smaller and more goose-like in flight, however, with a comparatively shorter-looking neck and a more rounded head; Whooper Swans tend to have a flatter profile of the bill and head. The easiest way to tell the two species apart, at least for yearlings and adult birds is by the distribution of the yellow and black markings on the bill. In Bewick's Swans the yellow markings at the base of the bill end behind the nostrils, whereas in Whooper Swans the yellow extends to a point beyond the nostrils, giving the yellow pattern a wedge-like appearance. Bill patterns cannot usually be used to differentiate between Bewick's and Whooper Swans for birds in their first winter, although emerging bill markings may be seen at close quarters towards the end of the winter season, but differences in the overall body structure and head shape are apparent after fledging. The grey plumage of Whooper Swan cygnets wintering in Britain tends to become whiter more rapidly than that of Bewick's Swan cygnets, perhaps indicating that they mature more rapidly. Large variation in cygnet plumage between broods and the darker grey feathering observed in Finnish-bred Whooper Swans seen wintering in Britain, however, indicate that other factors (e.g. diet) may also be involved. Some Bewick's Swans retain traces of the juvenile grey feathering on the head and neck into their third summer (J. Beekman pers. comm.), whereas many Whooper Swans have all-white adult plumage at one year old (second summer). The calls of the two species also differ, although both are highly vociferous. Bewick's

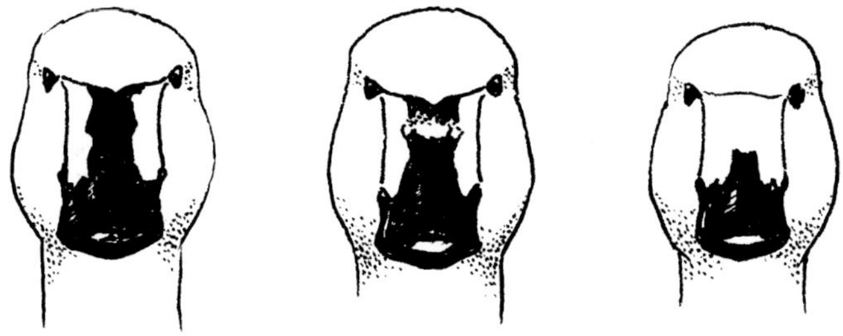

Figure 1.1 Illustration of variation in bill patterns of Bewick's Swans, showing the three main types: blackneb (Riddler), pennyface (Lille) and yellowneb (Dealer).

Swans generally make higher pitched and more musical notes than the more sonorous whooping calls of the Whooper Swan.

1.1.3 Separation of Bewick's Swans from Whistling Swans

Although Bewick's Swans are slightly smaller on average than Whistling Swans, the most reliable field characteristic for distinguishing between adults of the two subspecies is the proportion of yellow on their bills. Their body shape and posture are very similar, as are their calls (the position of the trachea within the sternum being the same for both), but experienced aviculturalists can differentiate between their voices at close range (N. Jarrett and M. Roberts pers. comm.). In Whistling Swans the bill is almost entirely black in colour, but usually with a small and variable patch of yellow in front of the eye. In Bewick's Swans the proportion of yellow is much larger, in some individuals (known as 'yellownebs') stretching in a continuous band across the bill to link the yellow patches on either side. In other Bewick's Swans (known as 'blacknebs') the centre-line of the upper mandible is black from the feathering to the tip of the forehead, and in a third group ('pennyfaces') the front of the bill is black from the brow-line to the tip, but a patch of yellow (the 'penny') occurs in the middle (Scott 1966; Evans 1977a; Rees 1981; Figure 1.1).

A comparison of photographs taken of the right bill profiles for Whistling Swans caught in the eastern part of their North American range (where hybridisation with Bewick's Swans is least likely) and for Bewick's Swans caught at Slimbridge showed that the average proportions of yellow to the whole bill profile were 3.1% for Whistling Swans ($n = 300$) and 31.5% for Bewick's Swans with blackneb bill patterns ($n = 104$) (Evans & Sladen 1980). Nine Whistling Swans (3% of the sample) had no yellow patch at all. The largest proportion of yellow recorded for the Whistling Swan bills was 15.8%, and the smallest proportion of yellow recorded for the Bewick's Swan bill was 22.9% (Figure 1.2). Thus the ranges appeared to be discrete, giving greater credibility to out-of-range sightings for the two species and also to records of intergrades.

The amount of yellow on the bill therefore has been used to support reports of Bewick's and Whistling Swans seen outside their normal ranges (reviewed in Evans & Sladen 1980). Generally, Whistling Swans breed on the tundras of North America and migrate south each autumn to spend the winter in California (the western population) and in Delaware, New Jersey, Maryland and Virginia (the eastern population), whereas Bewick's Swans breed in the Russian Arctic and winter mainly in northwest Europe and east Asia (see Chapter 2). Photographs of the bill patterns confirmed that three Whistling Swans or Whistling x Bewick's Swan intergrades occurred in the normal range of Bewick's Swans, and that

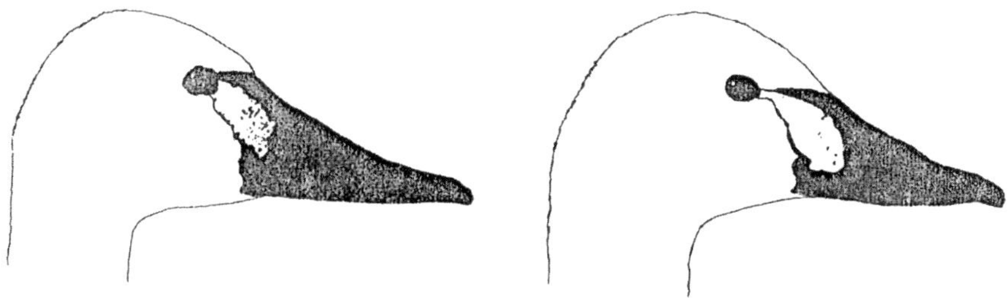

Figure 1.2 Comparison of the proportions of yellow to black on the bills of Bewick's and Whistling Swans. The Whistling Swan shown was the one with most yellow (15.8%) on its bill (left), and the Bewick's Swan was the one with the least yellow (22.9%). Bill patterns traced from photographs. From Evans & Sladen 1980.

four Bewick's Swans or Whistling x Bewick's intergrades occurred in the normal range of Whistling Swans during the 1980s. These consisted of two Whistling Swans seen in Japan, one Whistling Swan (or possibly an intergrade; 15.3% of yellow on the bill) in England, three Bewick's Swans in North America (in Maryland, Oregon and Saskatchewan) and a Whistling x Bewick's Swan intergrade (with 17.8% yellow on the bill), also seen in Canada. One of the Bewick's Swans seen in North America (in the Lower Klamath National Wildlife Refuge at Worden in Oregon) had a Whistling Swan mate and a family of two cygnets. A group consisting of a Bewick's Swan with Whistling Swan mate, three immatures (thought to be Bewick's or Bewick's x Whistling intergrades) and two other adults (possibly intergrades) seen on Hog Lake, near Red Bluff, California, the following winter may have been the same family party, but unfortunately photographs of the birds were not detailed enough to confirm their identities. A further out-of-range report was supplied in December 1977 when a Bewick's Swan marked with an orange neck-band was found dead at Adak Island in the Aleutian Islands, Alaska. The bird had been caught and marked as a cygnet in Chaunskaya Bay in northeast Siberia by A. Y. Kondratiev and other Soviet scientists as part of the USA–USSR Environmental Protection Agreement (Evans & Sladen 1980).

Most out-of-range reports of Bewick's and Whistling Swans have not had the benefit of photographic evidence, particularly the earlier sightings. Although photographs are an extremely useful diagnostic tool, the proportion of yellow on the bill can only be determined for certain by photographing the bill at right angles to the camera, in conditions free from shadow and when there is no mud (or, in one case, oil) on the bill, requirements that are difficult to achieve for swans in the field. Nevertheless, Whistling Swans appear to have occurred as vagrants in northeast Siberia since the second half of the 19th century, with sightings from Bering Island as early as 1882 and Novomarinsk (currently Anadyr) in 1897 (Dementiev & Gladkov 1952). More recently, a pair with three cygnets and five other Whistling Swans were discovered at Kolyuchin Bay on the Chukotka Peninsula in July 1974 (Kishchinski *et al.* 1975). There was a general increase in Whistling Swan numbers at Kolyuchin Bay during the mid 1970s to 1980s, and some 600–1,000 birds are now thought to occur in Chukotka in summer (Syroechkovski 2002; Chapter 2). A mixed Bewick's x Whistling Swan pair was seen on the Ekviatap estuary (179°E) in 1990, although a nest was not found (M. S. Stishov pers. comm., in Syroechkovski 2002).

In the southern part of the flyway, between 10 and 29 Whistling Swans have been observed at Bewick's Swan wintering sites in Japan each year since 1990 (Environment Agency of Japan 1998). A mixed Bewick's x Whistling Swan pair was seen in Japan in the 1970s (T. Ohmori pers. comm., in Evans and Sladen 1980), and there were a further four reports of mixed pairs in Japan in the late 1980s, including a family with two cygnets (Mikami 1989). In addition to the mixed family party seen for one or two years in Oregon and California, an adult Bewick's Swan accompanied by an immature was seen in California in February 1975 (Stallcup & Winter 1975), three possible Bewick's x Whistling Swan intergrades on Victoria Island, California, in the 1977–78 winter and another possible intergrade near Benton, California. A Bewick's Swan (identity confirmed by photographs) seen in Saskatchewan, Canada, had a Whistling Swan mate.

Although most out-of-range sightings have been of Whistling Swans in east Asia or of Bewick's x Whistling Swan intergrades along the Pacific flyways, a swan with yellow patches small enough to suggest that it was a Whistling Swan was spotted among the Bewick's Swans at Slimbridge in the 1968–69 and 1969–70 winters (the same bird) and a second individual in 1975–76. A swan named 'Serbia' that also had very little yellow on the bill, a possible Whistling x Bewick's Swan hybrid, visited Slimbridge in both 1996–97 and 1997–98, but there have been no other sightings of possible Whistling Swans at Slimbridge in recent years. For sites elsewhere in Britain, the British Birds Rarities Committee confirmed that an adult Whistling Swan was present at Nocton Fen, Lincolnshire, on 22 January 1998 (Rogers & the Rarities Committee 1999). The Committee describes the bird as 'only the second British Whistling Swan, with previous occurrences from 1986 to 1990 in Hampshire and (mainly) Somerset referring to a returning individual'. Whistling Swans are very rare visitors to the Netherlands. The Dutch rarities committee (the Commissie Dwaalgasten Nederlandse Avifauna), which has considered reports of rare birds in the Netherlands annually since 1979, and historical records dating back to 1924, has accepted fewer than 10 sightings of Whistling Swans in the country (van den Berg & Bosman 1999; van der Vliet *et al.* 2002). Most recently, a reported Whistling Swan in the area south of Den Oever (Noord-Holland), probably the same bird seen in northeast Holland earlier in the 2003–04 winter, turned out to be an aberrant Bewick's Swan. There have been five records to date of Whistling Swans in Ireland, and each record is believed (based on bill pattern) to refer to different individuals. Additionally, three of these five birds are thought to have returned to the same place in more than one winter. The most recent record was of a bird on the North Slob, Co. Wexford, in December 1990 that returned in March 1991 and December 1991 (O'Sullivan & Smiddy 1991, 1992; Irish Rare Birds Committee 1998; P. Smiddy pers. comm.). The first and second records of Whistling Swans in Estonia were of birds at Haeska, Martna, in November 1990 and April 1995 (Davies 2001).

The occasional records of interbreeding in the wild help to reinforce the case for conspecificity of Bewick's and Whistling Swans (Evans & Sladen 1980). The closer proximity of Whistling Swans in North America to the far eastern Bewick's Swan population means that mixing of the two populations is more likely to occur across the Bering Straits than across the Atlantic Ocean. This and an increase in the number of Whistling Swans nesting in east Russia since the 1970s (Chapter 2) probably accounts for the higher frequency of Bewick's and intergrade sightings in North America, and of Whistling Swans in Russia and Japan, than of Whistling Swans in northwest Europe. Despite the much more extensive overlap of the ranges of Bewick's and Whooper Swans, there are still no definite cases of Bewick's and Whooper Swans interbreeding in the wild, although it has been known in captivity. Since there are records of Whooper Swans breeding with Mute Swans in the wild, it seems likely that Bewick's × Whooper Swan pairings may occasionally occur. The remoteness of the swans' main breeding grounds in the Russian Arctic would reduce the likelihood of a mixed pair being observed during the summer, but it is perhaps surprising that Bewick's × Whooper Swan pairs are not occasionally reported in the winter months. Any hybrid offspring also would again be difficult to recognise, because of the similarity in the Bewick's and Whooper Swans' morphology and bill markings.

1.1.4 Genetics studies

The classification of plants and animals tends to be a movable feast, because taxonomists are concerned not only with grouping similar organisms into the same category, but also with describing the proximity of species in evolutionary terms. Traditionally this was based on studies of morphology, physiology, plumage and behaviour (particularly courtship displays), sometimes linked with fossil evidence. All extinct swans for which there is fossil evidence are from the northern hemisphere, with the exception of *Cygnus sumnerensis* from New Zealand (Brodkorb 1964). The earliest swans now attributed to Cygnini are *Cygnus mariae* and *Paracygnus plattensis* from the late Miocene of North America (Bickart 1990; Callaghan *et al.* 2005). More recent developments in avian genetics have enabled us to determine similarities between species at a molecular level, and thus to deduce their relationships over time, but even the detailed molecular studies give varying results depending on the types of analyses undertaken (Callaghan *et al.* 2005). In particular, the evolutionary status of the Coscoroba Swan, and also whether Bewick's Swans and Whistling Swans are genetically discrete, have given rise to some debate. Certainly the hybridisation of Bewick's Swans and Whistling Swans in the wild, and the cline in their bill patterns, lent some preliminary support to the view that they should be treated as a single species.

A recent PhD study at Nottingham University on the population and evolutionary genetics of swans (Harvey 1999) indicated that the Coscoroba Swan forms a sister group to the swans and geese, together with the Cape Barren Goose *Cereopsis novaehollandiae*, to which it is most closely related. This study suggested that the Black-necked Swan is the most ancient of the swan species, followed by the Black Swan, the Mute Swan, then the northern swans grouped together. However, morphological studies (Livezey 1996) and molecular studies using longer sequences but fewer species than the Nottingham study (Harshman 1996) suggested a sister relationship between the Black-necked Swans and the Black Swans. Molecular studies of the commonly sequenced cytochrome *b* gene have confirmed the difficulties of resolving relationships between the four northern migratory swans. This is probably because the four lineages separated at about the same time, but there is agreement that these swans are more closely related to each other than to the other swan species (review in Callaghan *et al.* 2005). Although there is no reason to separate Bewick's Swans and Whistling Swans on the basis of sequence divergence amongst the eight swan species and subspecies, preliminary population level analyses (using mitochondrial D-loop sequencing) suggest that they are genetically distinct. This indicates that speciation is underway and may continue if the level of interbreeding remains low (Harvey 1999).

1.2 BEWICK'S SWANS IN AVICULTURE

Bewick's Swans have long been kept in captivity. These are usually injured birds incorporated into waterfowl collections in Europe, but swans from the eastern population have also been sent from China to America. A wounded male survived for many years in Anjou during the latter half of the 19th century (Rogeron 1883). Blaauw (1904) described plumage changes in a young bird, winged on the Zuiderzee in the Netherlands, which was brought to him to keep. The Duke of Bedford reported that Bewick's Swans bred occasionally in captivity, the first record being one at Woburn, Bedfordshire, before 1914, but no dates or details were given (Delacour 1954). A hybrid between an eastern Bewick's Swan female and a male Whistling Swan was reared in Connecticut in 1942 (Delacour 1954). More generally, although most swan species regularly rear young in zoological gardens, breeding the Arctic-nesting Bewick's and Whistling Swans at lower latitudes has proved difficult. A wild male Bewick's Swan, thought to be in its second winter, was caught at Slimbridge in 1948 and was paired with a wild-caught adult female ('Mrs Noah') obtained from Holland in 1950. On 2 June 1956 they began nest building; two eggs hatched at the end of the month and one chick fledged (Johnstone 1957; Evans 1975). Thereafter the pair bred regularly until the death of the male in 1962. The female did not breed again for several years, but subsequently paired with one of her 1961 offspring to raise seven cygnets to fledging between 1966 and 1969 and, following his death in 1969, reared seven more cygnets when paired to one of her 1962 offspring from 1973 to 1982. The first report of Whistling Swans breeding in captivity was near Winnipeg, Canada, in 1945 (Delacour 1954). The species did not breed in Europe until 1976 when two cygnets were hatched and raised at the Flamingo Gardens and Zoological Park at Olney, Buckinghamshire. A pair of Whistling Swans at Slimbridge also bred successfully in the same year; two offspring were reared in 1976 and 10 from two clutches in 1977 (Evans 1977b; Kear 1977, 1978).

'Mrs Noah' remains by far the most successful breeding Bewick's Swan to have been kept in captivity. By the time of her death in 1982, aged at least 33 years, she had laid 165 eggs and reared 27 cygnets to fledging. Only 13 other Bewick's Swans have bred in Wildfowl & Wetlands Trust collections, of which nine were her offspring and two her grand-offspring. Other records of Bewick's Swans breeding in captivity are at Takamatsu, Japan, in 1962; at Moscow in several years since 1968; and at Askaniya-Nova, in the former USSR in 1971 (Evans 1975). The reason why Bewick's and Whistling Swans are more difficult to breed in captivity than other swan species may be linked with their breeding distribution in the wild. Photoperiod is thought to regulate the timing of the migratory and reproductive cycles (Murton & Westwood 1977; see Chapter 3) and the day-length even in mid-summer at most zoological gardens may be too short to stimulate laying in these High Arctic nesting species. Certainly captive Bewick's Swans breed late compared with other wildfowl at the Wildfowl & Wetlands Trust; some 15 hours of daylight appear to be necessary before they start to lay (Kear 1972; Murton & Kear 1973).

1.3 BEWICK'S SWANS: A CLOSER LOOK

1.3.1 Sex and age differences

Like other swan species, there is no major difference in physical appearance between male and female Bewick's Swans. Adult swans, aged two years or more, have white plumage, black legs and feet, and the highly variable black-and-yellow bill markings. Some individuals may have rusty stains on the head, neck and underparts if they have been frequenting iron-rich waters, but this usually wears off upon moving to a new site. The yellow markings on the bill are usually cadmium yellow in adults; paler shades may be indicative of ill health, particularly if the yellow is blotchy. Less common is a strong orange coloration on the bill, sometimes in birds with normal bill colour in the preceding or subsequent winter (Evans 1977a). The reason for this is still unclear but may be due to changes in diet, since swans with orange or pale bills may return with a normal yellow bill colour in subsequent years. Similarly, although most Bewick's Swans have a dark brown iris, a small proportion has grey-blue eyes. Eye colour seems to be consistent from one year to the next.

Juveniles can be distinguished quite easily during their first winter by their grey plumage and grey-pink bill. The grey plumage becomes whiter as the winter progresses, but yearlings can usually be aged in their second winter by traces of grey juvenile feathering in the white adult plumage. Grey feathering is rarely seen in third-winter birds, although some third-summer birds retain grey feathers on the head and neck (J. Beekman pers. comm.; Rees *et al.* 1997a). Bill colour also develops during the first winter, darkening from the tip as the young birds lose their reddish-pink bill markings. The outline of the adult bill pattern, albeit ill-defined, appears in chalk white tinged with yellow by early spring. Most swans have developed their adult bill markings by one year of age, although pink patches may remain on the bill into the second winter. Except in leucistic birds (see p. 28), these have usually turned black by the third year. The stage of development reached by the young, in terms both of whitening of the plumage and definition of the bill pattern, varies between broods. This may be due to differences in age, to the quantity and quality of food at different sites and, perhaps, also to their ability to gain access to the food available. The calls of juveniles are also diagnostic in that first-winter birds make higher pitched wheezing calls, rather than the musical bugling of the adults.

Downy young Bewick's Swans are very similar to the newly hatched young of other swan species, and cannot readily be distinguished from Whistling Swans of the same age. The down is pale greyish-white in colour with slightly darker patterning on the head, neck and back, and lighter underparts. The bill is flesh pink, grey at the tip and along the sides. It is relatively smaller than that of Whooper Swan and Trumpeter Swan cygnets and with less down extending along from the base (Boyd 1972). Fledging occurs at 60–70 days, usually in mid- to late September in the wild (Chapter 5).

Although there is no difference in the plumages of the two sexes, male and female Bewick's Swans can often be separated in the field by differences in body size. This is easiest for paired birds, since the two members of a pair can be compared directly. The sex of each individual can also be determined by cloacal examination when the birds are caught for ringing. Measurements made of Bewick's Swans' wing length, skull length, bill length and tarsus length confirmed that males are generally larger than females in every age category (as adults, yearlings or cygnets). However, there was substantial overlap in the range of measurements recorded for the two sexes (Evans & Kear 1978; Rees *et al.* 1997a; Appendix 1). The number of instances where the female has proved to be larger than her mate is surprisingly low, although one female, misnamed 'Dougie', was not only bigger than her mate 'Estralita' but was also the more aggressive member of the pair. There is also a significant difference in the sizes recorded for each of the age categories, with yearlings being distinct from adults as well as from cygnets, but again there was some overlap in the sizes recorded across the age categories. Weight changes have been recorded only for the winter months, when they increase from a low level upon arrival at the wintering site in late October or early November to a maximum in late December or early January. The most rapid weight gain is in the first few weeks after arrival (i.e. by mid-November). A decrease in body mass in late January and February, mirroring similar weight loss at this time recorded for other species of waterfowl, is followed by marked pre-migratory fattening in March, prior to departure on spring migration. Females have consistently higher abdominal profile scores (an indicator of body condition) than males throughout the winter, particularly if they have a mate, and this difference is most marked just before migration (Bowler 1994). The exception is single females, which are generally in poorer condition than single males, probably because they do not have a mate to protect them whilst feeding. Paired males have significantly lower abdominal profile scores than single males, indicating that protection of the female is achieved at the cost of reducing their own food intake. This is particularly evident for dominant males with mates and cygnets, which are in the poorest condition of all the social classes by March, whereas their cygnets showed a significant improvement in condition over the winter season (Bowler 1994, 1996).

1.3.2 Geographic variation

The extension of the Whistling Swan breeding range into east Russia, sightings of Bewick's Swan on the Pacific coast of North America, and reports of Bewick's × Whistling Swan pairs suggest that there may be a gene flow between the western Whistling Swans and eastern Bewick's Swans, which could lead to geographic variation within the Bewick's Swan population. Bewick's Swans in the eastern part of the range are reputed to have larger bills than those further west (Boyd 1972), but the wide range in the measurements recorded for swans wintering in northwest Europe and the lack of biometric data available for swans in China and Japan means that there is currently no evidence of an increase in body size towards the eastern end of the range. There is strong evidence for a clinal variation across the Arctic in the amount of yellow on the Bewick's Swans' bills, however, with those in the east having a smaller proportion of yellow than those in the west. A comparison of the bill markings of Bewick's Swans showed that the proportion of black-neb bill patterns was higher at five sites on Honshu, the main island of Japan, than it was amongst those wintering in Britain. The proportion of yellownebs was correspondingly lower at the Honshu sites (Scott 1981; Table 1.1). Interestingly, while there was no difference in the bill pattern categories recorded at Slimbridge and Welney, there was a marked difference in the proportions of bill pattern types between different sites in Japan, which may perhaps be attributable to a combination of site fidelity and heritability of bill patterns from one generation to the next. Studies have shown that individual Bewick's Swans, accompanied by their mates and cygnets, return to the same wintering grounds in successive years, and that the offspring frequently continue using the site when they reach adulthood (Scott 1966). The presence of one or two particularly successful breeding pairs, therefore, may result in a high proportion of related birds being present

Table 1.1 Proportion of different bill pattern types in Bewick's Swan flocks in Japan and England. From Scott 1981.

Site	Number of birds checked	% Blackneb bill pattern	% Pennyface bill pattern	% Yellowneb bill pattern
Slimbridge	2,400	17.9	19.1	63.0
Welney	300	18.0	18.0	64.0
Japan	312	35.9	13.1	51.0

in the wintering flock. In contrast to Bewick's Swans, Whooper Swans in the eastern part of the range have yellower bill markings than those in the west (Brazil 1981), indicating that there is no parallel gene flow between the eastern Whooper Swan population and the Trumpeter Swans of North America.

1.3.3 Leucistic Bewick's Swans

During the 19th century Swinhoe (1870) described *Cygnus davidi*, an unusual specimen seen in a Peking museum. The mystery deepened since the specimen was lost and no other records reached the west, making it difficult to confirm its identity. The Peking swan was reported as being smaller than a Bewick's Swan with all-white plumage, a vermilion bill tipped by a black nail, and orange-yellow legs and feet. The area between the bill and eye was feathered, leading some to suggest that it was a Coscoroba Swan, although there was no explanation of how it got from South America to China (Evans & Lebret 1973). Meanwhile, Dorogostaiskiy (1913) received a specimen from the ornithological collection of the Irkutsk Museum of the Russian Imperial Geographical Society (Eastern Department) of an adult male swan killed by peasant Filipov on the River Irkutsk in 1902, which upon examination Dorogostaiskiy judged to be *Cygnus davidi*. The Irkutsk bird had most of the physical characteristics of the swan described by Swinhoe, particularly extensive 'yellow' (rather than vermilion) markings on the bill except for a black nail, a feathered cere, all-white plumage, and brown-orange (not black) legs and feet. The measurements were again smaller than for an adult Bewick's Swan. Janet Kear (1972) pointed out that the degree of feathering described by Swinhoe, which was also noted for the Irkutsk specimen, is found in juvenile Bewick's Swans, that the size would be about right for a subadult bird, and that *Cygnus davidi* therefore might be a leucistic variety of *Cygnus columbianus bewickii*.

Leucism (from the Greek *leukos*, meaning 'white, bright, light', or in this case 'lack of colour') denotes pigment deficiency, which in turn is thought to be genetically determined. Leucism is well known in the Mute Swan, where the white phase, pale-footed individuals are also known as 'Polish Swans', following the import of these birds from the Baltic. A sexual imbalance of 'Polish Swans' on Rhode Island (10% of males and 26% of females) led to a genetic study that showed that colour inheritance is sex-linked in Mute Swans (Munro *et al.* 1968). A leucistic Bewick's Swan, named 'Needham', seen at Welney in the 1971–72 winter and at Slimbridge in 1972–73, was accompanied at Slimbridge by a normal mate and a completely white cygnet with flesh/chalk-grey legs.

Birds have also been reported with yellow legs (Hanby 1986), orange legs (Merne & Walsh 1991) and pink-red legs (Ogilvie 1986), with or without a predominantly pink bill. It is possible that leg colour changes over time, perhaps the bill darkens, or alternatively the different combinations may reflect different genetic coding (Evans & Lebret 1973), but since these individuals are seen for only one or two years, it has not been possible to monitor changes over time or from one generation to the next. Bewick's Swans with yellow legs and normal bill colour seem to be the most common variant; they are noted in flocks in the Netherlands in most winters, and there is some evidence for this being an inherited characteristic. In a flock of 821 swans in the Wieringermeerpolder, Netherlands, in December 2001, there were eight yellow-legged birds, which appeared to be related. They were close together in the flock and one adult (with a blackneb bill pattern) was accompanied by a yellow-legged yearling (a pennyface) and at least one cygnet also with yellow legs (W. Tijsen and H. Schobben pers. comm.).

1.3.4 Bill pattern variation

Studies of individual animals are greatly facilitated if they can be identified by natural characteristics, particularly since it is often difficult to catch and mark all the animals artificially and undue stress can be caused in the process. The ability of trained observers to recognise individual Bewick's Swans by the variations in the swans' black-and-yellow bill markings, and thus identify every individual in the flock, is one of the main strengths of the Bewick's Swan study at Slimbridge. Differences in Bewick's Swan bill patterns were first recorded by Clem Acland in 1923. The artist Charles Tunnicliffe was clearly aware of this since he sketched Bewick's Swans with different bill patterns in the 1940s (see Tunnicliffe 1979). Géroudet (1962, 1963) and Sermet (1963) made similar observations in Switzerland, despite that country not being a major wintering haunt for the birds. Sir Peter Scott, watching the Bewick's Swans using the lake in front of his studio window, also noticed the bill-pattern variations, particularly in the intricate patterns over the culmen and behind the nares, and realised that the presence or absence of particular swans on the lake could be monitored from day to day. Moreover, Sir Peter noted that of the 24 birds identified in the winter of 1963–64, 16 returned the following winter, indicating that this method could be used to track individual swans over several years. He quickly appreciated the value of this technique for a detailed study of the species and the Slimbridge Bewick's Swan study commenced (Scott 1966). The more extended black-to-yellow interface on the Whooper Swans' bill markings, although also variable, is less pronounced than in Bewick's Swans, making it more difficult to differentiate between individuals in the field (Brazil 1981).

From the outset, the bill patterns of adult and yearling swans wintering at Slimbridge were drawn to facilitate subsequent recognition. Initially, Sir Peter, his wife Philippa and daughter Dafila Scott drew the bill patterns, and in due course other scientists were also involved in the study. The three main categories of yellowneb, blackneb and pennyface bill patterns (Section 1.1.3) were soon identified. Within each category there are numerous variations, ranging from broad differences in patterning on the forehead, in the shape of the yellow protrusion towards the nostril (known as the 'tooth'), in the lower forward quadrant on the side of the bill and at the gape. More subtle nicks and knobs at the black-yellow interface, or spots and mush (of yellow on black or black on yellow) are also useful at close quarters. With other cues such as under-bill colouring and pattern, eye and eyelid colour, body size, head shape, bill profile and behaviour (including frequenting a particular part of the swan lake at Slimbridge) also available to the observer, the level of variation is such that the swans seem more different over time rather than less so. At the start of the winter, when memory is a little rusty and swans are arriving en masse, the presence of the same mate helps to recall the identity of a pair, although of course this is not to be relied on because Bewick's Swans do re-pair following the death of a mate. A 'face book' is compiled for swans seen at Slimbridge each winter, which includes not only the drawings of their bills but also their names and associations, and this is also very useful for confirming who's who in subsequent years.

As the number of swans identified increased (the bill patterns of some 7,600 adults and second-winter birds have been recorded in the 40 years between 1963–64 and 2003–04), more systematic methods for recording and retrieving the swans' bill patterns were developed so that individuals could be recalled in subsequent years without relying totally on human memory and the face book (Rees 1981 gives a review). Many of the patterns proved to be asymmetrical, which helped to tell birds apart but made it more difficult to develop a system for filing the bill patterns of different swans. Pennycuick (1978) demonstrated that only 29% of swans have sufficient information in their markings on the right side alone for accurate identification. Nevertheless, a code was developed to describe the whole bill pattern in numerical terms, which could then be stored on computer. Now when an unknown swan arrives, its bill pattern code can be typed into the computer and compared with the codes of swans already incorporated on the database. The computer then lists the swans that approximate most closely to the new bird (Rees 1981). The coding and computerisation of Bewick's Swan bill patterns was used during the 1980s since it proved effective in retrieving the identity of swans not immediately recognised, but it remains a time-consuming system. Only a small number of birds were found to have been missed each winter (one or two birds not identified), so identification by computer has been temporarily dropped, pending the development of a less labour-intensive process. The mushy grey-pink bill patterns of cygnets are too ill-defined for identification purposes even within a winter season, although broad likenesses may be noted between members of the same brood. Cygnet bill patterns therefore are not recorded, but their identity in subsequent years can be determined by ringing and through association with their parents on returning as yearlings.

The accuracy of the bill-pattern recognition system has been tested and proven on several occasions. For instance, when experienced swan watcher Dafila Scott was asked to identify swans from slides taken two weeks earlier she correctly named the birds in 29 out of 30 good quality slides and 23 out of 30 of those less clearly portrayed (Bateson 1977). Recognition tests have also shown that swans can be identified reliably over a period of years; a comparison of photographs showed that although small changes in bill pattern may occur, usually in the upper part of the culmen, observers familiar with Bewick's Swans were able to identify the birds without difficulty, and even inexperienced observers could tell the birds apart (Evans 1977a). The greatest changes in bill patterns (excepting cygnets) occur between the second and third winters, but birds first identified as yearlings can usually be recognised upon returning as adults. One factor that helps immensely for observers learning to identify swans for the first time is that the birds arrive in stages during the autumn. It is therefore possible to draw and become familiar with a relatively small number of individuals when they first arrive, so that these are instantly recognisable by the time the next birds appear at the wintering site.

1.3.5 Heritability of bill patterns

Since the faces of individual Bewick's Swans are highly distinctive, it might be expected that offspring would inherit their bill pattern from their parents. Moreover, since humans are able to identify individual swans at a glance, it is possible that the birds also identify each other (particularly their mates and offspring) by sight. Certainly, paired birds that become separated during migration and spend the winter in different parts of Europe usually return together the following year if both individuals survive. This is not solely due to them returning to the same breeding territory. Members of a pair that arrive separately at a wintering site, sometimes weeks apart, recognise each other almost at once and reunite in a flock of more than 300 birds. Also, it is not unusual for a swan to peck at a bird feeding with its head under water only to discover, when the startled bird looks up, that it has pecked its own mate. On these occasions the pair may greet each other with head bobs or courtship displays, usually without vocalising, and then resume feeding side by side. Yet although Bewick's Swans have variable bill patterns that can be used to distinguish between individuals, it is unlikely that this feature is of crucial importance for recognising a mate since other monogamous birds, including other swan species, do not have this facility. In the Trumpeter Swan the bill is entirely black with no yellow markings. The precise mechanism by which individuals recognise each other therefore remains uncertain, but it seems likely to be based on a combination of sight and voice, supplemented by other visual cues such as head and body shape, and perhaps also smell. Certainly the repeated calls of birds separated from their families, and the regular alternating contact calls made by paired birds prior to take off, emphasise the importance of auditory cues to the birds for ensuring that paired birds and family members remain together.

A general similarity in the bill patterns of parents and their offspring, originally noted by Sir Peter Scott (1966), was confirmed in a more systematic attempt to quantify this similarity by coding different features of each individual's face. A comparison of 12 pairs and 20 of their offspring showed that the latter resembled both their fathers and mothers more closely than would be expected by chance (Bateson *et al.* 1980). Family resemblances are particularly striking in the upper part of the bill on the brow line. The study also indicated, but with less certainty, that the bill patterns of mates are less similar than would be expected if swans paired at random. These results remain tentative but, if true, could mean that swans actively seek mates that look slightly different from their parents and siblings, having learnt the features of their immediate kin through imprinting at a young age (Bateson 1978). Alternatively, active dispersion by one sex may occur in immatures returning to the breeding grounds that, since there appears to be a clinal variation in bill pattern across Arctic Russia, would result in individuals being more likely to pair with birds having bill patterns different to their own. It seems likely that active dispersion does occur in Bewick's Swans. In other species of waterfowl immature males generally disperse away from their natal areas, whereas young females show a relatively high level of natal site fidelity upon returning to breed for the first time (Greenwood 1980). However, the one Bewick's Swan that was ringed at its natal site and subsequently seen in the same area as a breeding bird happened to be male (WWT/All-Russia Research Institute unpublished data). The extent to which Bewick's Swans of either sex disperse prior to pair formation and breeding therefore needs to be ascertained, before the likelihood of yellowneb males from the western part of the range moving far enough east to increase the probability of pairing with a blackneb mate can be determined.

Other studies suggest that fluctuating asymmetry in secondary sexual characteristics provides a measure of sensitivity to environmental or genetic stress, and as such may be a true signal of the environmental conditions encountered during growth and development, or alternatively of the ability of individuals to cope with environmental stress. For instance, asymmetry in tail size has an adverse effect on flight manoeuvrability in Barn Swallows *Hirundo rustica*, and females tend to select males with long and more symmetrical tails (Møller 1990). It is therefore possible that there is some evolutionary advantage to the Bewick's Swans in having variable bill patterns, if the markings help to indicate the quality of a bird, which can be used as a cue in mate choice. A preliminary analysis indicated that symmetrical markings are not significantly associated with life expectancy or breeding success, but further analyses are needed to control for other confounding variables (S. Choudhury pers. comm.). Most swan species do not have bill-pattern variation available as a cue for assessing potential mates, however, indicating that they use other characteristics in mate selection (including age, body size, fighting ability or breeding experience, see Chapter 7), and these factors are also likely to be of overriding importance to Bewick's Swans in choosing a mate.

1.3.6 Voice

Like the other northern migratory swans, Bewick's Swans of both sexes are very vocal throughout the year. There is no obvious difference between the two sexes, although the female generally has a higher-pitched voice than the male. They use a range of musical hooting or crooning calls in a wide range of contexts. Perhaps the most frequently heard call is the loud and rapid *oo-oo-oo*, similar but less sonorous than the whooping calls of the Whooper Swan, invariably emitted during aggressive encounters and the subsequent triumph ceremonies (see Chapter 6). They may also hiss in threat. Gentler, single-note contact calls are used to ensure that pairs and family members remain together. More urgent contact calls are particularly important before flight, when birds are most likely to be accidentally separated from their mates or offspring. Thus swans indicate their intention to depart by head-bobbing and calling repeatedly, paired birds calling alternately, with the calls growing louder prior to take off. A loud monosyllabic honking note is used to maintain contact during flight. Lost birds also give fairly loud single-note calls, repeated at rather infrequent intervals. Otherwise, the contact calls of a resting flock result in a quiet babble of musical notes, which becomes a more persistent 'singing' in spring immediately prior to migration. The calls of juveniles in their first winter are readily distinguished from adult birds by their higher-pitched wheezy notes. Downy young have high-pitched squeaking calls when distressed and softer, less urgent contact calls at other times (Bowler 2005b; Rees *et al.* 1997a).

1.4 BEWICK'S SWAN STUDIES

1.4.1 Early work (pre 1970)

Bewick's Swans were first described as a separate species in 1830, but their movements, ecology and life-history strategies have only been studied in detail with the substantial growth in ornithological research over the last 50 years. The initial aim, as for many migratory species, was to ascertain their abundance (or rarity) and describe their migratory range. The earliest Bewick's Swan studies therefore were concerned mainly with their numbers and distribution. National wildfowl count programmes have been initiated in several countries within the swans' migratory range since the 1950s. Before then, observations or sightings were reported on a more *ad hoc* basis – for instance, Bewick's Swans were thought to have a more northerly distribution in the UK during the 19th and early 20th centuries, occurring rarely in England and Wales but present in substantial numbers in northwest Scotland, particularly in the Outer Hebrides and Tiree (Owen *et al.* 1986). The extent to which this apparent shift in distribution was due to confusion with Whooper Swans has been much debated over the years, but detailed records of the swans' use of Tiree suggest that there is no reason to doubt the early observations (J. Bowler pers. comm.; Chapter 2). Information on the migration routes also indicates that the previous more northerly distribution is genuine, with the swans passing over the Shetlands and western Scotland into Ireland (Baxter & Rintoul 1953).

Initial reports from the Netherlands show that Bewick's Swans were common winter visitors to the shallow coastal waters of the eastern Zuiderzee before this inland sea was dammed in 1932 (Ten Kate 1930; Poorter 1991). The waters desalinated rapidly and the new freshwater lake became known as Lake IJsselmeer. No complete swan counts were made in the region before 1932, so it is not known for certain whether the enclosure affected the number of Bewick's Swans using the area. However, 1,900 to 3,200 birds were present on the newly created Lake IJsselmeer in the 1936–37 and 1937–38 winters, respectively (Brouwer & Tinbergen 1939). Brouwer & Tinbergen also state that large numbers of Bewick's Swans appeared on the shallow coast of Gelderland (where Lake Drontermeer and Lake Veluwemeer are now situated), and that the birds used a larger part of the former Zuiderzee after 1932. Extensive aquatic vegetation (Fennel Pondweed *Potamogeton pectinatus*) attracted the swans to these sites. It therefore seems likely that damming the Zuiderzee influenced the swans' distribution and increased the numbers wintering in this part of the Netherlands.

More regular counts of wildfowl in Great Britain commenced with the establishment of the Wildfowl Count network in 1947 by the International Wildfowl Inquiry Committee (a precursor to the International Waterfowl Research Bureau), with the aim of covering as many water bodies as possible on a monthly basis during the winter. Following fierce debate during the passage through parliament of the 1954 Protection of Birds Act, the Wildfowl Trust was asked by the Nature Conservancy Council to coordinate the National Wildfowl Counts in order to provide a sound basis for determining key sites for conservation. The work was supported by the Nature Conservancy Council (more recently the Joint Nature Conservation Committee), the British Trust for Ornithology (BTO) and the Royal Society for the Protection of Birds (RSPB); in 1993 the National Wildfowl Counts amalgamated with the Birds of Estuaries Enquiry organised by the BTO to form the Wetlands Birds Survey (WeBS), a BTO/WWT/RSPB/JNCC partnership. WeBS continues to monitor changes in population trends and distribution of wildfowl and waders from counts made at key sites by a large network of volunteer counters. The sites used by Bewick's Swans are well covered by the count network, so the monthly counts are thought to give a good indication not only of long-term population trends but also of total numbers present in the country each season.

In other countries wildfowl counts were undertaken less systematically during the 1950s and 1960s, although waterbirds were counted at several important Dutch wetlands as early as 1947, and a large-scale count of Bewick's Swans in the Netherlands was undertaken in autumn 1955 (Poorter 1991; K. Koffijberg pers. comm.). A lack of coordination between the national censuses would limit their value in monitoring changes in the whole population, particularly since annual variation in weather conditions affected the timing of migration and the distribution of birds in winter. Therefore, the International Wildfowl Census (IWC) was instigated by the International Waterfowl Research Bureau (IWRB; now Wetlands International) in the mid-1960s, with the aim of coordinating wildfowl censuses across northwest Europe, the Mediterranean and North Africa, concentrating on the mid-January counts. The first coordinated coverage of IWC sites was undertaken in 1967, with coordinators of the national programmes forwarding the results of the January counts to the IWRB. With countries including the Netherlands, Germany, Denmark and Estonia, as well as Britain, undertaking annual counts for the IWCs, Bewick's Swan population trends were monitored across most of its wintering range from the mid-1960s onwards. Population estimates based on IWC counts between 1967 and 1973 were of less than 7,000 birds (Atkinson-Willes 1976), although more recent estimates have been much higher (see Chapter 2). The level of monitoring in the Netherlands increased further in 1993–94, as SOVON took over the coordination of the national waterbird census scheme, and Bewick's Swans were counted monthly from October to March inclusive. Similarly, detailed monitoring of swans migrating through Estonia has been undertaken every third winter since 1995 (A. Kuresoo pers. comm.). Complete winter censuses of both Bewick's and Whooper Swans in Denmark were initiated in winter 1990–91 for Bjarke Laubek's study of Whooper Swan ecology (Laubek 1998), and were adopted by the Danish National Environment Research Institute (NERI) as standard practice thereafter (B. Laubek pers. comm.). The first major census of Bewick's Swans in Ireland was undertaken in 1975–76 (Merne 1977).

1.4.2 Long-term studies at the Wildfowl & Wetlands Trust

When the Wildfowl & Wetlands Trust (then known as the Severn Wildfowl Trust) was established at the New Grounds, Slimbridge, in 1946, only small numbers of Bewick's Swans were found in Gloucestershire. During the 1950s no more than 30 individuals were counted in the area on a single day (Ogilvie 1969). Those at Slimbridge were usually seen on the Dumbles, 65 hectares of raised saltmarsh between the sea wall and the Severn estuary that occasionally floods during high tide. The long-term Bewick's Swan study at Slimbridge therefore did not start until February 1964, when the wild Bewick's Swans were first attracted to a lake in the Rushy Pen, and the first bill patterns were drawn (Scott 1966). The Trust's founder and director, Sir Peter Scott, took a close personal interest in the swans and, with his family, he was responsible for the early years of the study. In addition to appreciating that bill pattern recognition could form the basis for a detailed study of the species, they also found that Bewick's Swans have a high level of winter site fidelity, and that paired birds generally remain with their mates and offspring throughout the winter. Thus, it was possible to gather information on the swans' lifetime pairing and breeding success without relying on artificial markings. The ability of trained observers to identify all the swans present, rather than just a proportion of the flock, is extremely valuable in helping to ensure that the studies do not concentrate on a biased sample of dominant or site-faithful individuals. There is a limit to the number of swans that one person can identify in a day (*c.*400–600, depending on viewing conditions and the percentage of new birds at the site), but on most days the number of swans at Slimbridge falls within this range. Bill pattern recognition also provides a facile and rapid means of identification particularly valuable for behavioural studies, when instant recognition of opponents in aggressive encounters is necessary.

Ever since the study started, adult and yearling swans recorded at Slimbridge have each been given a name, making it easier to remember the birds thereafter. The most famous swan, called 'Lancelot', was named by the painter Keith Shackleton after a character – Lancelot Sage from *Yachting World* magazine – who was a thinly disguised caricature of Peter Scott. 'Lancelot' was amongst the 24 Bewick's Swans identified during the first year of the study in 1963–64 and returned each year until the 1985–86 season. Since he was first recorded as an adult bird (at least two years old) he must have been at least 25 years old when he was last seen. Only nine swans have since equalled (and two exceeded) Lancelot's record for longevity. 'Casino' was 26 when she was last seen at Slimbridge in the 1997–98 winter, while 'Brimstone' was at least 29 when found freshly dead in Russia in 1993. With annual survival rates estimated at around 86–87% for Bewick's Swans (Evans 1979a; Scott 1988), however, most swans do not reach 'old age'. Only 68 Bewick's Swans are known to have reached 20 years or older since the Slimbridge study started, of which 37 (54%) are female.

A register has been kept of all Bewick's Swans present at Slimbridge since 1965, recording the names and dates of all swans seen at the site on a near daily basis. The timing of arrival and departure was noted, plus attendance levels during the winter season. The number of

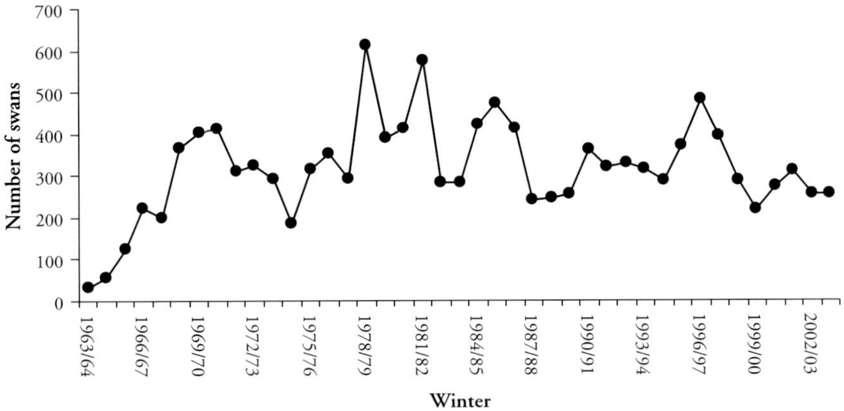

Figure 1.3 Maximum numbers of Bewick's Swans counted at Slimbridge, Gloucestershire. From Rees and Bowler 1996, WWT unpubl. data.

swans identified positively each winter rose annually from 24 in 1963–64 to 626 in 1970–71, with daily counts reaching 411 swans present on one day during the 1970–71 season (Evans 1979a; Figure 1.3). Thereafter, numbers fluctuated with the annual variation in weather conditions, reaching a maximum of 610 counted on one day when cold weather spread across northwest Europe in January 1979, when a total of 722 individuals were identified throughout the 1978–79 winter. The period of recruitment to the Slimbridge flock appeared to reach a plateau in 1969–70, since the proportion of 'experienced' swan units (i.e. those where the individual, or at least one member of a pair or family, had been recorded at Slimbridge in a previous winter) exceeded the proportion of new swan units from that season onwards (Evans 1979a; Rees 1988; Bowler 1996). Some 40% to 50% of all adults and yearlings recorded in any one year are birds which have visited Slimbridge in one or more previous seasons, with the proportion of new birds reflecting annual variation in hard weather movements of Bewick's Swans from wintering sites on the continental mainland.

In the early years, the swan research programme focused on making detailed observations of individuals wintering at Slimbridge, which described subtleties of the swans' social behaviour and provided the long-term data needed to analyse their life-history strategies (for example, mate fidelity, longevity, pairing success and breeding success) not easily detected by broader short-term projects. This yielded valuable information about the behaviour and ecology of the swans in winter, but formed only part of the picture for a species with an extensive migratory range. Moreover, data on survival and breeding success was being lost if the swans transferred their allegiance to other wintering sites, since only scientists familiar with the bill patterns could identify birds from the Slimbridge flock by their natural markings. Therefore from the 1966–67 winter onwards, some of the Bewick's Swans at Slimbridge were caught and marked with plastic leg rings each winter, each ring engraved with a unique code readable with a telescope at distances of up to 200 metres (Ogilvie 1972a; Rees *et al.* 1991), which could be used to identify the bird. Sightings of ringed birds reported from other parts of Europe made it possible to identify their migratory routes and staging areas, and occasionally to determine their breeding grounds, as well as ensuring that the progress of a bird could be monitored even if it wintered outside the Slimbridge area. Thus individual birds were monitored throughout their lifetime to determine factors affecting their pairing success, breeding success, survival rates and site selection. Researchers can use this information to explain changes in population size and distribution, and to identify the swans' ecological requirements, which in turn helps towards the conservation of the species.

The ringing programme had the further benefit of attracting the attention of both amateur and professional ornithologists to the swans, and partly helped to stimulate new studies of the species in the Netherlands, Estonia and Russia. Meanwhile the Trust's own swan research programme expanded to include fieldwork at a national and international level. The Bewick's Swan ringing programme within Britain was extended by the building of swan-pipes at Welney (in 1980) and at Martin Mere, Lancashire (in 1990). The development of a long-term Whooper Swan study, in collaboration with Icelandic and Irish colleagues, following the building of a swan-pipe at Caerlaverock, Dumfries and Galloway, in 1980, produced a valuable opportunity for interspecific comparisons. So far, these have dwelt on the different constraints of the swans' migratory and reproductive cycles, due to Bewick's Swans having to undertake a 2,500-mile (4,000km) migration to breed in Arctic Russia, whereas British-wintering Whooper Swans have a shorter 500-mile (800km) flight to their Icelandic nest sites (Rees *et al.* 1996). Information gained on the swans' natural history has also been disseminated through WWT's education programmes, through interest taken in the studies by the media and through WWT's Swan Adoption Scheme.

1.4.3 Changing numbers and distribution in the vicinity of Wildfowl and Wetlands Trust reserves

Regular records of the distribution of Bewick's Swans wintering in Gloucestershire since 1951 show that Bewick's Swans frequent only a limited number of sites in the area. The *Gloucestershire Bird Reports* record just 13 main sites as being visited by swans in more than one winter, with incidental sightings at a further 26. Originally it was assumed that the build-up in the number of Bewick's Swans wintering at Slimbridge was due to the conditions created by WWT, with the regular provision of grain in the Rushy Pen and protection from disturbance on the refuge encouraging the birds to return to the area. The population was then augmented as birds familiar with the New Grounds brought their new mates and offspring to the site. Whilst these factors are likely to have reinforced the New Grounds as a regular wintering site for the birds, other studies have indicated that there has been a substantial increase in the number of Bewick's Swans wintering throughout northwest Europe since the 1960s. Fewer than 7,000 swans were recorded in the early 1970s (Atkinson-Willes 1976) rising to 16–17,000 by the mid-1980s and 29,300 in 1995 (Dirksen & Beekman 1991; Rose & Scott 1994; Beekman 1997; Chapter 2). It has been suggested that the earlier surveys underestimated the size of the total population (Beekman *et al.* 1985), but regular observations at the Ouse Washes, Norfolk/Cambridgeshire, the main wintering site for Bewick's Swans in Britain, reinforced the view that there was a genuine increase in the number wintering in Britain during most of this period (Owen *et al.* 1986; Figure 1.4). There may also have been a movement of birds from the Netherlands to Britain during the 1960s due to the decline in submerged aquatic vegetation in the IJsselmeer, which was an important food supply for the swans during the 1950s (Poorter 1991).

Although the increase in the number of migratory swans wintering at the New Grounds could be due to changing population levels, there was a similar increase in flock size

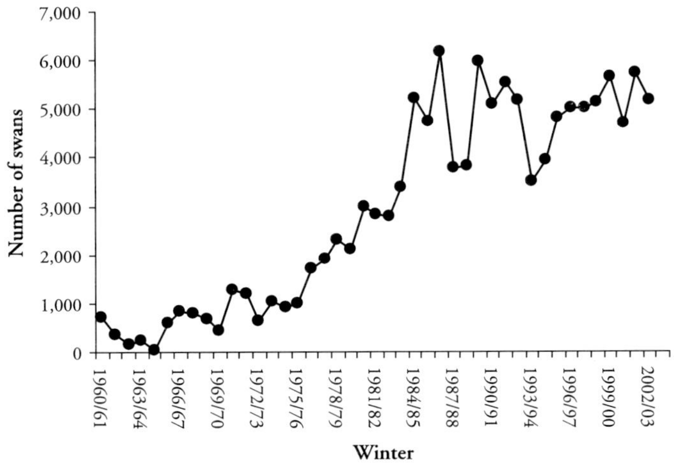

Figure 1.4 Maximum numbers of Bewick's Swans counted at the Ouse Washes, Norfolk. From Rees and Bowler 1996, WWT unpubl. data.

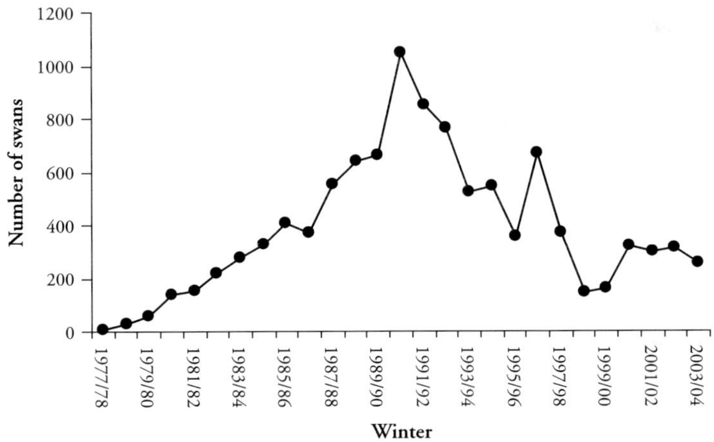

Figure 1.5 Maximum numbers of Bewick's Swans counted at Martin Mere, Lancashire. From Rees and Bowler 1996, WWT unpubl. data.

following the development of a reserve at Martin Mere in 1975 (Figure 1.5). Fewer than 10 birds were seen at the site each winter during the mid-1970s, but numbers rose dramatically during the 1980s and more than 1,000 birds were counted in the 1990–91 winter. The subsequent decline may perhaps be due to competition with the increasing numbers of Whooper Swans wintering at the site, although overt aggression between the two species is much less common than intraspecific encounters. The Ouse Washes was already an internationally important site for Bewick's Swans when the WWT Centre at Welney opened on the northern end of the Ouse Washes in 1970, and it was one of the sites designated when the UK joined the Ramsar Convention in 1976, but numbers have since continued to rise (Figure 1.4). A count of 6,164 Bewick's Swans on the Ouse Washes in January 1987 represented more than one-third of the northwest European population, then estimated at 16,500 birds. The current site record of 7,491 birds in January 2005 similarly accounted for a substantial proportion of the total mid-winter population. Welney and Martin Mere are both in areas of prime arable land, whereas pasture is more prevalent around Slimbridge. The provision of a secure roost (in the form of large lakes with shallow edges, which are protected from disturbance), in areas with a super-abundant food supply of root crops and winter cereals, is almost certainly the reason for migratory swans concentrating at Martin Mere and the Ouse Washes.

1.4.4 Dutch Bewick's Swan studies

With at least half the northwest European Bewick's Swan population wintering in the Netherlands, and more birds staging there en route to wintering sites in Britain and Ireland, it is critical for the conservation of the species that scientists in the Netherlands take an active interest in studying their ecological requirements and implementing their findings. In 1968 Ernst Poorter initiated a study of the breeding success of Bewick's Swans wintering in the Netherlands, since the disappearance of aquatic vegetation at many of their traditional feeding areas resulted in the birds switching to feed on pasture, then increasingly on arable land (Poorter 1991). He attributed the low proportions of juveniles recorded in swan herds in winter during the late 1960s to the birds feeding mainly on grass at this time, which has a lower metabolisable energy content than the starchy roots and tubers of pondweeds (*Potamogeton* species). Subsequent studies have found that the distribution of juveniles may be more complex, since the proportion of cygnets feeding on pondweeds compared with arable sites in the Netherlands was relatively low during the 1980s (Dirksen *et al.* 1991), and in the UK cygnet distribution is influenced not only by habitat but by flock size, time of year and geographic distribution (Chapters 2 and 4).

The Dutch Bewick's Swan Project 1982–1984 developed by Jan Beekman, Sjoerd Dirksen and Teus Slagboom further aimed to assess the importance of wintering sites in the Netherlands for the population as a whole. They also coordinated an international census of the species and used estimates of the percentage of juveniles present each winter to investigate differences in the distribution of breeding birds and non-breeders, both geographically and with respect to the food resources available in the Netherlands. The results of the 1982–1984 study were presented at the Third International Swan Symposium, held in Oxford in 1989 (Beekman *et al.* 1991; Dirksen & Beekman 1991; Dirksen *et al.* 1991), after which Jan Beekman embarked on a more detailed study of the swans' population dynamics, particularly in relation to how conditions on the wintering and staging sites affect breeding success.

Until the mid-1980s, most ringed Bewick's Swans were marked by the Wildfowl & Wetlands Trust, either at Slimbridge or at Welney, but in December 1985 Trinus Haitjema caught and ringed 22 birds at a wintering site in the Netherlands. He did not succeed again until 1989, but has caught Bewick's Swans regularly in the Netherlands since then, and also in Germany since March 1991, usually by cannon-net. In the first few years of the Dutch ringing programme, two plastic leg-rings were used (one on each leg), with a single digit inscription on each ring, which made them easy to distinguish from the single plastic leg-rings with two or three digit codes used in Britain. More than 100 swans were marked in this way. From winter 1990–91 onwards, the Dutch ringers switched to using yellow neck collars on Bewick's Swans caught in the Netherlands and Germany because of the high rate of ring loss, and to increase the number of sightings, particularly of birds feeding on aquatic vegetation. By the end of winter 2003–04, a total of 285 Bewick's Swans had been fitted with neck collars in the Netherlands and 91 in Germany (T. Haitjema, A. Degen and J. van Gils pers. comm.).

A recent development has been the founding of the Plant–Animal Interactions Department at the Centre for Limnology, Netherlands Institute of Ecology (NIOO), in 1994. The multi-disciplinary research group aims to describe the evolution of interactions between plant and animal species within aquatic ecosystems, and to investigate possible mechanisms underlying the evolutionary process. It initially concentrated on the interaction between Bewick's Swans and Fennel Pondweed, focusing particularly on the swans' intake of pondweed tubers when they arrive in the Netherlands in autumn (e.g. Nolet *et al.* 2001a, 2002) and how the swans' feeding activity affects the population dynamics and structure of the vegetation (e.g. Santamaría & Rodríguez-Gironés 2002). More recently, the NIOO team has been investigating distribution in relation to food availability (for example, sugar beet distribution) in more detail, both during migration and wintering in the Netherlands, with a view to understanding changes in their spatial behaviour (M. Klaassen and J. van Gils pers. comm.).

The development of the Bewick's Swan marking programme provided great potential for learning more about the swans' life-cycles, but this potential would not have been realised without efforts made to read the leg-rings and neck collars in the field. For this reason, the tail and wingtips of many birds were dyed yellow in the early 1970s to make them more obvious to amateur and professional ornithologists along the flyway and increase the chance of the ring being read (Chapter 3). This strategy worked; ring sightings from across Europe were regularly reported to the Wildfowl & Wetlands Trust during the 1970s and 1980s, and increased interest in Bewick's Swan studies in the Netherlands added a further substantial boost to the observer network during the 1990s. Indeed, having initiated the Dutch Bewick's Swan ringing programme, Trinus Haitjema started visiting the Ouse Washes in winter to look for missing birds not found in the Netherlands. The annual visit by Dutch ring-reading teams continued thereafter with different combinations of observers (Trinus and Miranda Haitjema; Henk Schobben and Wim Tijsen; Dico Tijsen; Reinhard Vohwinkel), and the now established tradition of mid-winter movement from the Netherlands to the fens (by both swans and people) provides not only valuable data but equally valuable international liaison.

1.4.5 International collaboration

Since the migratory Bewick's Swans traverse political boundaries, international communication is essential not only for improving our knowledge of the Bewick's Swan's life-history, but for promoting conservation measures along the flyway. International liaison was achieved initially through sightings of ringed birds reported by ornithologists in other parts of Europe, then through collaborative projects such as the co-ordinated international censuses, and most recently through joint expeditions to the swans' breeding grounds in the Russian Arctic. Peter Scott and his family made an exceptional visit to the Bewick's Swans' breeding grounds on the Yamal Peninsula in the former USSR in 1978, at a time when remote parts of Russia were largely closed to Westerners, at the invitation of Soviet scientists under the Anglo-Soviet Environmental Agreement. The invasion of Afghanistan in December 1979 brought this agreement to a close, so it was not until President Gorbachev introduced the concepts of *glasnost* and *perestroika* to Soviet politics that Western scientists again had the opportunity to work with their colleagues in Russia and the Baltic countries.

Towards the end of the Cold War, a Winston Churchill Travelling Fellowship enabled me to visit the USSR to observe the birds at migratory sites in Estonia, Karelia and the Gulf of Finland near Leningrad (now St Petersburg). Contact made in the Estonian capital of Tallinn in 1987 resulted in a formal invitation for a detailed study of the swans at Haeska on Matsalu Bay, Estonia, in 1988. This was an exciting time to be in Estonia, and not only for the excellent swan-watching in late April to early May. An Estonian film crew, also staying in the huts at Haeska, provided immediate insight into the historical changes underway; we had the benefit of a small portable television, and the Estonians provided instantaneous translations of the speeches demanding independence being broadcast by their political leaders. More systematic observations of migrating Bewick's Swans have been undertaken in Estonia since 1991 as part of a collaborative programme by Estonian, Dutch and Danish scientists (Luigujõe *et al.* 1996). The study confirmed the importance of Estonian wetlands as resting and feeding sites for the birds in both autumn and spring (Chapter 2). Estonian scientists aim to continue noting swan numbers at the staging sites and to monitor any changes in the swans' feeding habits attributable to changes in farming practice, with a view to introducing conservation measures if necessary.

After nearly a century of being unable to visit the Bewick's Swans' breeding range, following the Russian Revolution (1917) and during the Cold War (1945–91), the Russian Arctic again became accessible to Western scientists during the 1990s. Joint expeditions to study Bewick's Swans nesting in the Nenetskiy National District, Russia, have been undertaken by the Wildfowl & Wetlands Trust and the All-Russian Research Institute for Nature Protection in most years since 1991. In the early years we were joined by scientists from the Institute of Biology in Syktyvkar, Russia, the University of Groningen in the Netherlands and the Danish Ministry of the Environment, who were undertaking their own projects in the area, and more recently by staff from the Nenetskiy State Nature Reserve. In 1995, a collaboration between the Institute of Biology in Syktyvkar and RIZA (Institute for Inland Water Management and Waste Water Treatment) in the Netherlands resulted in a five-year (1995–1999) assessment of the Pechora Delta ecosystem. The relatively unspoilt nature of the delta meant that knowledge of its hydrology and wildlife would provide insight into ecosystem function that could be used in restoration programmes for areas where habitat had been degraded (van Eerden 2000). This work has evolved into the PRISM (Pechora Basin Integrated System Management) programme, which aims to provide support in developing policy to ensure the protection and wise use of the area.

Continued communication at an international level is of vital importance for the effective conservation of migratory waterbirds, including swans, since habitat loss in one area is likely to affect numbers and distribution elsewhere, particularly for the Bewick's Swan which is concentrated at comparatively few sites (Beekman *et al.* 1994a). Bewick's Swans have been protected from hunting for many years, but political changes within the former USSR make the situation more uncertain. The shooting of both Bewick's and Whooper Swans was legalised in Yakutsk, under regional law, for one year in the mid 1990s (Yu Shchadilov pers. comm.), and the debate continues on whether to reopen a hunting season at least for Whooper Swans in Yakutsk (Syroechkovski 2002). Also, although the effects of climate change associated with global warming are still being debated, there is a risk that changes in habitat and farming practice would increase the conflict with farmers and bring the birds closer to man-made hazards in a more closely managed environment. Swans, like other large birds, have a low reproductive rate, making them vulnerable to increases in mortality and decreases in breeding success. Bewick's Swans are also difficult to breed in captivity; the long days of the Arctic summer may be needed to trigger the start of the breeding programme (Murton & Kear 1978). It is therefore important to continue monitoring changes in population levels and distribution so that further conservation measures may be introduced before it is too late for them to be effective.

CHAPTER 2

Numbers and distribution

'More geese than swans now live, more fools than wise.'

Orlando Gibbons (*First Set of Madrigals and Motets in Five Parts*)

Bewick's Swans breed in the high Arctic regions of northern Russia, ranging from Chaun Bay, Chukotka, in the east to the coast of Cheshskaya Bay, Arkhangelsk District, in the west. Pairs have been reported nesting as far west as the Pechenga River, near the Fenno-Russian border (Scott & Rose 1996), and also on the eastern part of the Kanin peninsula (Mineyev 1991), but this information appears to be anecdotal. Unlike Whooper Swans, which nest in a broad band in the shrub tundra and taiga zones, the Bewick's Swans' breeding range is restricted to a relatively narrow belt of open Arctic tundra, mainly to the north of the Arctic Circle. Nesting pairs generally occur within 50–100 kilometres of the coastline of the Arctic Ocean, although in some regions such as the Yamal Peninsula and the Taimyr Peninsula they may breed up to 500 kilometres inland (Syroechkovski 2002).

The limited breeding distribution, combined with the severity of the Arctic winter, means that the birds cannot survive in their breeding range throughout the year. As a result, the Bewick's Swan is totally migratory. In most cases, the nearest wintering sites are at least 1,500km south of the breeding grounds, although a wild pair nested at fishponds in the Vilnius region of Lithuania in 1997, successfully raising two offspring (Stratford 1997). A pair continued nesting at the same site, albeit on different ponds, in 1998–2004 inclusive, and a brood of three offspring was again reared in 2004 (Švažas & Kozulin 2002; Kurlavicius 2003; S. Švažas pers. comm.).

The influence of photoperiod on the breeding season of wildfowl has been demonstrated by Murton & Kear (1978), who showed that a linear relationship exists between the mid-latitude of the breeding range and the onset of laying in many genera of Anatidae. Captive Bewick's Swans at several WWT centres follow this pattern in that few individuals breed, and those that do breed late compared with other species in the collection; some 15 hours of daylight seem necessary before they start to lay (Kear 1972; Murton & Kear 1973). Variation between individuals in the hours of daylight needed to stimulate breeding may therefore determine the extent to which the breeding range of the population is able to shift to more southerly latitudes. Since small groups of non-breeding Bewick's Swans have started spending the summer in Estonia and Lithuania in recent years (S. Švažas pers. comm.), it will be interesting to see whether they also start to breed more regularly and in greater numbers in the Baltic region.

Although Bewick's Swans breed across Arctic Russia, they follow two very different migration routes to winter at opposite sides of the Eurasian landmass. Those breeding west of the Urals migrate west along the Arctic coast to the White Sea, then head southwest across Karelia to winter in northwest Europe (the 'northwest European population'), mainly in Denmark, Germany, the Netherlands, Britain and Ireland. Those nesting east of the Lena River fly southeast to winter in eastern Asia (the 'eastern population'), in Japan, China and Korea. Whether there is any overlap in the distribution of the two populations in the breeding range, and the point at which the flyways divide, is still not known.

Small numbers wintering in the Caspian region of Iran and Turkmenistan, occasionally Turkey, and formerly around the Aral Sea on the Kazakhstan/Uzbekistan border, are thought to comprise a third population, although the origin of these birds is not known for certain. This chapter considers the historical and current status of Bewick's Swan populations, including fluctuations in population size and distribution, and the reasons underlying these changes. Factors affecting dispersal and site selection in winter are also discussed.

2.1 POPULATION SIZE AND TRENDS

2.1.1 Western population

Little is known about the size of the northwest European Bewick's Swan population prior to the 1950s, but mid-winter counts in Britain, the Netherlands and Ireland during the 1955–56 season put the figure at around 7,000 birds (Nisbet 1959). This figure was upheld by the mid-winter International Waterfowl Censuses (IWC), co-ordinated by the International Waterfowl Research Bureau (now by Wetlands International) from January 1968 onwards (Atkinson-Willes 1976), but additional counts cited by Poorter (1991) indicate that the official totals quoted for the 1950s and 1960s were underestimates. A single observer recorded 6,400 to 7,000 on one day in just one part of Lake IJsselmeer, in the Netherlands, in November 1955 (Bannerman 1957). Dutch national wildfowl counts found 11,560 in the Lake IJsselmeer area in December 1958, and some 8,000 birds in the same area in November 1961 (Poorter 1991).

Regular counts of Bewick's Swans wintering on the Ouse Washes and throughout Britain showed a marked increase in numbers from the mid-1960s onwards (Figure 2.1). Estimates of the numbers of Bewick's Swans wintering in northwest Europe, based on IWC

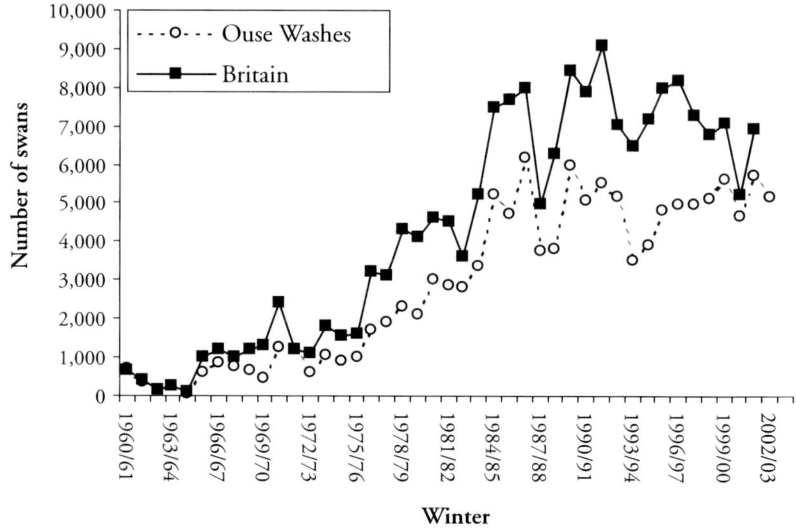

Figure 2.1 Counts of Bewick's Swans wintering on the Ouse Washes, and in Britain as a whole, since the mid-1960s.

counts, similarly found a substantial increase in the size of the population during the 1960s and 1970s. Numbers were put at 9,000–10,000 during the mid-1970s (Mullie & Poorter 1977), and at 18,000 birds by 1983 (Rüger *et al.* 1986), although the latter figure was revised downwards to 17,000 following the addition of a further three years' data in 1986 (Monval & Pirot 1989).

Since the early 1980s, the IWC has been augmented by five major coordinated censuses of the species across northwest Europe, made in mid-January in the years 1984, 1987, 1991, 1995 and 2000 (Beekman *et al.* 1985; Dirksen & Beekman 1991; Beekman 1997). These censuses aim to obtain accurate estimates of total population size through complete or near complete coverage of areas where Bewick's Swans are known to occur, in addition to the monitoring of trends at selected sites. Initially set up by the Dutch Bewick's Swan Project 1982–84 to assess the importance of Dutch wintering sites for the total population, the censuses are now scheduled to take place at five-yearly intervals as part of the activities of the Wetlands International Swan Specialist Group. Additional censuses, undertaken to monitor seasonal changes in distribution within northwest Europe, were made in November 1983 and March 1987 (Beekman *et al.* 1985, Dirksen & Beekman 1991). The November census proved unsatisfactory as many birds were moving westwards during the counting period. The census on 21–22 March 1987 found that numbers in Britain and Ireland were well below the January totals for those countries, because the swans had already headed east at the start of their spring migration, but only a few birds were recorded east of eastern Germany. Most were feeding on wet grassland in the Netherlands and West Germany, with 80–90% of the population thought still to be present in northwest Europe at this time.

The January censuses gave totals of 15,350–15,850 in 1982–83 (IWC plus supplements, Beekman *et al.* 1985), 16,243 in 1983–84 (Beekman *et al.* 1985), 16,910 in 1985–86 (IWC; Monval & Pirot 1989) and 14,600–15,950 in 1986-87, suggesting a rather stable population of about 16,000–17,000 birds in the mid-1980s (Dirksen & Beekman 1991). There was a dramatic increase in population numbers during the late 1980s and early 1990s, however, with 25,838 Bewick's Swans counted during the January 1990 census, and 29,277 in January 1995 (Beekman 1997). The Netherlands, Great Britain and Ireland held over 90% of the population on each occasion, with the Netherlands alone holding at least half of the Bewick's Swans wintering in northwest Europe during the 1990 and 1995 counts (Figure 2.2). These figures suggest an 83% increase in population size between 1987 and 1995.

Most of the increase was felt in the Netherlands, where numbers rose from nearly 9,000 in 1984 to more than 19,000 in 1995 (Beekman 1997; Figure 2.2). The consistent counting effort and good coverage of sites in the Netherlands throughout this period reinforces the view that the increase is genuine, and trends analyses for sites across northwest Europe derived from the IWC data similarly found a definite increase in population size between 1974 and 1994 (Delany *et al.* 1999; Figure 2.3a). Good breeding seasons in 1989 and 1990, when about 20% of swans in the wintering flocks were cygnets, may account for at least some of the increase. Yet several poor breeding years from 1992 onwards, when the percentage of cygnets was estimated at less than 15%, combined with an average annual mortality rate of around 16% (Evans 1979a; Scott 1988; Chapter 7), makes it difficult to explain how such a major change in population size has occurred. It is possible that there has been a shift in the swans' migration routes, with birds nesting further east in Russia following the European flyway rather than migrating to southeast Asia to winter. The eastern population also seems to have increased since the 1970s (see below), however, so there was no comparable decline in numbers. The lack of a ringing programme across the central part of the swans' breeding range means that any historical shift in

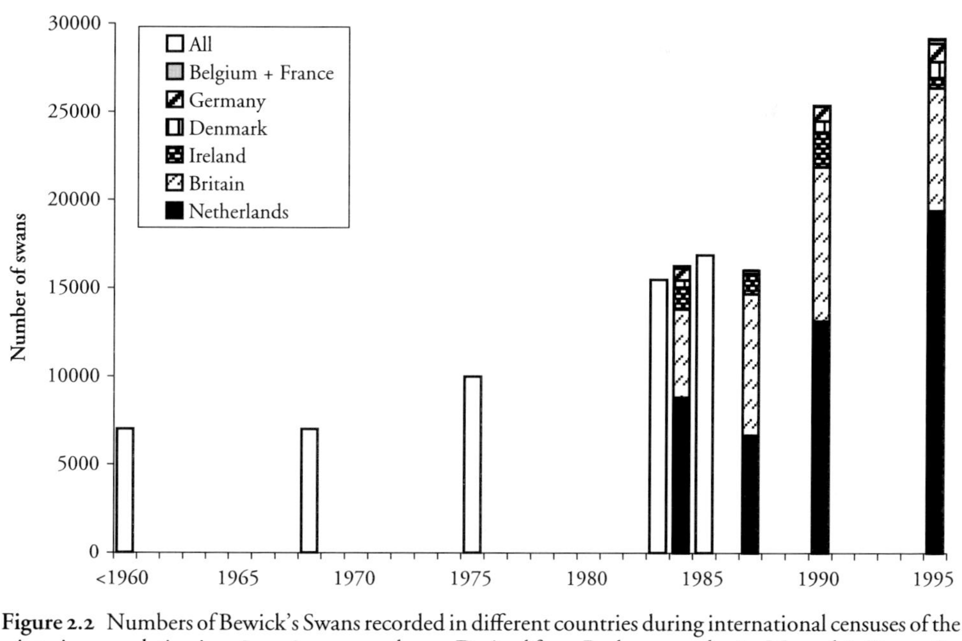

Figure 2.2 Numbers of Bewick's Swans recorded in different countries during international censuses of the wintering population in 1984, 1987, 1990 and 1995. Derived from Beekman *et al.* 1985; Monval & Pirot 1989; Dirksen & Beekman 1991; Beekman 1997.

distribution cannot now be detected, but the movement of birds between the flyways may be addressed in future years.

More recently, the IWC trends analyses indicate a worryingly sharp decline in the number of Bewick's Swans using key sites from 1996 onwards (Wetlands International unpubl. data; Figure 2.3a), with the pattern being evident both in the Netherlands (Figure 2.3b) and in Britain (Figure 2.3c). Regional variation in trends has been noted within the Netherlands, with a strong decrease in numbers recorded in the eastern and western parts of the country during the mid- to late 1990s, especially in those areas where numbers increased during the population increase in the 1980s. In the northern part of the country, where the swans feed on aquatic vegetation as well as the remains of the harvest, no overall trend could be detected (van Roomen *et al.* 2002). The extent to which the decline is attributable to the consistently poor breeding success, an increase in mortality levels, or to a redistribution of birds within the wintering range has yet to be investigated. The trends analyses (and particularly the IWC figures) are derived mainly from counts made at selected wetland sites, so it is possible that at least part of the decline is due to changes in distribution. For instance, the swans may have moved to new non-wetland habitats, or be short-stopping and remaining further east in the milder winters of recent years. Count data and ring resightings data both indicate that the mid- to late winter movement of swans from the Netherlands to Germany has been getting earlier since the mid-1990s (K. Koffijberg and J. van Gils, pers. comm.). The international Bewick's Swan census of January 2005 therefore will be particularly useful in helping to confirm the current status of the northwest European population.

2.1.2 Eastern population

Less is known about population trends for the eastern Bewick's Swan population than for those in the west. Indeed, in Japan, where the swans are monitored most carefully, it was not recognised as part of the Japanese avifauna until the late 1940s (Austin 1949; Austin & Kuroda 1953; Albertsen & Kanazawa 2002). Regular mid-January censuses, organised by the Environment Agency of Japan, have been made in Japan since 1970. Annual variation in numbers may be partly attributable to changes in distribution between Japan and China, although a lack of coordinated counts and of sightings of marked birds means that there is no concrete evidence to date for swans moving between the two countries. The annual censuses found 542 Bewick's Swans in Japan in January 1970 and 1,954 in January 1980 (Horiuchi 1981; Albertsen & Kanazawa 2002). Numbers continued to increase substantially during the 1980s and early 1990s to a peak of 31,198 in 1996, but the rate of increase was slower from the early 1990s onwards (Albertsen & Kanazawa 2002; Figure 2.4).

Following a decline in the late 1990s, numbers recovered to an average of about 40,000 birds in January 2001–2005 (Nature Conservation Bureau, Ministry of the Environment, Japan 2003, Y. Miyabayashi *in litt.*), at a time when large numbers were also reported in China (see below).

The level of increase up to the mid-1980s should be treated with caution, because the number of people involved in the counts and the number of swan sites monitored increased at this time, suggesting that the swan counts reflected improved coverage rather than an increase in total numbers (Albertsen & Kanazawa 2002). Nevertheless, an increase in the number of Bewick's Swans in eastern Siberia (Lena Delta, Kolyma Delta and Chaun Bay) during the 1970–1980s (Mineyev & Kondratiev 2001; Syroechkovski 2002), corresponded

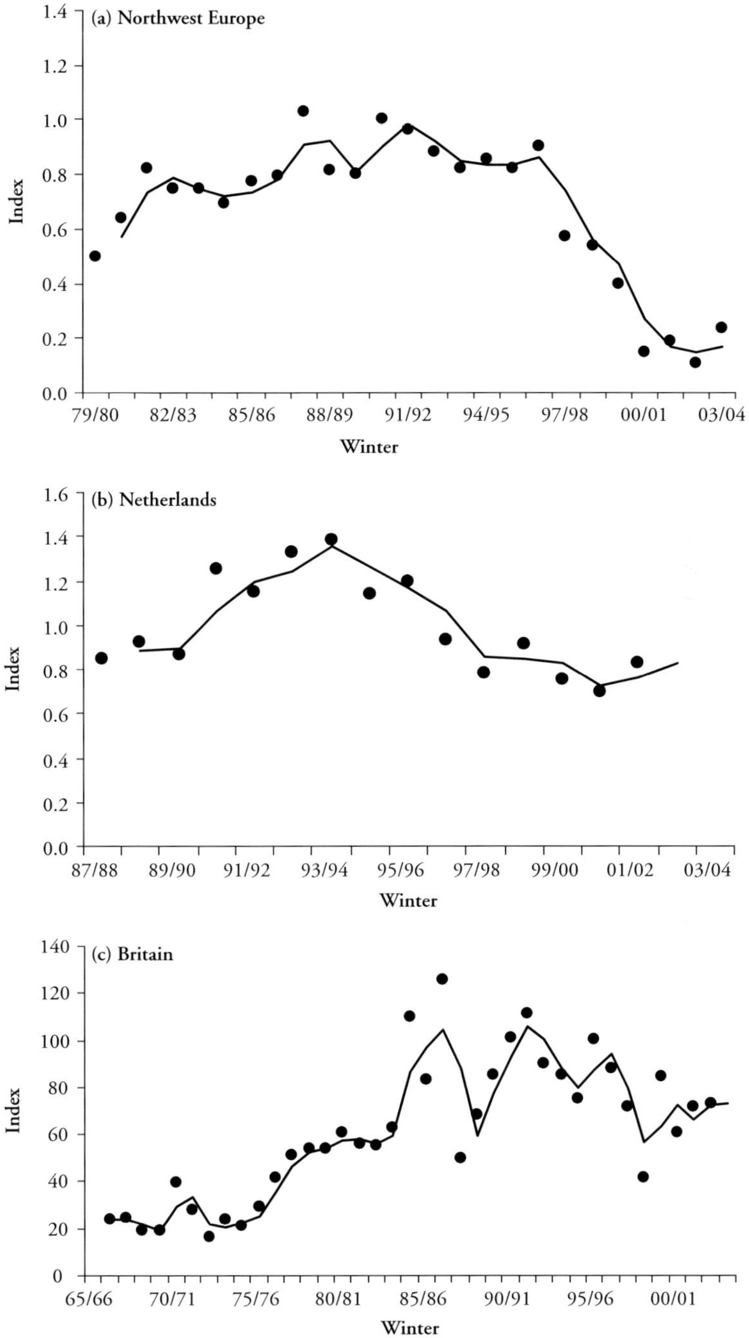

Figure 2.3 Population trends for wintering Bewick's Swans (a) in northwest Europe, (b) The Netherlands, and (c) Britain. The index shows changes in abundance from one year to the next at sites covered by the waterbird monitoring programmes. From Delany *et al.* 1999; S. Delany pers. comm.; van Roomen *et al.* 2003; Cranswick *et al.* 2005; A. Musgrove pers. comm.

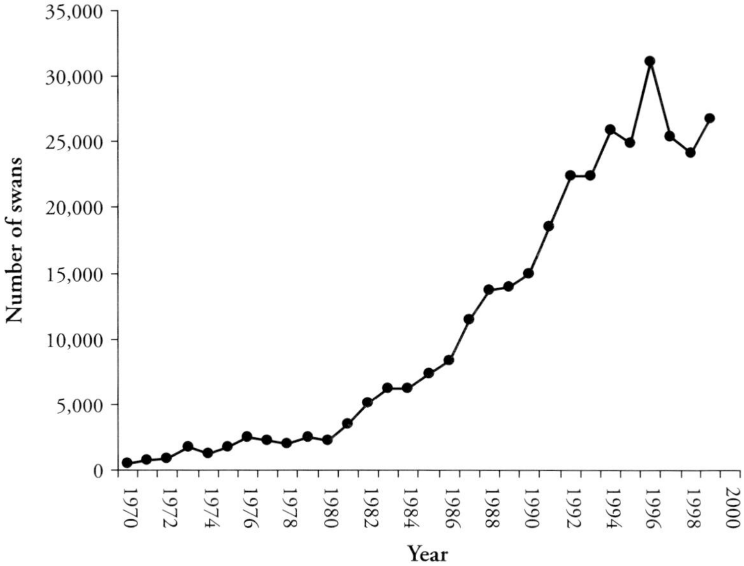

Figure 2.4 Increase in the numbers of Bewick's Swans wintering in Japan. From Albertsen & Kanazawa 2002; Environment Agency of Japan 1999.

with the increase in numbers recorded wintering in Japan over the same period (Environment Agency of Japan 1999; Albertsen and Kanazawa 2002). Breeding density in the Chaun lowlands of eastern Russia is thought to have increased by 50% between 1984 and 1994 (Kondratiev 1998; Syroechkovski 2002), but Kondratiev's (1991a) estimate of 1,200 Bewick's Swans between the Kolyma River valley and the Bering Strait during the 1980s seems rather low given the numbers wintering in Japan at this time. Although Bewick's Swans breeding further west on the Lena Delta probably also winter in eastern Asia, an aerial survey of 6,000–7,000 swans in the Chaun Bay area (including Ayon Island) in July 1993 suggests that discrepancies between summer and winter counts for the eastern population are due mainly to the difficulty in counting swans dispersed over the remote regions of Arctic Russia (Poyarkov *et al.* 2000; Syroechkovski 2002).

A second problem in determining population trends for the eastern population is the historic lack of regular systematic counts from wintering grounds in China or Korea. Numbers wintering in China were put at 10,000 during the 1990s, with no information on trends (Lu 1995; Zheng & Wang 1998; Ma & Cai 2000). In the Republic of Korea, numbers appear to fluctuate wildly, with a high of 1,300 nationwide in January 1992 and a reported low of 156 in January 1999 (Park pers. comm., in Miyabayashi & Mundkur 1999), although the latter can be considered an underestimate as a high proportion of 1,660 swans seen in the Nakdong estuary in mid-January 1999 were Bewick's Swans (N. Moores pers.comm). Thus, the total population was estimated at 40,000 birds by the end of the 1990s (Miyabayashi & Mundkur 1999), a 33% increase on Wetlands International's estimate of 30,000 birds wintering in China, Japan and Korea at the start of the decade (Perennou *et al.* 1994; Rose & Scott 1997).

Recent exceptional counts of more than 60,000 Bewick's Swans in China, reported first during an aerial survey of Poyang Lake in 2001 (Ji *et al.* 2001), and subsequently in a World Wide Fund for Nature (WWF-China) survey of the middle and lower reaches of the

Yangtze in February 2005, mean that a substantial proportion of the population may have been missed during the earlier censuses. If these latest figures are correct, the eastern population may now number more than 86,000 birds (Delany & Scott 2002).

The WWF surveys of the Yangtze River floodplain, undertaken in 2004 and 2005 (and scheduled for 2006) are made by 14 teams on the ground over a 2-week period, to minimise the risk of double-counting (Barter *et al.* 2004). The total of 65,114 Bewick's Swans recorded in 2005 is therefore thought unlikely to be an overestimate. Further information on midwinter changes in distribution between Japan (counted in January) and China (counted in February) is required to determine whether the national totals can be combined to give a reasonable estimate of total population size. Numbers in the eastern population certainly exceed the 40,000 estimated in the late 1990s, but coordinated counts across the wintering range would help confirm the true level.

2.1.3 Caspian/west Siberian population and out-of-range sightings

The counting of Bewick's Swans in northern Iran has been somewhat sporadic over the years, but results indicate a total wintering population in the south Caspian region of less than 1,000 birds. The latest estimate is of around 500 individuals (Delany *et al.* 1999). Some 843 birds were counted in Iran alone in 1969, since when counts submitted to the IWCs have not exceeded 322 (in 1994) (Rose 1995; Scott & Rose 1996). The 1994 counts included two birds seen on a lake at the Nad al Shiba golf course in the United Arab Emirates, the first record for the Arabian Peninsula for many years. The remainder were counted in Iran (Rose 1995). Small numbers also winter in Turkmenistan, on the eastern coast of the Caspian Sea, with most recently 22 birds seen in the Khazarskiy reserve in 2002 (V. Vassiliev and E. Rustamov pers. comm.). On the western side, up to 250 were counted in Armenia between 1990 and 2000, but only a handful (5-10 birds) were recorded in Azerbaijan (BirdLife International 2004a). The population estimate of 6,500 given by Rose & Scott (1994) was a printing error; it was intended to be 500 (Scott & Rose 1996).

It has been suggested that Bewick's Swans previously wintered on the Aral Sea and on lakes further to the east (Dementiev & Gladkov 1952; Delacour 1954). The destruction of the Aral Sea, its desertification, salination and pollution, following the diversion of its water supply from the Oxus River (now known as the Amu Darya) into irrigation programmes for the cotton industry, makes it unlikely that swans would now find suitable habitats along its shores. More recent population trend analyses, based on IWC data for the Black Sea/east Mediterranean and the southwest Asian regions suggest a population increase in the early 1990s, with a more recent decline, but variable coverage and the small number of swans using these sites means that population changes are difficult to ascertain (Delany *et al.* 1999).

The origins of the swans wintering on the Caspian Sea remain a mystery (Morozov 1996), although the barrier presented by the Ural Mountains suggests that these birds may nest on the Yamal, or even as far east as the Taimyr Peninsula, using the Urals as a navigational aid to help them find their winter quarters. Observations of Bewick's Swans during the spring migration (late April) in 1986 have shown that the swans fly from the northern Caspian Sea eastwards, skirting the Urals, then head north along the eastern slopes of the Ural mountains and along the Tobol, Ishim and Ob Rivers. Spring staging sites in the Volga–Ural interfluve were identified, most notably Kambak Lake in the Kamysh–Samara lake system, and also Edelsor Lake in the Kushum River basin (Belik *et al.* 1997). Further observations, from the Voykar River, supported the view that the Siberian breeding population (from the Yamal Peninsula and, probably, the Gydanskiy Peninsula) follows two migratory routes through western Siberia, with some birds heading to the Caspian/central Asia region to winter and others using unknown wintering sites further east in central to southeast Asia (Golovatin & Paskhalny 1997).

Recoveries in Astrakhan of two Bewick's Swans ringed in Britain indicate that the population wintering on the Caspian Sea is not totally separate from those wintering in northwest Europe (Rees 1991; Rees *et al.* 1997a). Moreover, a third bird ringed in Britain was found near Krasnovishersk, Perm, in the western foothills of the Ural Mountains, suggesting that the birds may migrate along the western side of the Urals, as well as the east. Occasional observations of small numbers migrating north through the Komi Republic, along the Urals east of Syktyvkar, supports this view (Y. Mineyev pers. comm.). One of the marked birds recovered outside the 'normal' flyway for the northwest European population was recorded as a cygnet when ringed, the second as a yearling and the third as a single adult. Two, if not all three, of the swans therefore were subadult when wintering in Britain. The greater likelihood of young swans transferring to another wintering area (see Chapter 3) thus appears to extend to their switching flyways.

Small numbers of birds occasionally reported in central or southeast Europe could be from either the Caspian or the northwest European wintering populations. Between 50 and 250 are recorded during severe winters in the Black Sea area of the Ukraine (Ardamatskaya & Korzyukov 1991; BirdLife International 2004a), with smaller herds turning up in Austria (maximum 6), Bulgaria (maximum 76), Cyprus (14 in 2002), Greece (maximum 46), Turkey (80–160 in 1991–2001), Romania (maximum 6) and Switzerland (maximum 10) (Beekman *et al.* 1994a; Scott & Rose 1996; BirdLife International 2004a). Reports of birds seen occasionally in Israel, North Africa (Algeria and Libya) and the Arabian Peninsula (Oman and the United Arab Emirates) may also be from this group. Similarly, the origins of two Bewick's Swans at Karavasta Lagoon, Albania, in January 1996, and of six swans (a pair with four cygnets) at the same site in January 2005 are unclear (M. Smart, pers. comm.). A bird ringed in the Pechora Delta region of European Russia was found shot in northeast Iraq (J. Beekman pers. comm. in Rees *et al.* 1997a).

Vagrants have also been seen in other parts of central Europe (including Hungary, the Czech Republic, Slovakia, Bulgaria, the Balkans and Italy), Svalbard, Bear Island and Iceland. A bird ringed in the Netherlands was subsequently found in a small flock of ten birds on the Palukia Lagoon/Evros, Alexanderupoli, Greece in 1998 (T. Haitjema and J. van Gils pers. comm.).

Some of the out-of-range sightings may be due to individuals joining larger herds of Whooper Swans on migration. This is particularly likely for the occasional Bewick's Swans seen in Iceland, since this is the main breeding area for Whooper Swans wintering in Britain and Ireland. For instance, one of the Icelandic Bewick's (a female yearling) was caught and marked with a neck collar during a round-up of Whooper Swans in summer 1984. A second bird, which had been ringed at Welney in February 1981, was seen at Jokuldalsheidi in eastern Iceland (an established Whooper Swan breeding area) in May the same year. Both swans appear to have followed more traditional migration routes in subsequent years, with midwinter records from the Netherlands and (in the case of the second bird) Welney, and spring sightings from staging areas in Germany. Except for those joining Whooper Swans on migration, adverse weather is likely to be the most common reason behind out-of-range sightings, particularly in more southerly latitudes.

2.2 REASONS UNDERLYING CHANGES IN POPULATION SIZE

It is well established that changes in the size of a population are due to one or a combination of four key factors: the recruitment of young, mortality, immigration and emigration. In closed populations, which have no interchange with others, variation in population size occurs only through mortality and recruitment. Recoveries of swans marked in Britain from the Caspian region suggest that at least limited emigration and immigration to and from the Caspian/west Siberian population, and perhaps also the eastern population, does occur. Nevertheless, although a Bewick's Swan marked with a neck collar on the eastern flyway was subsequently found in Estonia (Kondratiev & Kishchinski 1987), it seems likely that this was an exceptional vagrant rather than there being a significant movement of birds between the eastern and northwest European populations.

One of the remaining mysteries concerning Bewick's Swan populations is the extent to which birds nesting in the more central parts of the breeding range switch flyways and which factors stimulate a change in migration route. Meanwhile, the biological reasons underlying the rapid increase in the northwest European population between the mid-1970s and 1995 are not immediately obvious since, although there were several good breeding years in the late 1980s (particularly in 1989), there were some extremely poor years in the early to mid-1990s (Chapter 7). Variation in annual mortality rates, and even immigration from part of the eastern population, may therefore be relevant factors, although the eastern population also appears to have increased (rather than losing birds) over the same period.

Weather conditions in the Russian Arctic during the swans' breeding season have a major effect on Bewick's Swans' breeding success each year (Poorter 1991; Chapter 5). Within the wintering range, however, it is not always easy to determine whether it has been a good breeding year for the population as a whole because there is substantial regional variation in the distribution of juveniles. This is partly seasonal, since non-breeders generally migrate earlier than birds with young (Dirksen *et al.* 1991; Rees *et al.* 1997b), and partly related to previous experience of the wintering site, with swans returning more rapidly to a familiar site (Evans 1980). Nevertheless, by mid-winter the distribution of cygnets is still far from homogenous. For instance, only 8.3% young were recorded amongst 12,152 Bewick's Swans that were aged in the Netherlands in November 1995 (SOVON Ganzen- en Zwanenwerkgroep 1997), compared with 11.0%–14.7% young seen at sites in the UK (Cranswick *et al.* 1997).

In Britain and Ireland, variation in the percentage of young recorded has been attributed to flock size, with the proportion of cygnets being higher in smaller flocks, and also to geographic location. There is no major shift in the distribution of juveniles between regions during the winter, but variation in productivity remains evident, with southwest England usually recording the most young, due to the consistently high breeding success of birds wintering at Slimbridge (26% in southwest England over the 1990–91 winter, compared with 16% in northwest England, 17% in east England and 15% in Northern Ireland; Rees *et al.* 1997b). The provision of grain at Slimbridge may have contributed to the breeding success of birds wintering at this site, although it is clear that the distribution of juveniles is also related to habitat in other parts of the range. In the Netherlands, the proportion of cygnets seen feeding on pondweeds was low, and the proportion on winter cereals high (although this did not reflect its importance in terms of absolute numbers), but most cygnets fed on grassland from late November onwards (Dirksen *et al.* 1991).

Long-term studies have estimated annual mortality rates, with the percentage of Bewick's Swans surviving from one year to the next put at around 84% for adult birds and 66% for juveniles (Evans 1979a; Scott 1988; Chapter 7). Less is known about annual variation in mortality levels, however, or possible reasons for the fluctuations. *Post mortem* examination of birds recovered in the vicinity of WWT centres, and reports from ringing centres for swans found dead in other parts of the range, suggest that flying accidents, particularly with power lines, are the most common cause of death (Chapter 8). Since power lines are relatively constant in the swans' environment, it seems unlikely that they would be responsible for any annual fluctuation in survival for the population, although a general increase in man-made structures could affect the frequency of flying accidents in the longer term. Whether mortality rates vary with the time of year has yet to be investigated, but the energy demands of migration suggest that weaker birds may be more likely to succumb at this time. Weather conditions during the migration season, and food availability at the staging and wintering sites, therefore may result in higher survival in some years than in others, and this would affect the size of the population. Whether swans using some breeding, migratory or wintering sites have higher survival rates than those frequenting other areas also has yet to be determined.

2.3 BREEDING DISTRIBUTION

2.3.1 European Russia

Much of our current knowledge of Bewick's Swans' breeding distribution in the western part of their range is derived from the work of Yuri Mineyev, who monitored their nesting density and productivity on the Bolshezemelskaya tundra, the Malozemelskaya tundra and the Yugorsky Peninsula from the 1970s (summarised in Mineyev 1991, 2003). He found that breeding distribution across the Arctic in European Russia varies widely between tundras in different regions, despite the apparent similarities of these habitats. Moreover, numbers present in each region fluctuate substantially between years.

The westernmost location for Bewick's Swans in Arctic Russia is thought to lie between the Chesha and Shoina Rivers, on the eastern side of the Kanin Peninsula, where small numbers moult in mid-summer (Mineyev 1991). Although Mineyev (1991) also suggests that there might be some nesting pairs in this region, he was not working there himself, and an earlier analysis of the literature found no evidence for Bewick's Swans breeding this far west (Bianki & Shutova 1987). In the Red Data Book of USSR, Shchadilov (1984) excluded Kanin from the swans' breeding distribution, and put the breeding range as starting on the west coast of Cheshskaya Bay. The Timan tundras between the Indiga and Velt Rivers are thought to be used mainly during migration and for moulting, although autumn gatherings of swans west of Kolokolkova Bay have not been recorded in detail. The main breeding sites for the western population is 400 kilometres further east, in the Sengyakha basin, south of Kolokolkova Bay and at Zakharin Bereg (sometimes erroneously called the Russki Zavorot Peninsula; see Chapter 5) on the coast of the Pechora Sea. The main breeding regions in the Bolshezemelskaya tundra are the maritime lowlands of Bolvanskaya Bay (Neruta River) and west of Chornaya River, Varandeiskaya Lapta south of the Medynski Zavorot Peninsula and the southern coasts of Khaipudyrskaya Bay on the Barents Sea. The main sites on the Yugorsky Peninsula include the maritime tundra of the Barents Sea coast and the area west of Kara Bay (Figure 2.5).

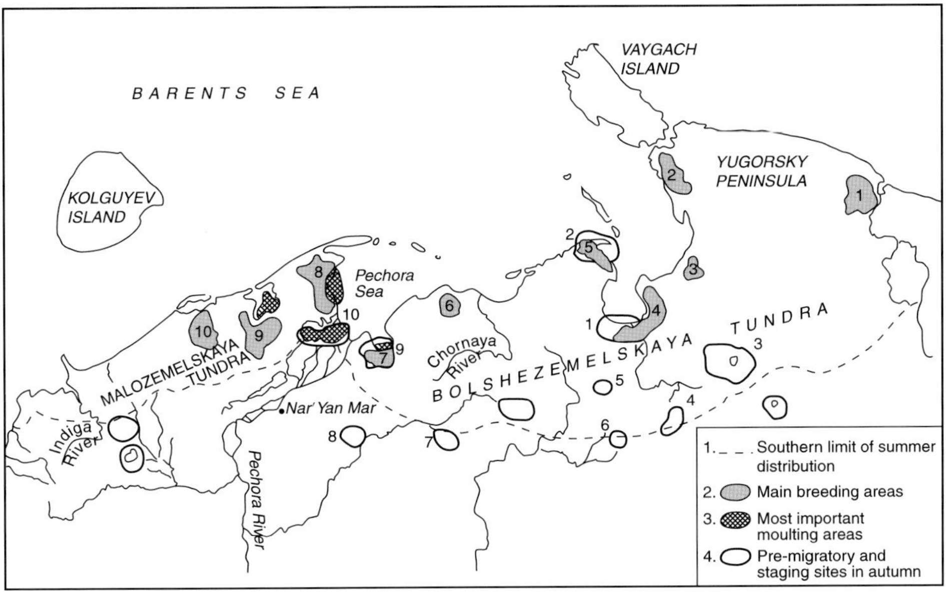

Figure 2.5 Distribution of Bewick's Swan breeding, moulting and pre-migratory sites in European Arctic Russia. Derived from Mineyev 1991, 2003.

Main breeding areas: 1 = Kara Bay tundra; 2 = areas between Rivers Bolshaya Oyu (Velikaya) – Lymbadayakha – Sedyakha; 3 = Kaipudyr Bay coast; 4 = Lower Morye-Yu; 5 = Medynski Zavorot Peninsula (Varandeiskaya lapta); 6 = Barents Sea coast between Rivers Chornaya and Dresvyanka; 7 = Bolvan Bay; 8 = Zakharin Bereg (south of Russki Zavorot peninsula); 9 = Kolokolkova Bay tundra; 10 = Senyakha Basin.

Pre-migratory and autumn staging sites: 1 = Khaipudyr Bay coast; 2 = Perevoznaya Bay; 3 = Vashutkiny Lakes; 4 = Lake Vatyarto; 5 = vicinity of Porchto and Syryerto Lakes; 6 = Lake Kolvato; 7 = Lake Beluga-ty; 8 = Lake Oshkoto; 9 = Bolvan Bay; 10 = Korovinskaya Bay and Pechora Delta (north).

The extent to which Bewick's Swans frequented Kolguyev Island was unclear for a number of years (at least to Western scientists), although a Bewick's Swan fitted with a satellite transmitter at Slimbridge spent the following summer there (see Chapter 3). Russian scientists have confirmed that the swans commonly nest on the island (Morozov & Syroechkovki in press), but an estimate of 6,500 birds present in July (Kalyakin 1995) seems too high. The latter may be due to counts being extrapolated from short aerial surveys.

Bewick's Swans are also known to breed on southern Novaya Zemlya. Two Slimbridge-ringed birds have been recovered there, at the southern island (Yuzhny) in August 1975 and Propaschaya Bay in September 1989. Early records suggest that both the northern and southern islands of Novaya Zemlya were used; small numbers were reported as far north as Cape Zhelaniya (Antipin 1938; Dementiev & Gladkov 1952), and the birds were commonly found on Yuzhny Island in the first half of the 20th century (Dementiev & Gladkov 1952). Recent observations confirm that the Novaya Zemlya archipelago is still an important breeding area for the species, with the swans nesting at latitudes of above 74°N, as far as the southern shore of Krestovaya Bay on the west coast of Severny Island (Kalyakin 1995). The main breeding areas are on Mezhdusharskiy Island and the southern part of Yuzhny Island, with swans also reputed (from questionnaire data) to nest along the east coast of Yuzhny Island (Kalyakin 1999).

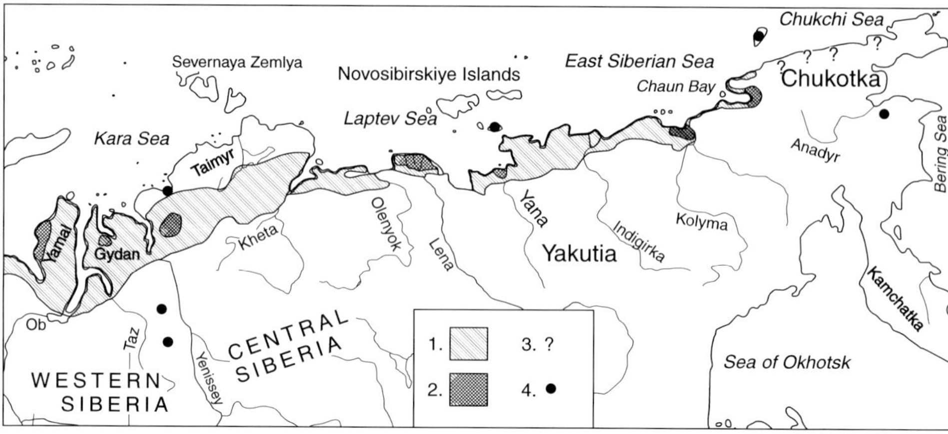

Figure 2.6 Bewick's Swan breeding range in the Siberian Arctic. 1 = breeding range; 2 = areas of high breeding density and non-breeding concentrations; 3 = areas where Bewick's Swans probably started breeding during the 1990s, but this has yet to be confirmed by ornithologists; 4 = breeding recorded outside of the main range. From Syroechkovski 2002.

2.3.2 Siberian Arctic

Further east, the breeding distributions of all swan species, including the Bewick's Swan, in the Siberian Arctic have been studied by scientists in the Goose, Swan and Duck Study Group of Northern Eurasia since 1988. Observations have been made at five or six sites each year ranging from the Gydanskiy Peninsula to Chukotka (Figure 2.6; Syroechkovski 2002). Additionally, Russian scientists have undertaken long-term aerial survey work in western and central Taimyr from 1966 to 2000 (Kokorev 1995, Kolpashikov 2001), and on the Lena Delta in Yakutia between 1963 and 1994 (Pozdnyakov 2002). This has provided valuable information on swan distribution in these vast and largely inaccessible regions (reviewed by Syroechkovski 2002). High breeding density recorded on Vaygach Island during the mid-1980s, particularly in the central region which averaged more than two birds (one pair) per square kilometre (Kalyakin 1987), was confirmed by subsequent ground-based monitoring, which found 0.12 to 0.36 nests per square kilometre in northwest Vaygach Island in 1986–1988 and 1995–1997 (Syroechkovsky *et al.* 2002). A count of 7,500 Bewick's Swans on Vaygach Island in 1984 was three times higher than numbers recorded in 1960 (Karpovich & Kokhanov 1967; Kalyakin 1987). In 1991, Bewick's Swan density was 0.75 birds per square kilometre, and there were an estimated 4,000 birds on the whole of the island at the end of the breeding season (Morozov 2001).

Comparatively little is known about the distribution of Bewick's Swans in the central parts of the breeding range, between Kara Bay (on the eastern side of the Yugorskiy Peninsula) in the west and the Kolyma lowlands to the east. Up to 87,000 Bewick's Swans have been recorded wintering in China and Japan in recent years, and it seems that most of these birds nest on Arctic tundras ranging from the Taimyr Peninsula to the Kolyma lowlands. Indeed, their breeding range may extend as far west as the Yamal Peninsula, since only 1,200 have been recorded breeding in the easternmost part of the range (Kondratiev 1991a), although the population size may have increased since Kondratiev's study. Certainly, Kondratiev (1991a) suggests that stable but extensive breeding by Bewick's Swans is mainly to the west of the Chaun lowlands. This is supported by reports that Bewick's Swans breed across the whole of the Yakutia tundra (albeit based on observations made between 1957 and

1975), extending into the northern taiga belt of the lower Lena and middle Kolyma Rivers (Perfiliev 1987). The swans are thought to be most common on the Lena Delta (reputedly 3.3 birds/100 sq. km in 1974) and the tundra of the Khroma–Indigirka watershed (2.5 birds/100 sq. km in 1961); less numerous on the Alazeya tundra (0.3 birds/100 sq. km in 1963) (Perfiliev 1987). Only small numbers of Bewick's Swans breed and moult between the Kara River (in the foothills of the Arctic Urals) and Baidaratskaya Bay, west of the Yamal Peninsula (Mineyev 1991), which suggests that this may be the point at which swans wintering in northwest Europe separate from those wintering in the Caspian region, or from those of the eastern population wintering in China.

Yamal and Gydanskiy Peninsulas. Information on the numbers and distribution of swans breeding on the Yamal Peninsula, and particularly on the Gydanskiy Peninsula, is especially patchy because of difficulties in gaining access to this remote region and in covering the terrain. Nevertheless, extrapolation of counts made during aerial surveys in the late 1960s estimated up to 5,600 Bewick's Swans in western Siberia, of which 2,600 were on the Yamal Peninsula and 3,000 on the Gydanskiy Peninsula (Uspenski & Kishchinski 1972; Syroechkovski 2002). It is known that swans still breed occasionally in the Upper Tanlova-yakha area of the southern Yamal Peninsula, although earlier publications indicate that they had ceased nesting along channels of the Shchuchya Delta where, a century ago, they were common (Finsh & Brem 1882, in Kalyakin & Vinogradov 1987). They were still nesting in the vicinity of the Shchuchya trading station in the 1930s (Kucheruk *et al.* 1975).

An expedition to the Ust Yuribey region, in the southwestern part of the Yamal Peninsula, in 1978, undertaken by Sir Peter Scott and his family with a party of Soviet scientists (led by Professor Vladimir Flint) under the Anglo-Soviet Environmental Agreement, found only one nest in the vicinity of their camp beside the Yuribey River, indicating a low breeding density in the vicinity. The party did see several small groups of swans during the flight from Salekhard, but again the largest herds did not exceed five birds, with some 83 birds being counted in total (Scott 1985). The Russian invasion of Afghanistan in 1979 brought the Anglo-Soviet Environmental Agreement to a close, precluding further joint expeditions to this or any other part of the Russian Arctic during the 1980s.

The current status of swans nesting on the Yamal Peninsula remains unclear, although a repeat of the aerial surveys in 1987 estimated a minimum of 5,460 birds for the Yamal Peninsula, 2,290 for the Gydanskiy Peninsula and 1,600 for the Tazovskiy Peninsula, indicating that some 10,000 Bewick's Swans breed in the arctic reaches of Western Siberia (Kalyakin & Molochaev 1990; Figure 2.7). Local ornithologists consider that these figures were overestimated and that many herds of non-breeding swans counted on the southern tundra were Whooper Swans (V.K. Ryabitsev pers. comm., in Syroechkovski 2002). Expeditions to study geese in the area (e.g. Paskhalny 1996), together with improved communication between research groups, may yield data over the next few years. Meanwhile, sources concur that the major Bewick's Swan concentrations in this region occur on the west Yamal lowlands and at the Yuribey basin on the Gydanskiy Peninsula (Syroechkovski 2002). Bird censuses throughout the west Siberian plain between 1959 and 1988 found only small numbers of Bewick's Swans to the south of the Yamal Peninsula, and the birds were recorded in the steppes only during spring migration (Ravkin 1991), indicating that the report of Bewick's Swans nesting in the Yenisey Basin (Ogilvie 1972b) was exceptional.

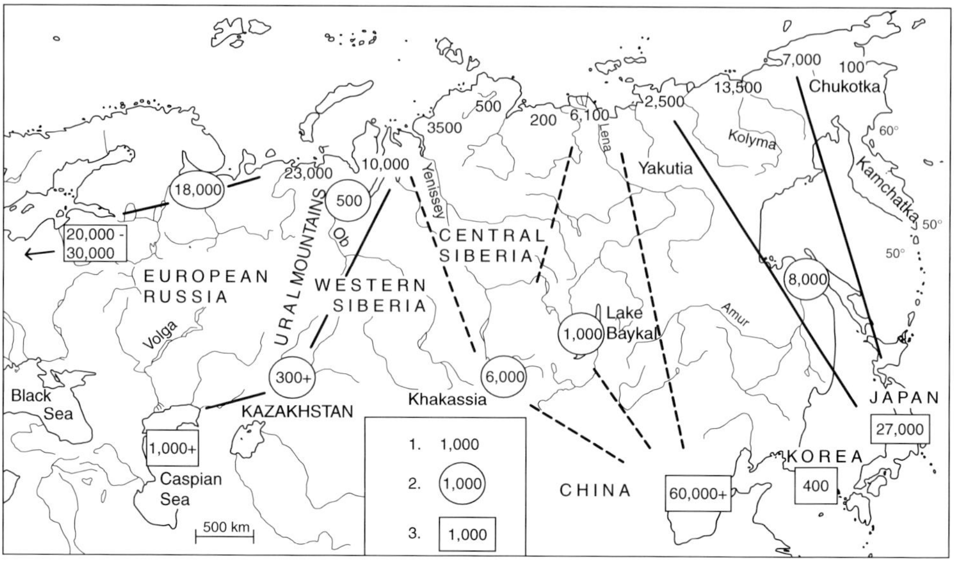

Figure 2.7 Bewick's Swan migration routes and population estimates in the late 20th century. 1 = numbers estimated in the breeding grounds; 2 = numbers estimated at stopover sites; 3 = numbers estimated in the wintering range. Solid lines indicate known migratory routes; dashed lines indicate possible migratory routes. From Syroechkovski (2002), who cites the sources of the count data, updated with recent population estimates from the wintering range described in the text.

Taimyr Peninsula. Numbers breeding on the Taimyr Peninsula declined by about one-third in the 12–13 year period between 1966 and 1978–79, from 0.57 birds (0.19 pairs)/100 sq km in 1966 to 0.24 birds (0.08 pairs)/100 sq km in 1978 and 0.17 birds (again 0.08 pairs)/100 sq km in 1979 (Vinokurov 1987). By the late 1970s, only 260 birds were thought to be present, including non-breeding birds as well as nesting pairs (Vinokurov 1987). Numbers for the whole of the Taimyr may have been underestimated, however, because eastern Taimyr and the inner delta of the Yenisey River were not surveyed regularly or extensively until the 1980s (Syroechkovski 2002). The disappearance of birds from the valleys of the Pyasina River and other large rivers on the peninsula has been attributed to the development of motorised water transport and to poaching, but the decline at more remote breeding and moulting sites was considered likely to be due to mortality during migration and in the winter (Vinokurov 1987).

Although the swans are no longer found in areas such as around Ust'-Tareya, where Bewick's Swans were seen regularly some 20–25 years ago (Rogacheva 1992), numbers started to increase during the 1990s especially in southwestern Taimyr, and are now estimated at about 3,500 birds for western and central Taimyr (Kolpashikov 2001), with a further 400–500 swans on the eastern Taimyr (Kolpashikov 2001; Figure 2.7). Moreover, fieldwork in the area from the 1990s onwards found swans at a number of sites in central and eastern Taimyr where birds were not reported during the 1980s, such as at Dudypta, Logata and on the lower reaches of the Balakhnya River (Russian Academy of Sciences unpubl. data, in Syroechkovski 2002). Main breeding concentrations are in southwest Taimyr (upper Pyasina River and Pura Lakes) and in central Taimyr in the Verkhnyaya Taimyra River basin. The inner delta of the Yenisey River (Brekhovskie Islands), which was an important breeding and moulting area at the beginning of the 20th century (Tugarinov

& Buturlin 1911), saw a substantial decrease in numbers due to heavy pressure from human activity in the 1960s–1980s, and now only about 100–300 swans occur regularly during the summer (A.A. Vinokurov, pers. comm., in Syroechkovski 2002).

Lena Delta. There has been detailed monitoring of Bewick's Swans on the Lena Delta, Yakutia, for almost 40 years. This was conducted by aerial surveys from 1963–1994 (Egorov 1965) and by ground transects using motor boats from 1994 to 2000 (Pozdnyakov 2002). The delta is low-lying, with over 30,000 lakes and 6,500 channels (Zalogin & Rodionov 1969); much of the area is polygonal marsh-tundra with 'low-centre' polygon pools providing a good nesting habitat for the swans (Pozdnyakov 2002). The aerial censuses found that numbers in the area decreased from 4,000 to 5,000 birds in the early 1960s to 700–800 birds by the mid-1980s, then increased to 6,100 birds by the mid-1990s (Poyarkov *et al.* 2000; Pozdnyakov 2002). The drop in numbers during the 1980s was not evident in counts in the wintering range, with numbers in Japan increasing at this time, but the long-term trends for China (likely wintering grounds for swans that spend the summer in the Lena Delta) are not known. The fluctuation in swan numbers in the Lena region has been ascribed to fluctuating numbers of wild Reindeer *Rangifer tarandus* in the area (which reached 35,200 animals on the delta in 1978), because the largest drop in swan counts occurred in years when Reindeer numbers were increasing, and swan counts subsequently increased after Reindeer numbers had declined. The Reindeer are thought to trample on the nests, and possibly displace swans from the area (Pozdnyakov 2002; Chapter 5). This hypothesis has yet to be confirmed, however, since there were gaps of several years between counts in the early years of the study, so other factors influencing swan numbers may have had an effect at this time.

The recent surveys have shown that swan distribution is uneven across the Lena Delta, with highest densities occurring in the northeastern part (up to 45 individuals/100 sq km in recent years), lower densities (maximum 5 individuals/100 sq km) in the southeast and lowest densities (0.5–0.9 individuals/100 sq km) in the northwest. This variation appears to be linked to habitat; the northeast is the lowest-lying part of the delta, rich with narrow, shallow channels in polygonal tundra. The northwest delta is an outlier of a marine terrace from the Upper Quaternary period, and the southeast is adjacent to the main shipping route separating the Lena River from the Laptev Sea (Pozdnyakov 2002).

East of the Lena Delta. Much of northeast Russia, from the Lena River to the Chukchi and Okhotsk Seas, is mountainous with associated mountain and boreal forest (taiga) vegetation, unsuitable nesting habitat for Bewick's Swans (Kondratiev 1991a). There are two extensive and geographically separate lowland belts, however, which serve as breeding areas: the Kolyma lowlands in Yakutia, and the Chaun lowlands in the Chukotka National District. The Yana Delta and its environs also holds reasonable numbers of Bewick's Swans (Poyarkov *et al.* 2000; Syroechkovski 2002; Figs 2.6 and 2.7). Aerial surveys of swan densities from the Yana Delta east to the Sellyakh River found, in 1982 and 1989, no change, or a very small increase in numbers, between these two years (Labutin *et al.* 1984; Pozdynakov & Degtyarev 2001), with some 400 birds estimated for the area. So a report of about 2,000 birds in 1994 was a substantial increase on the 1982 level (Poyarkov *et al.* 2000). Unfortunately, it has not been possible to determine whether this reported increase was due to rapid population growth during the 1990s, or to a variation in the methods used to count the swans and extrapolate the data recorded in the surveys (Syroechkovski 2002).

The Kolyma lowlands were recorded as holding some 700 Bewick's Swans during the 1980s (Kondratiev 1990), but this also may have been an underestimate because 13,000–14,000 swans were recorded by the mid-1990s (Poyarkov *et al.* 2000). The latter figure, taken together with counts from Chukotka, accords more closely with numbers wintering in China and Japan in the 1990s, although there remains a discrepancy with the large numbers recorded in China at the start of the 21st century. Elsewhere in Yakutia, the coastal lowlands between the Anabar and Olenyok estuaries held about 100 birds in 1983 (Labutin *et al.* 1984) and not more than 200 in the 1990s (Syroechkovski 2002), with 200 birds recorded in both the Indigirka Delta and the Khroma Estuary in 1994 (Poyarkov *et al.* 2000).

Chukotka. In Chukotka most breeding birds tend to frequent a narrow coastal belt just 10–15km wide, whereas in the Lower Kolyma region they disperse more widely across the flat tundras and may be found some 100–150km from the coast. In the 1970s and 1980s, breeding pairs were only occasionally recorded east of the Chaun lowlands, with few if any nests found in poor breeding years (Kondratiev 1991a), but there appears to have been a more recent expansion of the range eastwards along the north Chukotka coasts. Although this area is not well monitored, pairs were commonly seen nesting at Nolde Bay (173°30'W) in the early 1990s, with several breeding pairs as far east as the mouth of the Ekviatap River (179°E) (M. S. Stishov, pers. comm., in Syroechkovski 2002). The areas used by Bewick's Swans in Chukotka also seem to have extended further west in recent years; an aerial census of Chukotka in 1994 found that the last Bewick's Swans were close to the Pegtymel Estuary (174°W) (Poyarkov *et al.* 2000), whereas in earlier times their distribution stopped up to 500 km further east on the Amguema Estuary and Kolyuchin Bay (175°W) (Kondratiev 1998).

For at least the last two decades, the swans' main breeding area in Chukotka has been the lowlands at Chaun Bay (Figure 2.6). Numbers breeding there were estimated at 300 birds in the early 1980s (Kondratiev 1990), and although there was a 50% increase in breeding density between 1984 and 1994 (Kondratiev 1998), this figure seems rather low because an aerial survey in July 1993 gave 6,000–7,000 birds for the Chaun Bay area, including Ayon Island (Poyarkov *et al.* 2000). As with counts for the Kolyma lowlands, the higher estimates accord more closely with numbers recorded in the wintering range. The lack of regular ornithological surveys elsewhere on the Chukotka Peninsula means that sporadic breeding by Bewick's Swans in the region may have been overlooked. However, the good breeding habitats of the Kolyma and Chaun lowlands are thought to attract some 90% of Bewick's Swans nesting in the extreme northeast of Russia (Kondratiev 1991a).

The sub-Arctic Anadyr–Penzhina lowlands at the northern end of the Kamchatka Peninsula are frequented mainly by Whooper Swans (Kondratiev 1991a), which generally have a more southerly distribution than Bewick's Swans. Bewick's Swans also started spending the summer there during the 1990s, however, particularly in the lowlands of the Anadyr Estuary. A brood was seen on the southern coast of Onemen Bay in early August in the 1990s (M. Shulga, pers. comm., in Syroechkovski 2002), suggesting that this region may become a new breeding site (Syroechkovski 2002).

Recorded only as vagrants in Russia, following sightings in Chukotka during the 1950s and 1960s (Ptushenko 1952; Portenko 1972), Whistling Swans also have been seen nesting on the Russian tundras since 1974 (Lutsiuk & Sychev 1974; Kishchinski *et al.* 1975). Their breeding range has since expanded west and south of Kolyuchin Bay, along the north coast

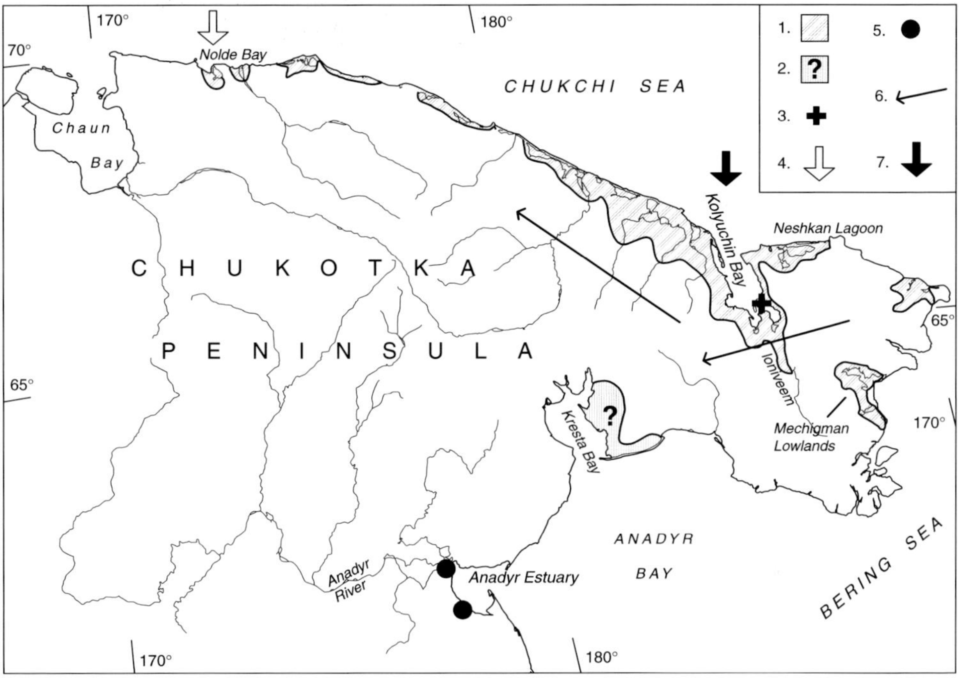

Figure 2.8 Whistling Swan breeding range in Chukotka, northeast Asia, in 2000. 1 = breeding range; 2 = breeding reported by local people, not yet confirmed by ornithologists; 3 = mixed Bewick's x Whistling Swan pair recorded; 4 = current western limit of breeding range; 5 = non-breeding Whistling Swan recorded outside main range; 6 = main direction of expansion of the breeding range in the last 20 years; and 7 = western limit of breeding distribution in the 1970s. From Syroechkovski 2002.

of Chukotka to Nolde Bay (174°E), the Mechigmen lowlands and, according to local people, part of the Ioniveem River basin (Poyarkov *et al.* 2000, Syroechkovski 2002, Figure 2.8). Local sources also report an expansion of the Whistling Swan breeding range southwest to Kresta Bay (65°30'N), but this has yet to be confirmed. The total number of Whistling Swans in Chukotka is not easy to determine due to incomplete coverage of the area by ornithologists, but it is estimated that some 600–1,000 birds may migrate to Chukotka for the summer (Syroechkovski 2002). Here their distribution overlaps with the eastern end of the Bewick's Swan's breeding range and, since they are conspecific, the potential for interbreeding ensues (see Chapter 1).

2.3.3 Historic changes in distribution across the breeding range

Longer-term (pre-1960) changes in the numbers and distribution of Bewick's Swans across the Eurasian breeding range are difficult to ascertain, although historical reviews state that the swans were commonly seen on the Russian tundras during the 19th and early 20th centuries (Ptushenko 1952; Dementiev & Gladkov 1952). In the western part of the range it seems that the breeding distribution has changed very little in recent times, except that the westernmost nest sites may previously have extended as far as the White Sea (Dementiev & Gladkov 1952), but numbers decreased substantially in many parts of the Siberian Arctic between the 1950s and the 1970s. During that period the breeding range became fragmented, with the species disappearing from many peripheral parts of the range and becoming rare even in some optimal habitats (Syroechkovski 2002). These declines have been attributed to a combination of hunting (illegal throughout the range since the 1950s), habitat loss and disturbance due to human activity, but conservation efforts both in the wintering grounds (Japan and Western Europe) and also in Russia saw an increase in numbers during the late 1980s and 1990s, together with a return of the swans to much of their former breeding range (Syroechkovski 2002).

In addition to determining key breeding areas for the species, the gathering body of data concerning the swans' current distribution and their habitat requirements in the breeding range is likely to be of major importance in predicting and ameliorating any future changes in population size and distribution that may result from the impact of global warming on the biosphere in northern Europe. A delay in the onset of the Arctic winter, or a more rapid thaw in spring, could also improve Whooper Swan breeding success in the more northern parts of their breeding range, since the slower-growing Whooper Swan cygnets would then have more time to fledge and migrate before the ice returns. Any northward shift in the Whooper Swans' breeding distribution could result in them competing with the Bewick's Swans for breeding territories, since the breeding ranges are close at present and even overlap in some areas. Whooper Swans, being larger and more dominant, might be expected to displace the Bewick's Swans from the prime territories, particularly since Whooper Swans monopolise key feeding and roost sites in places where both species occur in winter. It is also possible, however, that Whooper Swan pairs colonising the northern part of the range are subordinate birds displaced from traditional Whooper Swan breeding territories further south, and this may enable the Bewick's Swans to maintain their ground.

2.4 MOULTING DISTRIBUTION

Like most species of wildfowl, Bewick's Swans moult their primary and secondary flight feathers all at one time, in late July to early August, and are unable to fly until the new feathers have grown. Although non-breeding swans congregate in flocks during June and July, in areas with rich ground vegetation, such as maritime marshes, they may redistribute to safer areas to moult. Moult sites used by the non-breeders tend to be the same each year (Mineyev 1986), but do sometimes vary between years, and there may also be shifts in numbers using the different sites. Up to 1,000 swans moulting in the southern part of Vaygach Island in 1984 is the largest recorded concentration for the island (Kalyakin *et al.* 1987, K. Litvin pers. comm.), and some 3,000–3,500 birds moult on Korovinskaya Bay at the mouth of the Pechora River, north of Nar'Yan Mar (Mineyev 2003). Other major moulting sites in the European Arctic have been located on coastal waters and shallow bays in the vicinity of the main breeding areas, ranging from Chechskaya Bay to southern Vaygach Island, particularly in the Malozemelskaya tundra between Kolokolkova Bay, Pechora Sea and Korovinskaya Bay (i.e. the area sometimes referred to as 'Russki Zavorot Peninsula'; see Chapter 5) (Mineyev 1991, 2003). Territorial pairs usually remain on their territories for the flightless period, particularly if they succeed in raising a brood, but failed breeders may join non-breeders in nearby moult flocks.

Although up to 3,000–3,500 Bewick's Swans usually moult on Korovinskaya Bay (Mineyev 2003), a remarkable 15,000 individuals, representing three-quarters of the northwest European population, were seen around the shores of the bay in 1992 (Beekman *et al.* 1994a, 1996b; Rees *et al.* 1997a). The count was made in September, so may perhaps be classified as a pre-migratory gathering, when the swans gather in very large numbers as they shift south and west to feed in preparation for autumn migration. Nevertheless, the concentration of such a large proportion of the population in one place at one time, particularly if they are still flightless, makes them extremely vulnerable to sudden catastrophes, such as oil spills or disease. Fortunately, the major oil spill at Usinsk in October 1994, which resulted in oil leaking into the Pechora River, which flows into Korovinskaya Bay, did not appear to have an impact on the swans' breeding and moulting sites to the north (Bowler *et al.* 1995; see Chapter 8). After their moult, Bewick's Swans form even larger pre-migratory groups on the tundra water meadows (*c.* 50% of sightings), coastal marshes (28%) and shallow coastal waters (*c.* 20%) (Mineyev 1991), including several sites to the south of the main breeding and moulting grounds (Figure 2.5).

Many duck and goose populations undertake well established 'moult migrations' prior to becoming flightless, moving to safe areas with an abundant food supply because being grounded makes them more vulnerable to predators and restricts their ability to disperse in search of food. Ducks may make for milder climes, but moult migration in geese is almost invariably northwards, perhaps because the longer day-length at high latitudes enables the birds to detect and thus evade predators more easily (Ebbinge & Ebbinge-Dallmeijer 1977). Certainly there is a tendency for species breeding in temperate or sub-Arctic regions (for example, the larger Canada Geese *Branta canadensis*, Bean Geese *Anser fabalis* and the Central European population of Greylag Geese *Anser anser*) to show the longest and most well-defined moult migrations. Some species that undertake moult migrations breed north of the Arctic Circle, however, whereas others, such as the Lesser Snow Geese (*Chen caerulescens*) of Hudson Bay, Canada, moult south of the Arctic Circle despite there apparently being vast areas of suitable habitat further north. An alternative hypothesis therefore is that the birds seek out areas with better food quality, associated with the later spring flush of grass in Arctic regions (Owen & Ogilvie 1979). This would explain both the northerly trend in movement and the greater advantage to be gained by the southern populations.

The reason why Bewick's Swans, and indeed the other northern migratory swans, do not travel long distances on moult migrations remains unclear. A possible explanation is that their large body size makes the adults less prone than the smaller ducks and geese to predation by raptors or Arctic Foxes *Alopex lagopus.* The local redistribution and convergence of non-breeding Bewick's Swans to moult sites also may be analogous to the weaker movements shown by geese nesting at high latitudes, which similarly show less tendency to undertake long journeys prior to losing their wing feathers. In the case of the swans, distances travelled may be constrained by their narrow breeding distribution along the arctic coastline. Certainly the main moulting sites, both for Bewick's Swans and for Black Brant *Branta bernicla nigricans*, occur in the northern parts of their respective breeding ranges (Mineyev 1991; Palmer 1976; Owen 1980). The pre-moult movements of the Mute Swan, which has a more southerly distribution than the migratory swans, varies between populations and appears to depend on how far north and east the birds occur (Wieloch 1991; Spray *et al.* 2002). For example, non-breeding Mute Swans in the western Baltic area and from parts of southern Sweden congregate to moult in the shallow coastal waters of southwest Sweden (Mathiasson 1973), but the swans do not make such definite movements in other parts of their range, such as in Britain and Ireland (Spray *et al.* 2002).

2.5 DISTRIBUTION IN WINTER

The lack of comprehensive, regular censuses across Europe before the mid-1960s and in parts of the Far Eastern wintering grounds to this day means that information on earlier changes in the Bewick's Swans' geographic distribution is patchy and incomplete. Nevertheless, historic reports give some idea of major changes in the distribution of swans in the northwest European population. More recently, coordinated international censuses of the northwest European Bewick's Swan population have been undertaken by the Wetlands International (previously IWRB) Swan Specialist Group as part of the programme of providing base-line information on population size, trends and distribution to Wetlands International and national authorities. The results are used to identify sites of international importance for the species. Thus, from the mid-1980s, sites in northwest Europe averaging at least 170 Bewick's Swans *per annum* over a five-year period were considered to be of international importance, since this represented 1% of the total northwest European population at that time (17,000 birds; Monval & Pirot 1989; Dirksen & Beekman 1991). Following the continued increase in population trends into the 1990s, the threshold has been raised to 290 birds, pending the formal publication of the 1991 and 1996 international censuses and the results of the January 2000 census (Delany *et al.* 1999).

Mid-January counts in 1984 found some 54% (8,801) of the estimated population of 16,243 birds in the Netherlands, 31% (4,995) in Britain and 8% (1,244) in Ireland (Beekman *et al.* 1985; Figure 2.2). In January 1987 the situation was reversed, with 50% (8,018) of the population of up to 15,950 birds in Britain by mid-winter, 41% (6,560) in the Netherlands and 7% (1,148) in Ireland (Dirksen & Beekman 1991). The shifts in distribution are likely to be due mainly to the severe weather conditions in the Netherlands, although changes in farming practices, resulting in the availability of waste sugar beet for the birds, may also have had some effect on the timing of the swans' move to Britain and on the numbers involved.

The results of the 1990 and 1995 censuses have not yet been published in full, but there is evidence for a major increase in the size of the northwest European population during the late 1980s and early 1990s, with numbers reaching around 25,000 by January 1990 and 29,000 by January 1995 (Beekman 1997). Most of this increase was sustained in the Netherlands, with counts in other parts of the wintering range maintained at mid-1980s levels (Beekman 1997; Figure 2.2). An overview of sites used by Bewick's Swans in the Netherlands found that not only were numbers at traditional sites increasing, but the swans were also moving into new areas from the mid-1980s onwards (Koffijberg *et al.* 1997).

About 100 sites in Western Europe are known to have received internationally important numbers (>1% of the population) in winter since the early 1980s, but not all in the same winter season. Of these, the Ouse Washes consistently holds the largest numbers in midwinter (January), but the swans also congregate in very large numbers (over 1,000 birds) at sites in the Netherlands during the winter, and at staging areas in Denmark, Germany, Estonia and Russia during autumn and/or spring migration (Appendix 2). The eastern population was estimated at 40,000 at the end of the 20th century (Miyabayashi & Mundkur 1999), giving a slightly higher threshold (of 400 birds) for sites qualifying as being of international importance for this population at that time. Nevertheless, three staging sites and 23 wintering sites in Japan fulfilled this criterion (Miyabayashi & Mundkur 1999, Albertsen & Kanazawa 2002). Even with the higher population estimate of 87,000 birds, 16 sites in Japan and two in the Republic of Korea remain internationally important for the species (with >860 birds; see Appendix 3). Just four sites were listed as being important for wintering Bewick's Swans in China at the end of the 20th century (Miyabayashi & Mundkur 1999) but the more recent WWF-China surveys have described 38 sites in 10 wetland areas, including four in the Poyang Lake National Nature reserve (Barter *et al.* 2004, Appendix 3). There are further sites thought to be important for swans in China (Ma and Cai 2000; see also sections 2.5.2 and 2.6.2), but it seems that the numbers using these areas have not been published. Although Joonam Reservoir is listed as a separate site for Bewick's Swan in the Republic of Korea (Appendix 3), the Upo, Joonman and Dongpan Reservoirs and Nakdong Estuary all seem to be connected, with surveys showing that the waterbirds move between these sites depending on weather conditions and disturbance levels (N. Moores pers. comm.).

Less is known about key sites for Bewick's Swans in the Caspian/west Siberian population, where a group of 50 birds would be significant. The Gareh-Gheshlagh Marshes in Iran is possibly a regular wintering area; 39 birds were reported from there in 1991 (Rose & Scott 1997), rising to 90 in 1995 (Delany *et al.* 1999; Appendix 4). Up to 103 Bewick's Swans wintered at the Khazarskiy (formerly Krasnovodskiy) State Nature Reserve on the Turkmenbashi Gulf in Turkmenistan in 1977, and smaller numbers have been seen there in several years since then, including counts of 75 in 1979, 25 in 1991 and 22 in 2002 (V. Vassiliev and E. Rustamov pers. comm.; Appendix 4). Russian scientists believe that the number of Bewick's Swans wintering on the Caspian Sea may be underestimated, due to the difficulty of separating Bewick's Swans from Whooper Swans during aerial surveys (Morozov 1996; Syroechkovski 2002).

2.5.1 Winter distribution: northwest European population

The Netherlands. It seems likely that the Netherlands, which currently holds some 70% of the total northwest European population in autumn and early winter, has been an important wintering area at least since the first half of the 20th century. Certainly, Bewick's Swans were common winter visitors to the eastern part of the Zuiderzee before this inland sea was dammed in 1932 (Ten Kate 1930; Poorter 1991) and counts of 1,900 to 3,200 birds reported around the new freshwater lake, Lake IJsselmeer, were made during the winters of 1935–36 and 1936–37 (Brouwer & Tinbergen 1939). Numbers in the Netherlands increased substantially during the 1950s, coinciding with an increase in the numbers wintering in Britain, but then declined during the early 1960s, apparently due to a decrease in the amount of aquatic vegetation available to the birds, although this has not been substantiated (Poorter 1991).

The Frisian coast of Lake IJsselmeer was an important haunt for Bewick's Swans wintering in the Netherlands during the 1930s, late 1950s and early 1960s, but its carrying capacity diminished rapidly from 1967–68 onwards. The situation at the lakes near Kampen similarly deteriorated from the mid-1960s onwards, and submerged aquatic vegetation at Lake Eemmeer suddenly disappeared in 1966 following the construction of the southern dyke at the new polder of Zuidelijk Flevoland. This was due entirely to eutrophication, as the drop in water quality in the ditches and streams feeding into the lakes resulted in a turbid water system and blue-green algal blooms, where previously (up to and including the 1960s) there had been clear waters and large stands of aquatic vegetation (Noordhuis 2000). The biomass of pondweeds at Lake Drontermeer also declined substantially during the 1960s and 1970s. At Lake Veluwemeer, traditionally one of the outstanding areas in the

Netherlands for Bewick's Swans, the extensive stands of Perfoliate Pondweed *Potamogeton perfoliatus* had almost disappeared by 1970 (Poorter 1991). During the late 1960s, small groups of Bewick's Swans began to appear on drained grassland (polders) around Lake IJsselmeer, with up to 700 seen grazing on pasture during the 1968-69 winter. Thereafter, the numbers feeding in fields increased steadily, and the swans' distribution in the Netherlands increased considerably as the birds started feeding not only on grasslands but on waste potatoes, sugar beet and cereal crops. Meanwhile Lake Lauwersmeer has provided an important source of aquatic vegetation for the birds in autumn since 1973–74, following its closure from the Wadden Sea four years earlier.

The international surveys show that Bewick's Swans are more widespread in the Netherlands than in other parts of the northwest European wintering range (Figure 2.9). The country is the main stronghold for Bewick's Swans in late autumn and early winter, and internationally important numbers have been recorded at 62 sites since the mid-1980s (Appendix 2; Figure 2.10), including 9 used primarily on arrival in autumn (Appendix 2; Koffijberg *et al.* 1997). The decline since the late 1990s in the size of the northwest European population is becoming apparent at key sites in the Netherlands, however; 37 sites were recorded with peak counts averaging at least 190 birds in the five years to winter 2000–01 (van Roomen *et al.* 2002), but only 25 sites had an average of at least 190 birds in the five years to 2002–2003 (van Roomen *et al.* 2004; Table 2.1).

Of all the sites in the Netherlands, Lake Lauwersmeer, in the north of the country, usually receives the largest numbers in autumn, with up to 3,712 birds recorded in 1993 (van Roomen *et al.* 2003), nearly 7,000 in 1996 (SOVON Ganzen- en Zwanenwerkgroep 1999) and 6,461 in 1988 (Koffijberg *et al.* 1997). There is some annual variation in the autumn counts, since numbers fluctuate depending on the availability of pondweed stands, which form an important part of the Bewick's Swans' autumn diet. Nevertheless, the autumn average for the years 1985–2002 was an impressive 1,500 birds (van Roomen *et al.* 2004), and the site also regularly holds over 300 swans into the mid-winter period (van Roomen *et al.* 2003).

The Eempolders and Eemmeer make up the pre-eminent wintering site in the country, with more than 5,000 swans recorded in 1993, and peak annual counts averaging 3,632 between 1984–1993. More recently, numbers in the Eempolders have dropped to around 1,600 birds, but trends analyses for the winters 1985–86 to 2002–03 still make it consistently one of the most important sites for the species in the Netherlands (van Roomen *et al.* 2003, 2004; Table 2.1). Other sites also regularly receive over 1,000 swans, notably Oost-Flevoland-noord (average 1,700 birds in the years 1985–86 to 2002–03), Wieringermeer (1,300), Arkemheen en Putterpolder (1,200), the Harderwijk-Elburg polders in Veluwemear (1,200) and Het Bildt (1,100). Although Lake Veluwemeer and its surroundings (Harderwijk-Elburg) generally receive fewer swans than Lake Lauwersmeer, a few years with abundant pondweed and a new (albeit brief) period of feeding on stoneworts in the Netherlands boosted numbers to 6,200 birds in October 1996 and 11,500 birds in October 1997 (SOVON Ganzen-en zwanenwerkgroep 2000). Areas such as Het Bildt, Texel, Noord-Beveland, and the relatively new site at Wieringermeer have – despite a national decline in numbers – shown increasing trends in numbers in recent years (with the swans feeding on the remains of sugar beet following the harvest), whereas some traditional sites such as Noordoostpolder-west appear to be in decline (van Roomen *et al.* 2003, 2004). The latter is still more important than many other parts of the wintering range, however, with peak counts of some 210–500 swans recorded in the winters of 2000–2001 and 2002–2003,

1. Bewick's Swans flying to feed in fields near WWT Slimbridge, England (Paul Marshall).

2. A flock of Bewick's Swans rides out wintery weather in Japan (Mark Brazil).

3. A flock on mixed farmland in Friesland, the Netherlands, with modern Dutch windmills in the distance (Wim Tijsen).

4. Winter habitats from around the Palearctic: a) A few of the Bewick's Swans counted at Dahuchi, Poyang Lake reserve, China, in November 2004 (Mark Barter); b) On winter ice in northern Japan (Mark Brazil); c) Feeding in a stubble field of shooting rye cereals in Estonia during spring migration (Eileen Rees); d) On flooded pasture near WWT Caerlaverock, SW Scotland (Philippa Scott).

5. Swans coming in to land. Note the lowered legs and splayed feet, which act as air brakes (Paul Marshall).

6. Feeding on aquatic vegetation in the fishponds at Notec River, Poland (Mariusz Blank).

7. Tundra vegetation emerges as the snow melts at Khabuicka in the spring. This is Bewick's Swan country (John Bowler).

8. Bewick's Swan on her nest at Khabuicka in 1994 (Andrew Cooper).

9. A cygnet preens (Paul Marshall).

10. A Bewick's Swan family, comprising an adult pair and two cygnets (Paul Marshall).

11. Physical combat between rival swans (Philippa Scott).

12. An aggressive encounter; the wings are open and there is a flapping display (Paul Marshall).

13. A pair of swans performs a triumph ceremony (Paul Marshall).

14. A juvenile White-tailed Sea Eagle harasses a flock of Bewick's Swans. These eagles may prey on the cygnets when they are young (Mark Brazil).

15. Bewick's Swans, Bean Geese and White-fronted Geese feeding on arable land at the Waardpolder, Noord Holland, the Netherlands (Wim Tijsen).

16. Feeding under floodlights at WWT Slimbridge (Philippa Scott).

17. Grain is distributed for the swans and other wildfowl in the Rushy Pen at WWT Slimbridge (Philippa Scott).

18. The author measuring non-breeding Bewick's Swans caught in the Nenetskiy Nature Reserve in 2003 (Dave Paynter).

19. 'Abelhard' fitted with a satellite-transmitter (Mark Hulme).

20. Sir Peter Scott at his studio window, overlooking swan lake in the Rushy Pen (Philippa Scott).

21. The grace and elegance of a Bewick's Swan in flight (Philippa Scott).

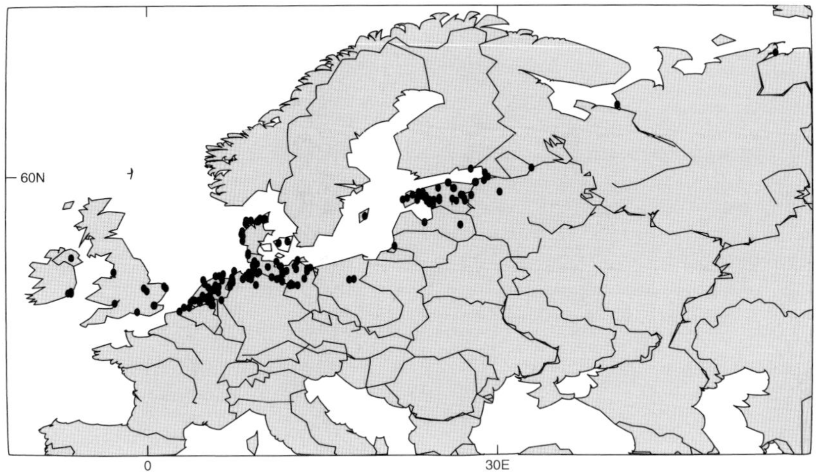

Figure 2.9 Wintering and staging sites of international importance for Bewick's Swans from the northwest European population. Based on data in Appendix 2.

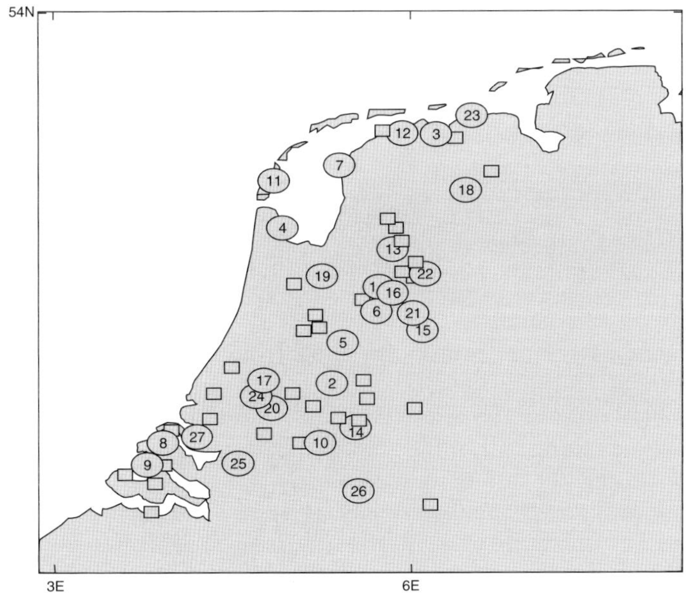

Figure 2.10 Key sites for Bewick's Swans in the Netherlands. Ovals = sites with internationally important numbers in 2001–02 to 2002–03; rectangles = sites with internationally important numbers since the mid-1980s. From van Roomen *et al.* 2003, 2004, Appendix 2. 1 = Oost-Flevoland-noord, 2 = Eemmeer, Nijkerkernauw, Nuldernauw e.o., 3 = Lauwersemeer, 4 = Wieringermeer, 5 = Arkemheen, Putterpolder en randmeren, 6 = Veluwemeer (Polders Harderwijk-Elburg), 7 = Het Bildt, 8 = Schouwen-Duiveland, 9 = Noord-Beveland, 10 = Vughtse Gement, 11 = Texel, 12 = Oost-en-Westsdongerdeel, 13 = Noordoostpolder-west, 14 = Maasland Den Bosch-Oss, 15 = IJssel: Zwolle-Deventer, 16 = Drontermeer (Kamperveen, Polder Oosterwolde), 17 = Reeuwijkse Plassen, 18 = Gronings-Drentse Veekolonien, 19 = Wonseradeel en Workum, 20 = Alblasserwaard, 21 = NW Overijssel, 22 = Staphorsterveld en Haerster- en Gennerbroek, 23 = Groningse Noordkust, 24 = Krimpenerwaard, 25 = Lopikerwaard, 26 = Maas Well-Mook, 27 = Overflakkee.

Table 2.1 Numbers and trends of Bewick's Swans at internationally important sites in the Netherlands for winters 1985–86 to 2002–03. Trend codes: + increasing (33%–100% in 10 years), - declining (<50% in 10 years), X lack of data, (f) fluctuating (no significant trend). From van Roomen *et al.* 2004.

Site	Average maximum	Maximum 2002–03	Month of max.	Total trend
Oost-Flevoland-noord	1,700	889	Nov	(f)
Eempolders en Eemmeer	1,600	2,224	Feb	(f)
Lauwersmeer, Anjumerkolken en Kollumerland	1,500	60	Jan	–
Wieringermeer	1,300	1,844	Dec	+
Arkemheen en Putterpolder	1,200	1,600	Feb	(f)
Harderwijk-Elburg polders en Veluwemeer	1,200	1,272	Nov	(f)
Het Bildt	1,100	250	Nov	X
Schouwen–Duiveland	990	415	Nov	(f)
Noord-Beveland	890	73	Oct	(f)
Vughtse Gement	690	1,977	Jan	+
Texel	680	825	Nov	+
Oost- en Westdongeradeel	670	2,198	Nov	+
Noordoostpolder-west	610	354	Nov	–
Maasland Den Bosch-Oss	560	696	Feb	(f)
IJsseldal Zwolle-Deventer	430	235	Dec	–
Kamperveen, Polder Oosterwolde en Drontermeer	410	806	Dec	–
Reeuwijkse Plassen e.o.	410	277	Feb	+
Gronings-Drentse Veenkoloniën	400	223	Dec	(f)
Wonseradeel en Workum	380	167	Oct	(f)
Alblasserwaard	370	362	Jan	(f)
Northwest Overijssel	370	227	Jan	–
Staphorsterveld en Haerster- en Gennerbroek	360	691	Jan	–
Noordkust Groningen	350	5	Nov	–
Krimpenerwaard	270	505	Jan	(f)
Lopikerwaard	200	260	Dec	(f)
Maas Mook-Ammerzoden	150	131	Jan	–
Overflakkee	98	128	Dec	–

compared with around 1,500 in the early to mid-1990s (Koffijberg *et al.* 1997; van Roomen *et al.* 2002, 2003, 2004).

The increase in the number of sites of international importance for Bewick's Swans in the Netherlands during the second half of the 20th century was quite remarkable. Only five such sites were cited during the mid-1970s (Atkinson-Willes 1976) and 13 during the 1980s (Monval & Pirot 1989). The change partly reflects more frequent coverage, since only January counts were available for many of the Dutch sites before the Dutch Bewick's Swan Project was initiated in 1982, whereas the distribution of the swans in the Netherlands changes during the winter season (van Roomen *et al.* 2002, 2003). The most recent estimate of 27 key sites in the Netherlands (van Roomen *et al.* 2004; Table 2.1) therefore is to some extent due to monthly counts monitoring sites used by the swans throughout the winter, but also can be attributed to the increase in population sustained mainly by Dutch wintering sites in recent years. With the recent decline, some areas which were colonised by the end of the 1990s, as the population expanded, are now being abandoned, particularly in the southwest of the country (K. Koffijberg pers. comm.).

Britain. Bewick's Swans appear to have been rare in England and Wales during the 19th century, but they were reputedly common in Scotland, especially on the islands of Tiree and North Uist (Baxter & Rintoul 1953, Ogilvie 1969). There were also reports of a regular autumn and spring passage through Shetland, although this seems a long way north for birds migrating from mainland Europe to Ireland. Although it is likely that there has been a southward shift in the distribution of Bewick's Swans wintering in Britain during the 20th century, it has been suggested that some of the historic reports may have been due to misidentification of Whooper Swans from the Icelandic breeding population, which do winter mainly in Scotland and Ireland (Cranswick *et al.* 1996; Rees *et al.* 1997b). However, in a recent reassessment of the literature, John Bowler (pers. comm.) concluded that the early records are accurate despite the current scarcity of the species in Scotland, because of the detailed nature of the reports (including descriptions of the later arrival dates for Bewick's compared with Whooper Swans; Anderson 1913), and particularly because specimens were collected from Ardnamurchan and Mull in 1879–80, a winter in which the swans were unusually abundant along both the west and east coasts of Scotland (Harvie-Brown & Buckley 1892). Bewick's Swans continued to occur in good numbers on Tiree until the Second World War, but they ceased visiting the island from the early 1940s onwards (Boyd 1958).

Elsewhere in Britain, some increase in numbers in the southern part of the country was observed during the 1920s, and they were recorded in Glamorgan in 1923 (Acland 1923). The first noticeable influx to England occurred during the 1938–39 winter, probably due to cold weather on the Continent encouraging the birds to move westwards. The first records for the fens of East Anglia occurred at this time, and this led to a pattern of the swans regularly wintering in southeast England (Witherby 1939; Ogilvie 1969). Numbers increased to only a few hundred during the 1950s, with the exception of the 1955–56 winter when cold weather elsewhere in Europe again resulted in an influx to Britain and numbers on the fens rose to around 1,600 birds. Peak counts in subsequent years usually remained under 500 until the 1965–66 season, when 1,300 to 1,500 were recorded in Britain, but rose steadily during the 1970s and 1980s at the time when the northwest European population was increasing substantially (Figure 2.1; Section 2.1.1).

In the middle of the 20th century, most swans wintering in Britain were found on the Ouse Washes, on the Cambridgeshire/Norfolk border, and this has since remained the single most important site for the species in the wintering range. Only 200 Bewick's Swans were seen at the site during the 1950s, but counts reached 1,000 in 1971, 2,000 in 1979, 5,000 in 1985 and peaked at 6,160 individuals in January 1987, representing some 33% of the northwest European population at that time. Consistently large numbers have used the Ouse Washes in mid-winter ever since. Although there appears to have been a decrease in population size in recent years, the five-year mean count for the Ouse Washes remained high at 5,277 birds in winters 1998–99 to 2002–03 inclusive (Cranswick *et al.* 2005; Figure 2.1; Table 2.2). Numbers again exceeded 6,000 birds in winter 2003–04, and reached 7,491 in January 2005 (D. Stevens pers. comm.).

Although Britain is at the southwest edge of the Bewick's Swan's range in Europe, the status of other sites of national and international importance for the species in the country has remained relatively constant over the years. Here the species congregates in large numbers at just a few lowland sites (Figure 2.11), which are covered routinely by the national Wetland Bird Survey (WeBS) (Table 2.2). Most have received Bewick's Swans since the 1960s and, although numbers may have fluctuated, these have been the main haunts for the species in Britain since the mid-20th century (Robinson *et al.* 2004). A main exception is Martin Mere WWT and the nearby Ribble Estuary in northwest

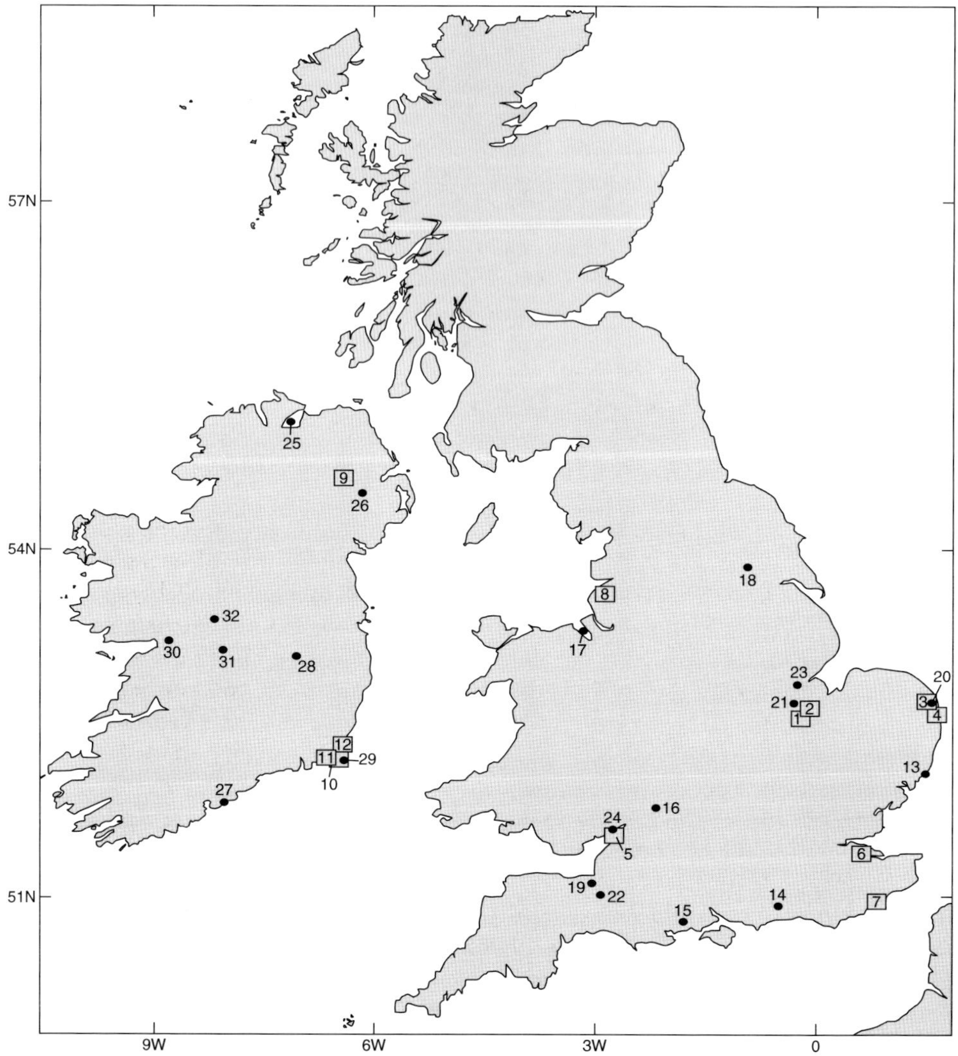

Figure 2.11 Key sites for Bewick's Swans wintering in Britain and Ireland. From Hunt *et al.* 2000, Robinson *et al.* 2004, Cranswick *et al.* 2005; Appendix 2. Rectangles = Sites of international or national importance in the five years to 2003–04; dots indicate additional sites of national importance during the late 1990s. 1 = Ouse Washes, 2 = Nene Washes, 3 = St. Benet's Levels, 4 = Breydon Water & Berney Marshes, 5 = Severn Estuary, 6 = Medway Estuary & Marshes, 7 = Walland Marsh & Dungeness Gravel Pits, 8 = Martin Mere & Ribble Estuary, 9 = Loughs Neagh & Beg, 10 = Tacumshin Lake, 11 = The Cull & Killag, 12 = Wexford Harbour & Slobs, 13 = Alde-Ore Complex, 14 = Arun Valley, 15 = Avon Valley, 16 = Coombe Hill Canal, 17 = Dee Estuary (England/Wales), 18 = Lower Derwent Ings, 19 = Durleigh Reservoir, 20 = Hickling Board, 21 = Ludham How Hill, 22 = Somerset Levels & Moors, 23 = The Wash, 24 = Walmore Common, 25 = The Lough Foyle/Swilly Complex, 26 = River Lagan, Flatfield, 27 = Ballycotton to Shanagarry, 28 = Barrow Upper Floodplain: Monasterevin to Athy, 29 = Lady's Island Lake.

Table 2.2 Maximum numbers of Bewick's Swans recorded during WeBS counts at sites of national and international importance for the species in Britain and Northern Ireland since 1996–97, with mean value for the five years to 2002–03. From Pollitt *et al.* 2003, Cranswick *et al.* 2005.

	96/97	97/98	98/99	99/00	00/01	01/02	02/03	5-year mean
Sites of international importance in the UK								
Ouse Washes	4,997	4,257	5,129	5,649	4,693	5,735	5,177	5,277
Nene Washes	863	2,585	723	327	1,100	347	1,068	713
Severn Estuary	555	393	287	216	272	310	345	286
Martin Mere/Ribble Estuary	669	368	144	163	322	296	315	248
St. Benet's Levels	286	161	126	209	206	147	289	195
Sites of national importance in the UK								
Medway Estuary	32	15	302	42	(0)	(0)	(0)	(172)
Breydon Water & Berney Marshes	476	231	210	132	186	85	240	171
Walland Marsh & Dungeness Gravel Pits	324	306	256	64	10	180	221	146
Somerset Levels	285	68	(120)	(117)	(146)	(108)	(69)	(146)
Dee Estuary	107	79	48	56	(118)	156	(70)	109
Lower Derwent Valley	139	81	21	61	?	?	?	76
Arun Valley	68	98	52	78	71	?	?	73
Site of all-Ireland importance in Northern Ireland								
Loughs Neagh & Beg (Northern Ireland)	117	77	53	16	102	19	6	39

England, which saw a substantial increase in Bewick's Swan numbers during the 1980s to a peak of 1,046 in autumn 1990, but numbers have since declined to a few hundred birds, possibly due to interspecific competition with the increasing numbers of Whooper Swans in the area (Rees & Bowler 1996). The Medway Estuary in Kent is included because of an exceptional count of 302 swans in 1999. However, the swans have used other sites in the county – Walland Marsh for feeding and Dungeness Gravel Pits to roost – since at least the 1970s. Varying numbers on the Nene Washes, Cambridgeshire (which usually receives over 1,000 birds), is attributable to birds moving to and from the Nene from the nearby Ouse Washes. The Severn Estuary (including Slimbridge) still receives internationally important numbers in mid-winter, with peak counts of around 300 birds in the early 21st century.

Ireland. Ireland is historically a major wintering area for the species but, as in Scotland, there was a big decline in numbers in the early part of the 20th century (Kennedy *et al.* 1954; Ogilvie 1969). The first major census of Bewick's Swans in Ireland was not made until the 1975–76 season, when 1,900 birds were recorded and the total wintering population was assessed at 2,000–2,250 (Merne 1977). This figure was higher than expected at the time, not only because it greatly exceeded earlier incomplete counts (Ogilvie 1969) but because weather conditions were dry and mild, which should have minimised movements of birds to Ireland (Merne 1977). Poorter (1991) suggests that the numbers wintering in Ireland may have been even higher during the late 1950s and early 1960s because at least 7,000 and 6,000 birds appeared to pass through the Netherlands in the autumns of 1958 and 1961 respectively, but numbers in Britain were thought to be below 1,000 at this time.

Regular monitoring of Bewick's Swans throughout Ireland has been undertaken annually since 1994–95, when the Irish Wetland Bird Survey (I-WeBS) started monitoring the waterbird populations wintering in the Republic of Ireland. I-WeBS is run concurrently with the UK's WeBS scheme, which has monitored waterbirds in Northern Ireland since 1986-87. Together they provide all-Ireland information on Bewick's Swan numbers and distribution in this westernmost part of their range. Although small flocks of less than 20 birds may be found across Ireland, the key sites (>100 birds) are mostly in the southeast, notably at the three internationally important sites of Tacumshin Lake, the Cull and Killag, and Wexford Harbour and Slobs in the Republic of Ireland (Colhoun 1998, 2000; Robinson *et al.* 2004; Figure 2.11). Historically, Tacumshin Lake was used by fewer birds than the other two sites. Around 150–200 birds were recorded there regularly between the mid-1960s and 1980–81, but up to 470 were counted annually in the mid-1980s and late 1990s. The numbers at the three sites are closely related, however, with the Cull and Killag becoming important for the Bewick's Swans when the Wexford Slobs flock redistributed in the late 1970s (Robinson *et al.* 2004). It is thought that a single flock now uses Cull/Killag and Tacumshin Lake, together with the River Bride and Lower Blackwater River (Colhoun 1998).

Most Bewick's Swans in Northern Ireland are usually to be found at Lough Neagh and Lough Beg, where peak counts of over 100 birds were recorded in the late 1990s (Pollitt *et al.* 2003), although only up to six swans were seen there in winter 2002–03 (Cranswick *et al.* 2005; Table 2.2). The surveys have also shown that the counts can be rather variable, both for individual sites and for the all-Ireland totals, which may reflect weather conditions and food supply, particularly for the majority of the population, which winters further east in Europe.

Even allowing for this variation, there has been a substantial decline in the number of Bewick's Swans in Ireland in recent years. Over 1,000 birds were recorded in 1995–96 and 1996–97, but counts returned to low levels (maximum of 520 in the Republic of Ireland; 133 in Northern Ireland) in 1997–98 (Colhoun 2000). Only 382 were found in Ireland (mostly in Wexford) during the international census in January 2000, and only flock sizes of less than 50 Bewick's Swans have been seen in Northern Ireland since then (G. McElwaine pers. comm.).

Denmark. Denmark is a major staging post for Bewick's Swans during the autumn migration, and sites in West Jutland are thought to have been important since the early part of the 20th century, especially Ringkøbing Fjord and the now largely reclaimed Lake Fiilsø (Poorter 1991). They were rarely recorded elsewhere in Denmark before the late 1930s, when they began to appear on passage in Tudea Lagoon and at Lake Tissø in West Sealand (Schlede 1961; Poorter 1991). More frequent censuses in recent years indicate large fluctuations in the numbers present in mid-winter, which can generally be explained by weather conditions. Bewick's Swans have been seen in North Jutland each spring since 1958, initially in very small numbers, but with at least 100 birds in the Osterrild and Lønnerup Fjords since 1962, and more than 300 birds at Lønnerup Fjord in 1994 (Rasmussen *et al.* 2000; Appendix 2).

It seems that the birds used Denmark increasingly during the 1960s, with birds also occurring in North Jutland in mid-winter from 1960–61, and regularly in autumn from 1964, but again mid-winter counts depend strongly on the weather conditions (Poorter 1991). Monthly countrywide counts undertaken in Denmark in the early 1990s found that

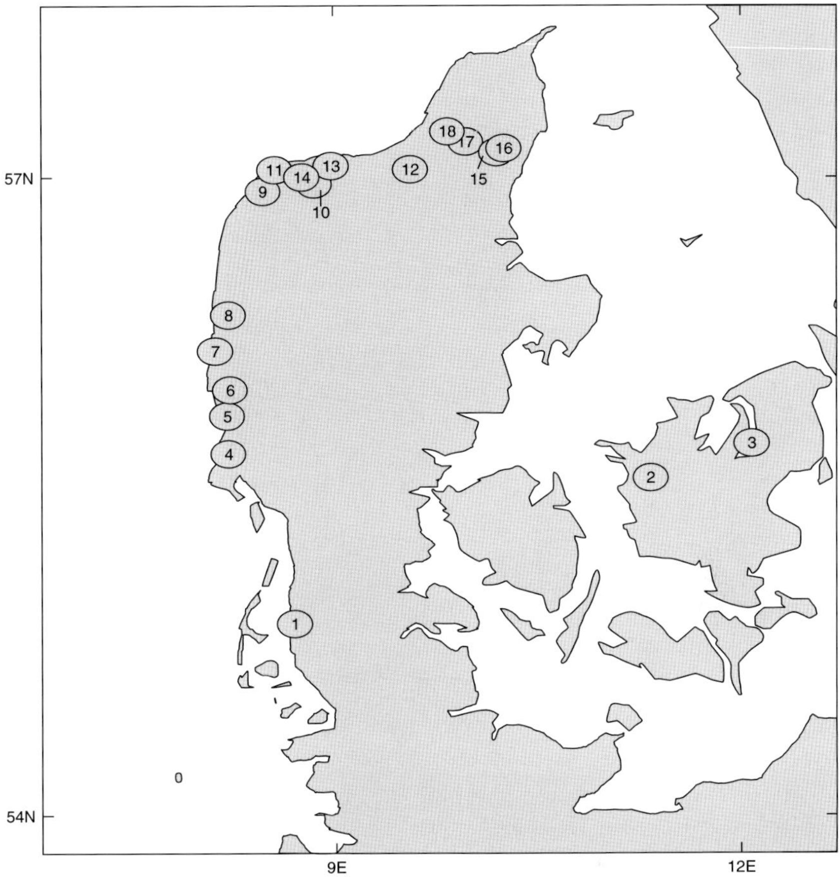

Figure 2.12 Key sites for Bewick's Swans in Denmark. From Appendix 2. 1 = Tøndermarsken Magisterkog and Rudbøl Sø, 2 = Tissø, Lille Åmose & Hallenslev Mose, 3 = Roskilde Fjord, Selsø & Kattinge Søerne, 4 = Fiilsø, 5 = Tipperne, 6 = Ringkøbing Fjord, 7 = Stadil Fjord & Vest Stadil Fjord, 8 = Nissum Fjord, 9 = Skjern Å, 10 = Vejlerne (west), 11 = Nors So & Vendlet So, 12 = Ulvedybet and Nibe Bredning, 13 = Vejlerne (east), 14 = Lønnerup Fjord, 15 = Bolle and Try meadows, 16 = Droninglund, 17 = Tylstrup, Arup Holm and Hovsør Røn, 18 = Store Vildmose, Ryå & Stavad Enge.

numbers range from 4,000 to 6,000 birds during peak passage periods in autumn (mid-October to mid-November) and spring (mid- to late March), with about 1,000 birds present in mid-January (Laubek 1995a). The major concentrations were again found in west and north Jutland, the most important areas being Fiilsø, Ringkøbing and Vest Stadil Fjords (more than 1,000 swans in 1994), Vejlerne (up to 400), Ulvedybet (more than 4,000), Store Vildmose (1,000) and Bolle Enge (2,500), with these last two sites being new staging areas for the species (Rasmussen *et al.* 2000; Appendix 2; Figure 2.12). In the 1960s, swans at Ringkøbing and Nissum Fjords were found mainly on aquatic habitats, but by the early 1990s they had shifted partly or completely onto farmland, perhaps due to eutrophication at the wetland sites (Laubek 1995a, b).

Germany. The eastern part of Germany has been an important staging area for Bewick's Swans since the 1950s, particularly the Baltic region in the northeast (Schubert 1974,

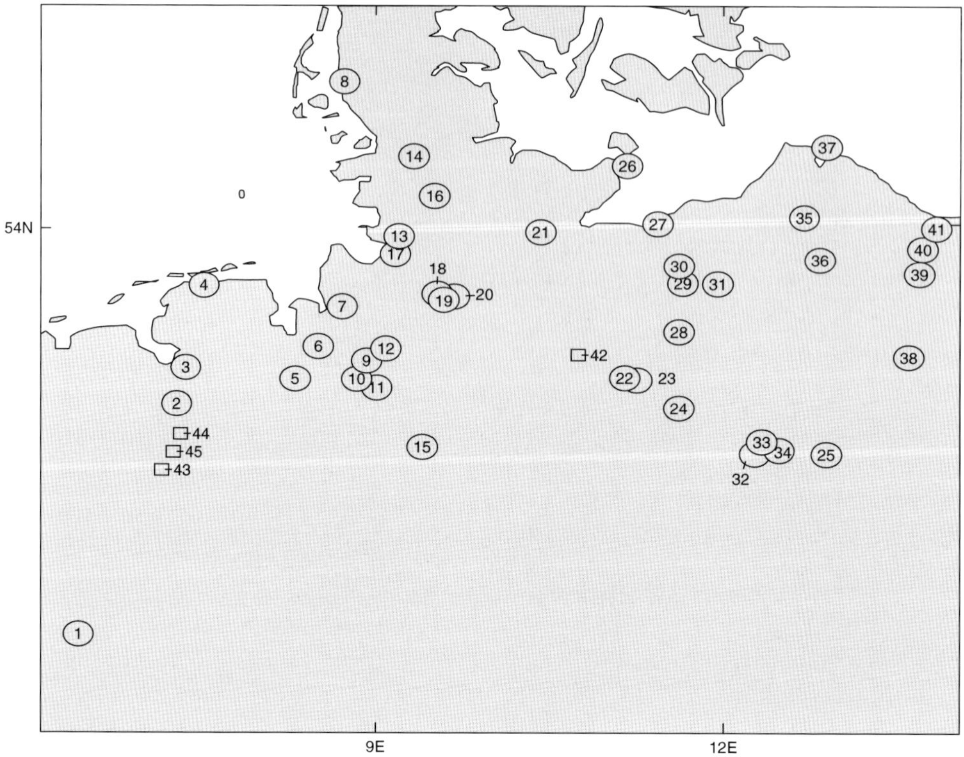

Figure 2.13 Key sites for Bewick's Swans in Germany. Ovals = used mainly as staging sites; rectangles = used mainly as wintering sites. From A. Degen pers. comm. and Appendix 2. 1 = Unterer Niederrhein (Lower Rhine area), 2 = Ems lowlands (Lingen - Papenburg), 3 = Rheiderland, 4 = Wadden Sea (Niedersachsen), 5 = Hunte lowlands, 6 = Unterweser (Lower Weser), 7 = Lowlands near Bramel, 8 = North of Nordfriesland, 9 = Hamme lowlands, 10 = Wümme lowlands/St. Jürgensland, 11 = Wümme lowlands/Fischerhude, 12 = Huvenhoopsmoor & Breddorfer Wiesen, 13 = Kudensee, 14 = Lowlands of the Rivers Eider, Treene & Sorge, 15 = Aller lowlands (Buchholz - Etelsen), 16 = Haaler-Au, 17 = Elbe from Stade to Otterndorf (lower tidal reaches of the River Elbe), 18 = Elbe, Bishorst to River Krückau, 19 = Elbe, Haseldorfer Marsh, 20 = Elbe, Wedeler Marsh, 21 = Warder See, 22 = Elbe from Schnackenburg to Hohnstorf (Niedersachsen), 23 = Elbe from Polz to Boizenburg (Mecklenburg-Vorpommern), 24 = Aland-Elbe lowlands, 25 = Elbe from Werben to Gaarz (Brandenburg), 26 = Ostküste Oldenburgs, 27 = Wismarbucht & Salzhaff, 28 = Lewitz fishponds, 29 = Obere and Mittlere Warnow with Glöwe and Mildenitz, 30 = Lake Schwerin, Lake Dambeck & Lake Neukloster-Warin, 31 = Mildenitz, Lake Gägelow and Lake Upahl, 32 = Lowlands of Havel, Lake Schollene & Lake Gülpe, 33 = Lowlands of Dosse and Jäglitz, 34 = Unteres Rhinluch, Lake Dreetz Lake, 35 = Recknitz and Trebel lowlands, 36 = Kummerower See, 37 = Vorpommersche Bodden, 38 = Greifswalder Bodden, 39 = Lake Putzar & Lake Galenbeck, 40 = River Peene lowlands, 41 = Island Usedom, 42 = Elbe floodplain (Schnackenburg - Lauenburg, including Amt Neuhaus), 43 = Groß Fullener marsh, 44 = River Ems lowlands near Laten, 45 = Wesuwer Brook.

Klafs & Stübs 1987), and remains so to this day. Several hundred swans use Greifswalder Bodden (a bay on the southern shore of the Baltic Sea) as a staging site, and the Lewitz fishponds with adjacent lowland wet meadows of the River Elde are also used regularly, particularly during spring migration (Scott & Rose 1996, Unselt *et al.* 2000, Appendix 2, Figure 2.13). Southwest of Lewitz, the Elbe floodplain from Schnackenburg to Lauenburg was recognised as an important wintering area for the birds in the 1980s, and a number of

Slimbridge-ringed birds were seen both staging and wintering in the area (H. Eggers pers. comm.). This part of the Elbe formed the border between the Federal Republic of Germany (FDR) and the former German Democratic Republic (GDR), so counting the swans and reading their rings on the GDR side of the river had to be undertaken with care during the Cold War period. Nevertheless, Helmut Eggers' efforts yielded nearly 200 sightings of over 100 ringed swans in the northern (GDR) part of the Elbe floodplain during the early 1980s and, following the unification of Germany, more than 2,000 Bewick's Swans were recorded wintering across the floodplain during the 1990s (Scott & Rose 1996; Unselt *et al.* 2000).

In western Germany, only small numbers wintered in Holstein and on the banks of the Lower Elbe River before 1964 (Schmidt 1965; Dahms & Grosskopf 1978), but there was some increase in wintering numbers during the 1970s, and a substantial increase in numbers present during the spring migration. Key wintering haunts are now mainly in the Niedersachsen region, including the tidal reaches of the Elbe River, the lowlands of the River Ems from Lingen to Papenburg, and the Lower Weser River. The River Ems is the most important area for the swans in Germany, with up to 1,000 present in winter, and 2,310 birds present during the spring migration. The Lower Rhine also receives a few hundred birds (Unselt *et al.* 2000), and a major study of the swans in Niedersachsen in January 1995 found 890 Bewick's Swans in the area at that time (Degen *et al.* 1996).

Schleswig-Holstein is still used by the swans during their northbound journey in spring, when they stage at the Warder See (up to 300 swans), the lowlands of the Rivers Eider, Treene and Sorge (1,500) and the Pinneberg Elbe lowlands (220). Other spring staging areas include sites in Niedersachsen (e.g. the valley of the River Ems and the Lower Aller lowlands), the Elbe and Aland lowlands and northeastern Germany (e.g. Greifswalder Bodden and the Trebel lowlands). Over 3,500 Bewick's Swans were counted in Niedersachsen alone during the March 1995 survey (Degen *et al.* 1996), and 3,000 Bewick's swans were counted in the Lower Elbe Valley in 1997 (Unselt *et al.* 2000), which highlights the importance of German staging areas for the species.

Belgium. Each winter, several hundred Bewick's Swans visit traditional wintering areas in northwest Belgium, regularly reaching more than 400 birds. The most important area is close to the border with the Netherlands, with up to 553 at Krekengebied Noord-Oost-Vlaanderen (a complex of former brackish and freshwater sea inlets in a polder landscape; Anselin *et al.* 2000) during winter 1999–2000. In most winters, quite large groups are also recorded in the Oostkustpolders between Oostende, Brugge and Knokke (with up to 200 in 2002–03) and the IJzervallei (up to 195 in 1999–2000) (K. Devos, E. Kuijken and C. Verscheure pers. comm.). Tens of birds are found at several other sites, such as the Scheldepolders near Antwerpen, and further inland at the Maasvallei and Harchies. Numbers of wintering Bewick's Swans in Belgium have clearly increased since 1990, with no evidence for the decline noted in other parts of Europe in the early 21st century (K. Devos pers. comm.; Figure 2.14).

France. Further south, a relatively isolated herd of some 80 swans (135 in 1979; 120 in 1997) has wintered in the Camargue in southern France since the late 1960s (Monval & Pirot 1989; Deceuninck *et al.* 2000). In recent years, a small but regular flock has also wintered on the three adjacent lakes of Lac Amance, Lac du Temple and Lac d'Orient, near Troyes in northeast France, with numbers reaching 71 in November 2003 (F. Croset pers. comm.).

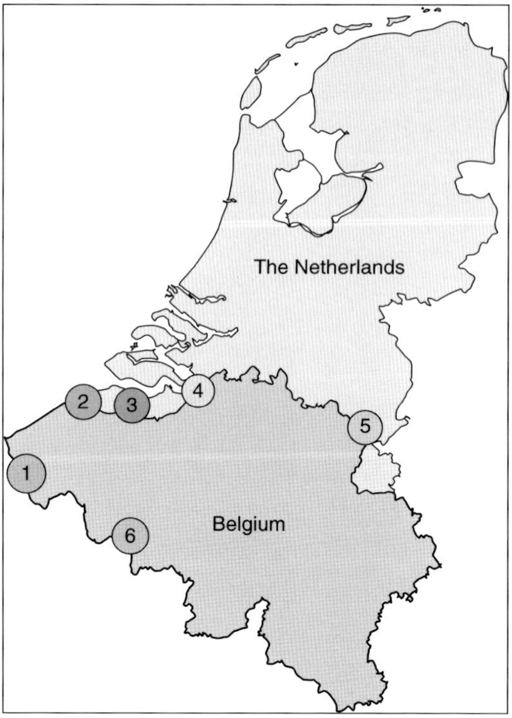

Figure 2.14 Key sites for Bewick's Swans in Belgium. From K. Devos pers. comm.; Appendix 2. 1 = Ijzervallei, 2 = Oostkustpolders, 3 = Krekengebied, 4 = Scheldepolders Antwerpen, 5 = Maasvallei, 6 = Harchies.

2.5.2 Winter distribution: eastern population

Bewick's Swans from the eastern parts of the breeding range migrate to wintering areas in China, Japan and Korea. They are sighted occasionally in Kutch district, northwest India, and Sind province, Pakistan (Ogilvie 1972b). Any changes in distribution over time are difficult to ascertain because until very recently swan censuses did not separate Bewick's from Whooper Swans.

Korea. Reports of a major decrease in the numbers and distribution of migratory swans wintering in Korea since the Second World War are likely to be true for both species, since they comprise most of the swans wintering in the country; the Mute Swan is a rare winter visitor to Korea (Won 1981). In the early part of the 20th century, thousands of yellow-billed swans wintered in Korea, using ponds, lakes and reservoirs along the northeastern and northwestern coasts upon arrival in early winter, moving south to winter on ponds and wetlands in Gyeongsangnam-do as the weather became colder. Some would move to the southern islands, such as Jindo, when waters froze elsewhere on the Korean peninsula.

The Korean War of 1950–53, when soldiers and civilians illegally hunted the local wildlife, had a disastrous effect on a number of species. Whether the swans were affected is unclear, because there are very few papers on birds in Korea from this or earlier times, but the limited information available suggests that they were not present in huge numbers

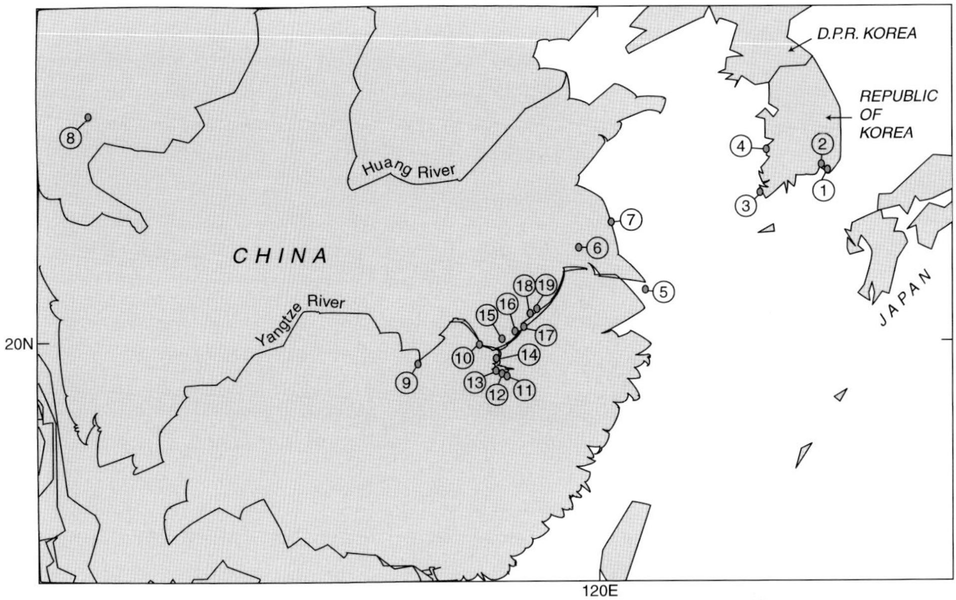

Figure 2.15 Sites of international importance for Bewick's Swans in China and Korea. From Appendix 3. 1 = Nakdong Estuary, 2 = Joonam reservoir, 3 = Jindo Island, 4 = Upper Geum River, 5 = Chongming Island, Shanghai, 6 = Gaoyou Lake, 7 = Yancheng Nature Reserve, 8 = Qinghai Lake, 9 = Donting Lake wetlands, 10 = Wang Lake, 11 = Nanchang City sites (Bei Shen Lake, Chang Lake, Poyang Lake and San Lake), 12 = Poyang Lake NNR sites (Can Dou Lake, Chang Hu Chi and Dahuchi), 13 = Shangrao City sites (Baishazhou, Kangshan Dyke, Nanjiang Lake-Dalian Lake, Raopoyangdalianzi Lake, Raopoyanglian Hujiaoxi, Raopoyangzhu Lake-Hanqi Lake-Qi Lake, Shi Lake and Yugankangshanei Lake), 14 = Jiujiang City sites (Jishan Lake, Saicheng Lake, Xin Miao Lake) 15 = Ergu Fan, Daguan Lake, 16 = Shengjin Lake region (Chi'an Xinshengyu, Shang Lake Bailianwei-Dayangou, Shang Lake Da Zhou, Shengjin Lake, Xia Lake Tan Bu Cha, Xia Lake Tang Tain Cha, Xia Lake, Xiaoluzui-chi'an and Xiaoxihu), 17 = Wuchang Lake region (Wuchang Lake, Xia Lake Guo He Cun and Xiahuliuhu), 18 = Baidang Lake region (Baidang Lake and Wu Jin Du), 19 = Fengsha Lake region (Fenlincun and Zhou Tan Zhen).

during or after the war (Park 2002, N. Moores pers. comm.). More recently, the development of agriculture during the 1960s and 1970s destroyed many coastal wetlands and inland waters or marshes, including those in the Hyopcheon and Changyong areas, which had been protected as National Monuments since the 1930s (Won 1981). By the early 1980s, most Bewick's and Whooper Swans wintering in Korea congregated in the southeast part of the Korean peninsula in the Nakdong Estuary, and also in smaller numbers at the Joonam Reservoir, with fewer than 200 yellow-billed swans visiting Jindo Island (Won 1981, BirdLife International 2004b, Figure 2.15). There are too few accurate counts to confirm current trends, but Whooper Swan numbers have increased substantially, at least in the Nakdong Estuary, during the late 1990s, with mid-winter counts exceeding 2,000 birds in recent years. Bewick's Swan counts have declined over the same period, not only in the Nakdong Estuary and wetlands of the Nakdong floodplain, but also on the Guem River where small numbers previously wintered (N. Moores *in litt.*).

China. In China, Bewick's Swans have been recorded as far south as Guangdong Province, and inland up the main river valleys. Poyang Lake on the Yangtze River was an important site historically (Ogilvie 1972b) and remained an important wintering area during the 1980s with 900–2,100 swans recorded in winter from 1989–90 to 1991–92 inclusive (Lu 1990). More recently, there was a phenomenal count of over 60,000 Bewick's Swans made during an aerial survey of the lake in 2001 (Ji *et al.* 2001; Section 2.1.2). Although fewer swans (42,843 Bewick's) were recorded in the Poyang Lake region during the WWF-China survey of the middle and lower reaches of the Yangtze floodplain in early 2005, a total of 65,114 Bewick's Swans was counted – almost all in Jiangxi and Anhui Provinces (Barter *et al.* 2004 and Barter *in litt*; Appendix 3; Fig. 2.15). Thus it seems that much higher numbers winter in China than had been previously recognised.

Chongming Island in Shanghai used to be an important Chinese wintering site, with 3,000–3,500 birds counted in consecutive winters between 1986 and 1989 (Yu 1991, 1992; Ma & Cai 2000) but none during the WWF-China surveys in 2004 and 2005. Several hundred were recorded at Qinghai Lake in the late 1980s (Miyabayashi & Mundkur 1999; Appendix 3). Other wintering grounds included Changxin Island and Hengsha Island in Shanghai, Yancheng, Gaoyou Lake and Dongtai in Jiangsu Province, Dongting Lake in Hunan Province, the Han River valley and Hong Lake in Hubei Province (Ma & Cai 2000; Figure 2.15). Swans are also sometimes seen in Zhejiang, Fujian, Guandong and Sichuan provinces and in Taiwan (Ma & Cai 2000).

Japan. Data collected on the distribution of Bewick's Swans in Japan during the 1970s showed that they were reasonably widespread in this country, occurring in five prefectures: Fukushima (33%), Shimane (23%), Miyagi (20%), Niigata (9%) and Akita (6%) (Horiuchi 1981). Numbers wintering in Japan were relatively consistent at this time with 1,000–2,000 birds recorded during the 1970s and 1,954 counted during the mid-January national census in 1980, although numbers have since increased substantially (see Section 2.1.2). More recent surveys have indicated that swans still frequent these five prefectures, and that they are now also found in good numbers elsewhere in the country. Twenty-six sites in Japan have been identified as being internationally important staging or wintering areas for eastern Bewick's Swans, in that they regularly receive at least 1% of the eastern population (Miyabayashi & Mundkur 1999; Albertsen & Kanazawa 2002; Figure 2.16a,b). Supplementary food is provided at a number of these sites (Chapter 8). The number of swans wintering at Lake Inawashiro in the northwest part of Honshu has risen markedly since 1965, when the Lake Inawashiro Wild Swans Conservation Association began protecting and feeding swans (mainly Bewick's) wintering in the vicinity (Ohmori 1981). The site is now of international importance for the species (Albertsen & Kanazawa 2002).

Sites on the northern island of Hokkaido are used regularly as staging posts for birds wintering on Honshu Island, during both the autumn and the spring migration (Matsui *et al.* 1981). Thus 20,000 swans may stage at Lake Kutcharo-ko and Lake Onuma in north Hokkaido, but ice formation forces them south in mid-winter. On average, only 2%–7% of the Bewick's Swans in Japan spend the winter in Hokkaido each year (Environment Agency of Japan 1999; Albertsen & Kanazawa 2002).

The main wintering areas for Bewick's Swans in Japan are Lake Hyoko in Niigata Prefecture (with over 3,000 Bewick's Swans in mid-winter), the Mogami-gawa River in Yamagata Prefecture (over 2,000 Bewick's and a further 5,000+ Whooper Swans), around Niigata City and Toyostato Town in Niigata Prefecture and the Hazama-gawa River in Miyaga Prefecture (Albertsen & Kanazawa 2002). Large numbers of swans can be found

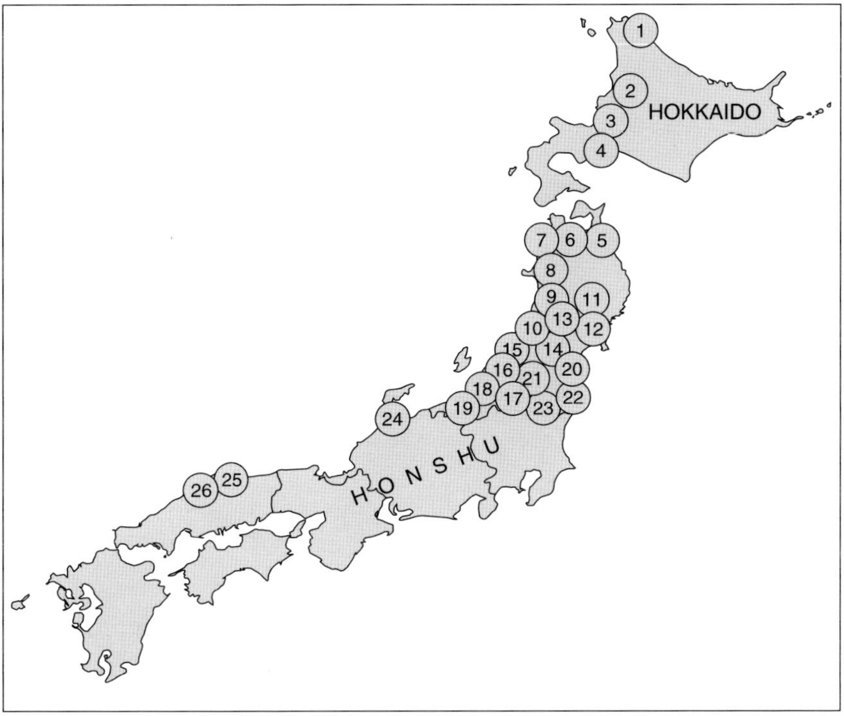

Figure 2.16a Sites of international importance for Bewick's Swans in Japan during migration or in mid-winter. From Miyabayashi & Mundkur 1999, Albertsen & Kanazawa 2002. 1 = Lake Kutcharo-ko, 2 = Lake Fukurojinuma, 3 = Lake Miyajimanuma, 4 = Lake Utonai-ko, 5 = Lake Ogawara-ko, 6 = Ezogatate Reservoir, 7 = Mawarizeki Reservoir, 8 = Otomonuma Reservoir, 9 = Hachirogata Rice Fields, 10 = River Mogami-gawa, 11 = River Hazama-gawa, Wakayanagi Town, 12 = River Hazama-gawa, Toyosato Town, 13 = Kitashinbo-ooike Reservoir, 14 = River Shiroishi-gawa, 15 = Lake Fukushimagata, 16 = Lake Toyanogata, 17 = Lake Hyoko, 18 = Lake Sakata, 19 = River Shinano-gawa, 20 = River Abukuma-gawa, 21 = Lake Inawashiro-ko, 22 = Kamishigeoka Reservoir, 23 = River Natsui-gawa, 24 = Ouchigata Rice Fields, 25 = Lake Nakaumi, 26 = Nougi Rice Fields.

even further south, with about 2,000 birds wintering in the Tottori–Shimane area of western Honshu (Environment Agency of Japan 1999; Albertsen & Kanazawa 2002).

2.6 STAGING AREAS AND NETWORKS OF KEY SITES

2.6.1 Western flyway

A study of the energy requirements of Bewick's Swans during their annual migratory cycles emphasises the importance to birds undertaking long-distance migrations of having established sites at regular intervals along the flyway to replenish their nutrient reserves (Beekman *et al.* 1996a). The limited ability of large birds to carry the body fat needed to fuel flapping flight, together with the high energy cost of flying (Pennycuick 1989; Norberg 1996), means that Bewick's Swans are able to fly only an estimated 2,000km before needing to refuel for the next stage of their journey (Beekman *et al.* 2002). Since the breeding sites are some 3,500km from the wintering grounds, it was thought that the

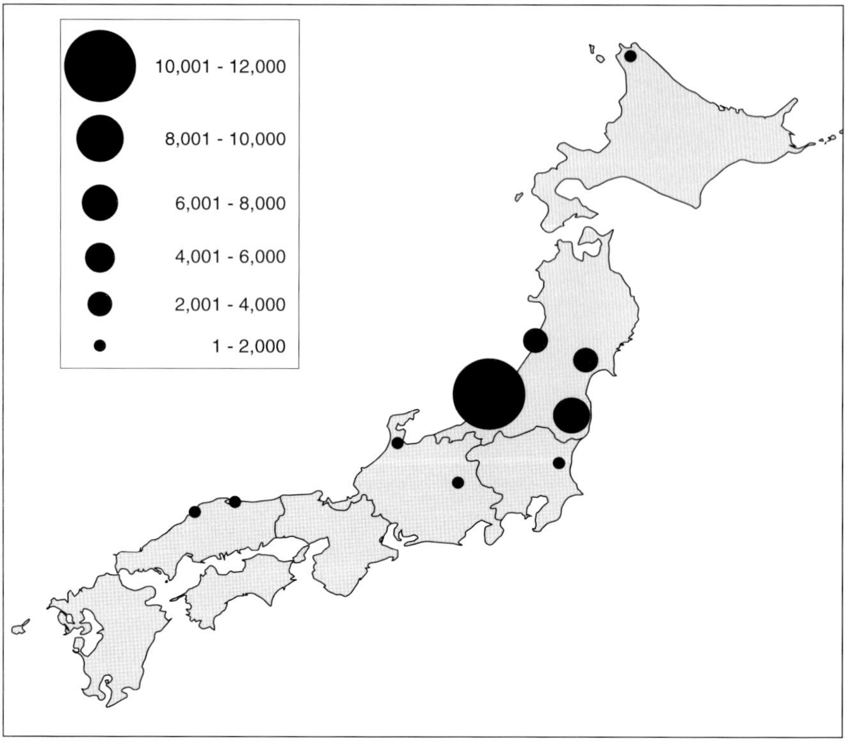

Figure 2.16b Bewick's Swan wintering sites in Japan. Size of dots indicates total number of birds recorded in each prefecture in mid-January 1999, while the location of the dots shows the most important site in each prefecture. From Albertsen & Kanazawa 2002.

swans need to stop at least twice to feed during the autumn and spring migrations (Beekman *et al.* 1996a; Chapter 3). Satellite-tracking studies have since confirmed that the swans make at least two stops during spring migration, but in autumn they may use just one migratory site (in the Baltic region) for feeding between the breeding grounds and wintering range (Beekman *et al.* 2002; WWT unpubl. data; see Chapter 3). More frequent stop-over sites, spaced at shorter distances, might be needed in certain circumstances, for example if the birds encounter severe weather conditions during the spring migration, or if a pair is accompanied by their offspring. Additional sites are also likely to be important for resting for a few hours during migration, although these may hold relatively few individuals and for only short periods of time.

Northern Russia. In the early stages of the autumn migration, swans from the European northeast of Russia congregate on shallow coastal waters, large lakes and river estuaries between the Pechora Sea and the Yugor Peninsula. One of the most important areas may be the Pechora Delta, where 15,000 have been recorded in premigratory flocks in autumn (Beekman *et al.* 1994a, 1996b). Most records of autumn gatherings of swans were to the east of Kolokolkova Bay (Mineyev 1991; Figure 2.5). It seems likely, however, that more important moulting, breeding and staging areas exist than have been identified so far, particularly for the eastern population, which has been studied in less detail. Major shifts in the distribution of moulting flocks may also occur between years in response to the availability of aquatic vegetation in the large areas of open water used as moulting sites.

Satellite-tracking has shown that the swans tend to pass over the White Sea in northwest Russia without stopping in autumn (Beekman *et al.* 2002; WWT unpubl. data), but the Dvina Delta and Dvina Bay, near Arkhangelsk on the White Sea, are very important to the birds in spring. Counts of swans passing through the area in 1996 estimated that some 18,000 birds used Dvina Bay in May, where thousands of Bewick's Swans were seen feeding on Fennel Pondweed *Potamogeton pectinatus* and eelgrass *Zostera* spp. (Nolet & Drent 1998; Nolet *et al.* 2001b). This site, together with Matsalu Bay in Estonia, can hold over 50% of the population at any one time, and it seems that almost the whole of the population may use Dvina Bay for refuelling during spring migration (also see Chapter 4).

Gulf of Finland. The easternmost point of the Finnish Gulf, near St Petersburg, has long been recognised as a potentially important staging area because this is where swans migrating north along the Baltic coast in spring must head overland to reach the breeding grounds to the northeast. The journey from the Gulf of Finland to the White Sea, via Lake Ladoga and Lake Onega, is the shortest and most direct route to the breeding areas, although the value of the two lakes to the birds is still uncertain. The recovery of a Slimbridge-ringed bird from the Finnish Gulf (Rees 1991) reinforced the view that this might be a key migratory site, and in May 1987 I found a pair of swans (named 'Mr Big' and 'Small') among 305 Bewick's Swans feeding and sleeping along the north shore of the Finnish Gulf, in the Primorsk region near St Petersburg (Rees 1991). 'Mr Big' and 'Small' had spent the 1986–87 winter at Slimbridge and departed on their spring migration on 12 March, before being identified by bill pattern (leg rings also seen) in the flock near St Petersburg on 8 May 1987. They returned to Slimbridge the following winter, arriving on 2 December 1987. Some 1,000–2,000 swans (both Bewick's and Whooper) were seen sitting on the ice in the Finnish Gulf in late April 1989, presumably waiting for weather conditions to improve before continuing their northward migration (Beekman *et al.* 1996a). More recently, large numbers (counts of up to 5,000 birds) have been recorded there in spring during the 1990s (Sviridova 2000; Appendix 2). A high turnover of birds passing through the area may mean that most of the northwest European population depends on this site.

Estonia. Some of the most important of the staging sites are on the west coast of Estonia, where shallow bays and their surrounding fields hold large numbers of swans, particularly in spring but also in autumn. Of these, Matsalu Bay has long been recognised as a key site for Bewick's Swans on migration, and has been protected as a nature reserve since 1957. The Soviet government designated it a wetland of international importance under the Ramsar Convention in December 1976 and, following the official reinstatement of Estonian independence in 1991, it was re-designated as a Ramsar site by the Estonian government in 1994. The reserve now covers an area of 518.8 square kilometres, including the bay itself and coastal pastures (Paakspuu & Kastepold 1985; Kalamees 2000). The maximum number of migratory swans recorded each spring has declined since a combined count of 40,000 Bewick's and Whooper Swans, made on one day in 1957 (T. Kastepold and V. Paakspuu pers. comm.) to between 8,000–10,000 in 1963 and 1964 (Paakspuu 1967), and up to 6,000 birds (2,000 Bewick's and 4,000 Whoopers) during the mid-1980s (Rees & Bowler 1991). There was an increase in Bewick's Swan numbers on Matsalu Bay during the 1990s (with counts ranging from 3,000–13,500 birds between 1990-99), which may reflect the increase in total population size at this time (Kalamees 2000).

The development of a Bewick's Swan study in Estonia since 1991 has yielded much more detailed information on the swans' use of Estonian migratory sites in recent years (Luigujõe *et al.* 1996). Counts made between 1992 and 1994 found 14,500 Bewick's Swans at Matsalu Bay in spring (3,000 in autumn), 7,000 at Parnu Bay and the Parnu rivershed (1,000 in autumn), 500 on the west shore of Lake Peipsi (2,000 in autumn) and at least 500–1,000 individuals in the Pandivere region, River Emajõgi, Haapsalu Bay and the Väike-Väin Strait (Luigujõe *et al.* 1996; Figure 2.17). Lake Peipsi is in the northeast of the country, which may be a reason why it is more important for the swans in autumn than in spring. Certainly three of five Bewick's Swans tracked by satellite in autumn 1998 and all three birds tracked in autumn 2003 stopped at Lake Peipsi en route to the wintering grounds (Beekman *et al.* 2002; WWT unpubl. data; Chapter 3). The lake is situated on the Estonian border with Russia, where it is known as Pskov-Chudskoye Lake. Russian ornithologists confirm that the east (Russian) side is also important for the swans in autumn in some years, but that the use of the site by both Bewick's and Whooper Swans is variable. They suggest that annual variation in the number of migratory swans seen on the lake is influenced by water levels, with swans more likely to occur in years when the water levels are low (Borisov *et al.* 1999).

Lithuania and staging areas to the south. There are internationally important Bewick's Swan staging areas in the coastal wetlands of Lithuania, with up to 3,200 birds counted in the seasonally flooded meadows of the Nemunas River delta area in 1992, and up to 700 swans observed in the northern part of the Curonian (Kursiu) Lagoon (Švažas *et al.* 1999, Švažas 2001). In recent years, an average of 200–300 birds was estimated for both sites (Raudonikis 2004), and inland lakes in southern Lithuania support up to 100 Bewick's Swans each year (S. Švažas pers. comm.).

South of Lithuania, there is no indication that the swans use the lagoons on the Kaliningrad coast, despite the southern part of Kursiu Lagoon falling within the borders of the enclave (i.e. there were no important birds areas [IBAs] reported for Bewick's Swans in Kaliningrad in the list of Russian IBAs; Sviridova 2000). Poland is an important staging and wintering area for Whooper Swans, but only two sites have been described as being important for Bewick's Swans on migration. Up to 190 birds were recorded at the Ostrówek and Smolgulec fishponds in 1993, and over 400 birds at the Ślesin and Występ fishponds in the mid-1990s (Gromadzki & Wieloch 2000). Nevertheless, Bewick's Swan migration over Poland has changed in recent years; in the mid 20th century they were seen only along the Baltic coast, but the swans now fly up to 200km into the country to feed at inland sites such as the fishponds on the Notec River (M. Blank, pers. comm.). Several sites in Denmark, Germany and the Netherlands are used as staging areas by swans wintering further south or west. These may also be the final winter destinations for smaller numbers of birds that remain at the sites all winter (Scott & Rose 1996; Appendix 2; J. van Gils pers. comm.).

2.6.2 Eastern flyway

The Bewick's Swan's migration and use of staging sites in the Far East has been studied by fitting birds with neck collars since 1974 (Kondratiev & Kishchinski 1987), and by the satellite tracking of individual birds caught in Japan since 1990 (Kamiya & Ozaki

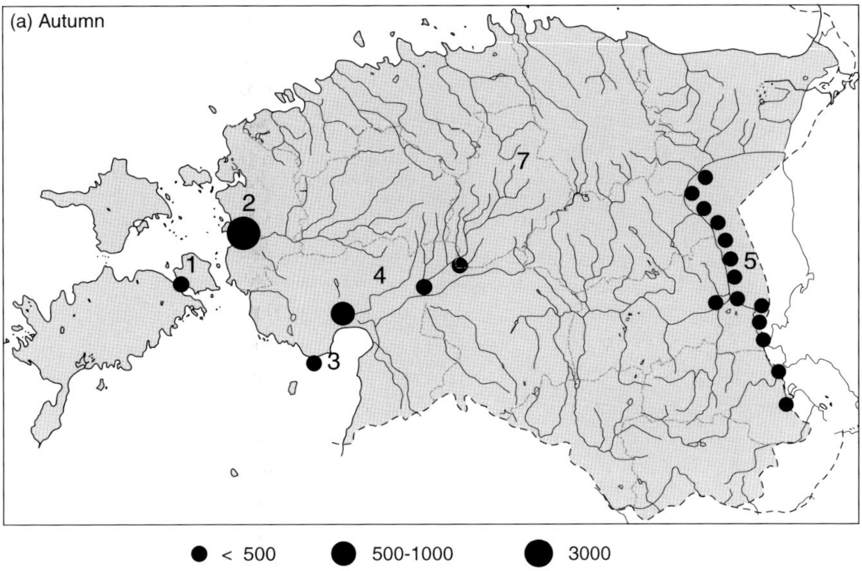

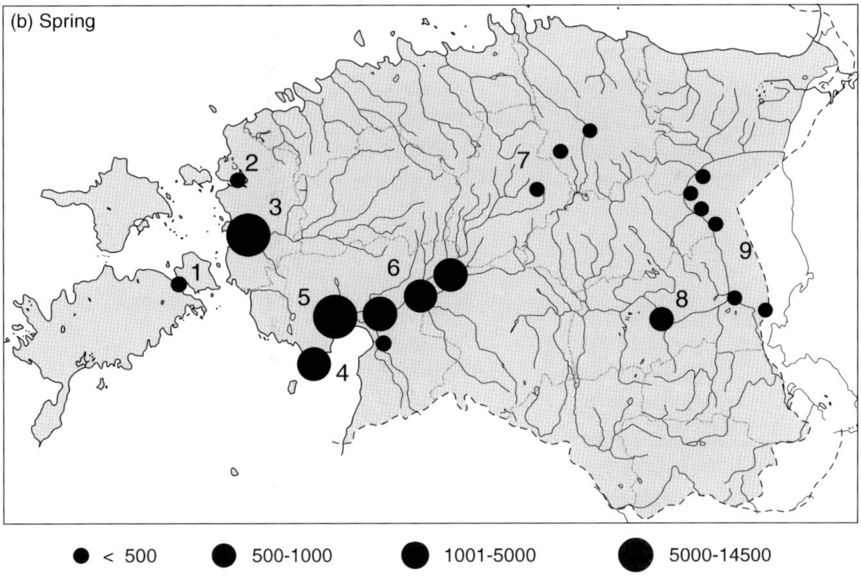

Figure 2.17 Key autumn and spring staging sites for Bewick's Swans in Estonia. Dots indicate maximum numbers recorded between 1992–1994. From Luigujõe *et al.* 1996.
(a) Autumn: 1 = Väike Väin Strait, 2 = Matsalu Bay, 3 = Lao-Lui coast/Pärnu Bay, 4 = Pärnu rivershed, 5 = Lake Peipsi.
(b) Spring: 1 = Väike Väin Strait, 2 = Haapsalu Bay, 3 = Matsalu Bay, 4 = Lao-Lui coast/Pärnu Bay, 5 = Audru Polders, 6 = Pärnu rivershed, 7 = Pandivere region, 8 = River Emajõgi, 9 = Lake Peipsi.

2002). A total of 43 birds were marked with neck collars near Chaun Bay in Far East Russia between 1974 and 1978, and yielded interesting results. Most were recorded (some individuals several times) at migration sites along the islands of Japan, but only three sightings have been recorded in Russia (Kondratiev 1984; Kondratiev & Kishchinski 1987; Bird Migration Research Centre 1987). Additionally, one young swan was found dead on Adak Island (in the Aleutian Islands) and another was found in Estonia (Kondratiev & Kishchinski 1987). This and more recent satellite-tracking studies all suggest that Bewick's Swans breeding in the far east of Russia winter mainly in Japan (Kishchinski 1979; Higuchi *et al.* 1991; Kamiya & Ozaki 2002), although it should be noted that only a few birds have been tracked by satellite, and these were birds already wintering in Japan.

The lack of ring/collar re-sighting effort in the more western breeding and wintering areas of the eastern population means that wider ranging movements may be undetected. Nevertheless, it seems likely that birds from the more western areas (Lena Delta and Eastern Taimyr) winter in China, with some perhaps migrating to Korea. A programme of marking Bewick's Swans caught on the Lena Delta with red neck-collars, initiated in 1999, should throw further light on this issue (a total of 28 birds have been marked to date; Pozdnyakov 2001). None have been seen in Japan to date, but a bird with a red neck-collar was seen on Poyang Lake, China, in 1999. In January 2002 ornithologists succeeded in reading the codes on three neck collars at Saicheng Lake, Jiujiang, China (29.41°N, 126.06°E), confirming that these birds originated from the Lena Delta (Bird Ringing Centre of Russia, unpubl. data).

China. Bewick's Swans arriving in China use sites in the north-central and northeastern parts of the country, and their wintering sites also tend to be more easterly, whereas Whooper Swans occur across the country. The main migration sites are in the provinces of Inner Mongolia (Dashai Lake, Dalinor, Wulagai River), Tianjin (Pohai Bay, North Big Bay, Jinghai), Hubei (Beihaihe), Liaoning (Wafandien, Panjin, Panchi, Zuangho, Dantong, Dalien), Jilin (Tongyu, Momoke, Changbeishan), Heilungjiang (Chalung, Xingkai Lake, Three River Flatland) and Shandong (Yungzhen, Qingdao, Wuli and the mouth of the Yellow River) (Ma and Cai 2000). Field observations indicate that the migration route into China follows the Ro River and the Hui River in Inner Mongolia, the Peony River, Chichiherh and Xingkai Lake in Heilungjiang Province, Xiang Lake and the Changbei Mountains in Jilin Province, Ying Ko and the marshes at Kangping on the upper reaches of the Liao River in Liaoning Province, the sand banks at Huayuanko in Henan Province, Yentai Long Island in Shandong Province, and also the Hubei, Gansu and Xinjiang Provinces (Ma and Cai 2000).

Japan. Detailed observations of the flyways used by Bewick's Swans arriving to winter in Japan show that the birds use three main routes on heading south through the Japanese islands (Matsui *et al.* 1981; see also Chapter 3). In autumn, most arrive from Point Klilion on Sakhalin to follow a central route from the Wakkanai district at the northern tip of Hokkaido, then follow the Teshio or Ishikari Rivers to Lake Utonai before flying to the Shimokita Peninsula on Honshu island. The Sarukotsu Marshes in northern Hokkaido and Lake Utonai may be used as staging sites. The Okhotsk-Pacific route, generally taken by birds arriving from Point Aniva on Sakhalin, is used mainly by Whoopers, but also by a few Bewick's Swans. These birds migrate along the Okhotsk coastline, alighting on Lake Komuke (in Monbetsu City), Lake Tofutsu (in Abashiri City) and on the Shiretoko and Shimokita peninsulas, before following the Pacific coast of Hokkaido to Point Erimo and Lake Utonai. Many of the Whooper Swans remain at Lake Utonai over winter, but the Bewick's continue south via the Shimokita Peninsula to wintering grounds on Honshu. The third route follows the coast of the Sea of Japan, starting downstream of the Teshio River and extending to the Tsugaru Peninsula on Honshu (Matsui *et al.* 1981).

There are at least two migration routes used in spring by swans wintering in the southern part of Japan. Satellite tracking of individual birds has shown that they either follow the coast of the Japanese islands, or that they cross the Sea of Japan to the Lower Tumangan River, before heading north via Lake Khanka on the China/Russia border (used mainly by Whooper Swans; Krivenko 1999) and the Sanjiang Plain to the mouth of the Amur River. Field observations suggest that those swans migrating along the Japanese islands in spring continue to the large Sakhalin Island, which is in Russia. There they follow both the east and west coasts of the island, and also the interior lowlands, before coming together in the northern part of Sakhalin, and perhaps moving a little east to join those that took the continental route on the Amur lowlands (Ostapenko 1991). Swans gather in large numbers at Pil'tun and Chayvo Bays in the north of Sakhalin, and at Salmon Bay (Bukhta Lososey) in the Gulf of Aniva in the south. The number of yellow-billed swans (both Bewick's and Whoopers) migrating across Sakhalin was put at 5,000 birds during the 1950s (Gizenko 1955), but some 30 years later it was estimated at not less than 15,000 birds, with Japanese ornithologists claiming an even higher figure of 30,000 birds (Ostapenko 1991). If this is accurate, it indicates a major increase in the Far Eastern Bewick's and Whooper Swan populations during the second half of the 20th century (most likely, given the increase in numbers of both species counted in Japan; Albertsen & Kanazawa 2002), or a change in their migratory distribution.

Eastern Russia. After Sakhalin, it is thought that only Whooper Swans migrate along the Kuril Islands to Kamchatka, but both Bewick's and Whoopers traverse northern Kamchatka, following the Penzhina River valley and Parapol'ski dol. The estuary of the Amur River, just across the strait from northern Sakhalin, is an important staging area for Bewick's Swans, with about 8,150 counted there in mid-May 1991. Local people report that they usually stay for about three weeks in May, arriving when the first ice-free areas appear on the estuary and departing after the ice breaks (Babenko 1995). Small numbers (200 birds) also use Bolon Lake, some 400km upstream from the estuary (Babenko & Poyarkov 1998).

The Penzhina Valley is probably a link route for Bewick's Swans migrating from Sakhalin and the Amur region into Chukotka, but the Kolyma Valley appears to be the main route for both species of swans (Ostapenko 1991). Those that winter in continental Asia but breed in the far eastern part of the range are thought to fly initially along the coast of the Sea of Okhotsk after leaving the wintering sites in spring, then cross the watershed mountain ranges to the valley of the Kolyma River (Kondratiev 1991a). They follow the river northwards and only disperse to the breeding grounds, including those visiting the most easterly limits of the range, upon reaching the Lower Kolyma region. Up to 2,253 Bewick's Swans have been counted on the Kava River (some 200km west of Magadan) during spring migration (between the third week of April and mid-May) (Krechmar & Krechmar 1997), so this may be an eastern staging post for swans migrating north along the Kolyma River.

2.7 FACTORS AFFECTING SITE SELECTION IN WINTER

Once the swans have arrived in the wintering range, they do not necessarily remain at the same site throughout the winter season but may transfer between habitats or between sites. Although a high proportion of the northwest European population congregate in the Netherlands in autumn, depletion of the food supply or freezing conditions there usually results in many birds moving further west to Britain and Ireland as the winter progresses. Thus, the final wintering sites for some individuals may be used only as staging areas by others.

2.7.1 Geographical location

A major survey of the distribution of all three swans species wintering in the United Kingdom was undertaken during the 1990–91 winter, in order to assess the importance of geographic location and habitat variables on site selection, and so determine the proportion of the wintering swans now feeding on farmland. It transpired that, although there was a within-winter shift in the Whooper Swans' distribution, with the birds using regions closest to their Icelandic breeding grounds upon arrival in the UK before dispersing southwards, there was little variation in the Bewick's Swans' distribution once the birds had arrived (Rees *et al.* 1997b). Moreover, as described earlier, sites with large numbers of Bewick's Swans were widely dispersed, occurring in northwest England, east-central England and southern Ireland (Rees *et al.* 1997b; Figure 2.11). Consistency in the distribution of Bewick's Swans during the winter, and the apparent tendency for birds to target sites in particular regions, may be due to Britain and Ireland being at the westernmost edge of their migratory range. Here winter site fidelity, reinforced in some cases by active management of sites for the birds, results in Bewick's Swans flying directly to traditional wintering sites at the start of the season. Seasonal shifts in distribution would be more evident if sites in Germany and the Netherlands were also included in the analysis, because these are closer to the swans' Arctic breeding grounds, and so are more likely to be used for staging than sites in the British Isles.

2.7.2 Habitat

Over the course of the 20th century there was a shift in the distribution of Bewick's Swans wintering in northwest Europe, not only geographically but also in relation to the habitats selected. Traditionally the swans frequented shallow tidal waters, coastal lagoons, inland freshwater lakes and marshes and flooded pasture, where they fed mainly on aquatic vegetation such as the rhizomes and tubers of pondweeds *Potamageton* spp., horned pondweeds *Zannichellia* spp. and stoneworts *Chara* spp. They also took the roots and rhizomes of eelgrass *Zostera* spp. in tidal basins and the leaves, stems, stolons and roots of grasses and herbs when in wet or flooded pastures (Poorter 1991 gives a review).

During the 1960s, the swans increasingly grazed on arable land and pasture in winter, the change being attributed mainly to wetland reclamation and drainage, and to the depletion of submerged macrophytes due to water pollution (Poorter 1991; Rees *et al.* 1997a). It is quite possible that the move was also a positive one, reflecting changes in farming practice, with the swans selecting root crops, cereals and fertilised swards because of their high nutritional value (see Chapter 4). Nevertheless, it is interesting to note that following an improvement in water quality and the recolonisation of the border lakes (especially Lake Veluwemeer) in the Netherlands by stoneworts (*Chara globularis,*

C. contraria and *C. aspera*), the maximum number of swans counted and the extent to which the swans used these sites during the winter season increased annually during the mid-1990s (at least from 1994–95 to 1997–98 inclusive), indicating that they still prefer aquatic vegetation when it is available (Dirksen *et al.* 1998). More recently, numbers have again declined, and few Bewick's Swans now feed on the lakes. It was initially thought that this might be due to competition with Mute Swans also feeding in the area, but it now seems that there are other (unknown) reasons for the move from the site (K. Koffijberg pers. comm.).

In the Netherlands, the swans still feed mainly on submerged macrophytes in autumn (October), before moving on to arable land in mid-winter and, increasingly, to grassland as the season progresses (Dirksen *et al.* 1991; Koffijberg *et al.* 1997). In Denmark, aquatic vegetation is taken mainly by Whooper Swans in October, with Bewick's Swans seen on stubble fields at this time, moving increasingly to winter cereals and Oilseed Rape *Brassica napus* from late November onwards (Laubeck 1995b). Bewick's Swans wintering in Britain similarly feed mainly on farmland, but their distribution varies from month to month as the birds shift between arable fields and improved pasture (Figure 2.18). They are relatively consistent in their use of permanent inland waters during the winter and are only occasionally seen on unimproved pasture or coastal sites. The total number of swans counted on arable land was highest in autumn, presumably because of the availability of spilt grain and waste root crops following the harvest, with the proportion recorded on improved pasture increasing each month from January onwards. More detailed consideration of the arable habitats used showed that stubble fields were selected upon arrival in autumn. The swans shifted to root crops in mid-winter and to cereals or Oilseed Rape thereafter (Rees *et al.* 1997b).

Although most Bewick's Swans in Britain were found on arable habitats, Bewick's Swan flocks were more commonly recorded on permanent inland waters than on arable land, indicating that the birds congregate in large numbers at comparatively few arable sites, and disperse over a wider range of inland waters. For instance, the average flock size for Bewick's Swans in January 1991 was 90.5 for birds on arable land, 41.6 on permanent

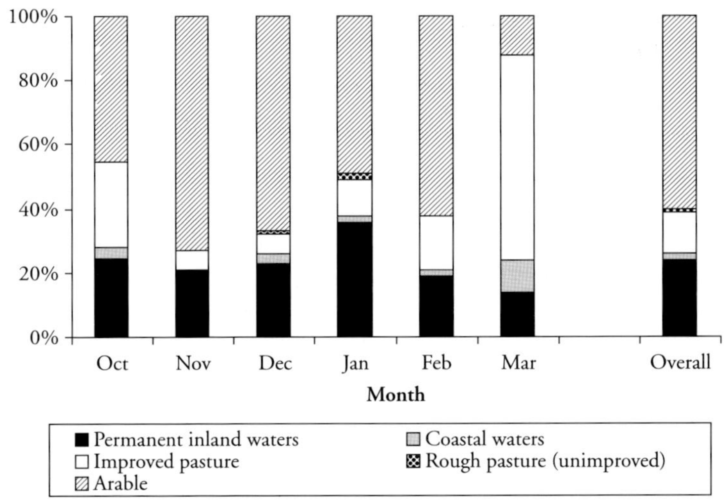

Figure 2.18 Habitat selection by Bewick's Swans wintering in Britain. From Rees *et al.* 1997b.

inland waters and 38.5 on improved pasture. The numbers of flocks recorded for each habitat category during the 1990–91 survey did not differ significantly from one month to the next, indicating that changes in the proportion of birds recorded in different habitats were due to fluctuating flock sizes, rather than flocks transferring totally from one habitat to another.

2.7.3 Interspecific competition with Whooper Swans

When two or more species with similar feeding habits spend time together in the same area they may compete for the available food resources, or adapt their behaviour in order to coexist; either way this may affect their dispersal patterns. A grazing succession has been recorded for White-fronted Geese *Anser albifrons* and Eurasian Wigeon *Anas penelope*, with Wigeon selecting fields that had been used intensively by geese during the previous month (Rees 1990). The grazing of fields by Bean Geese in the Yare Valley, Norfolk, also made the grass more suitable for Wigeon, but the Wigeon then rapidly depleted the vegetation to a level which could not sustain the geese. A change to sheep-grazing, and a resultant increase in the number of Wigeon in part of the site, is therefore thought to account for the shift of Bean Geese from Buckenham Marshes to Cantley Marshes within the Yare Valley during the 1980s (Sutherland and Allport 1994; Hearn 2004).

The observation that Whooper Swans dominate Bewick's Swans at sites where both species occur suggests that the Whoopers may displace the Bewick's in these areas (Black & Rees 1984). However, so far, there is no conclusive evidence that competition between the species is affecting their winter distribution, except at a local level. A comparison of the number of single-species and mixed Bewick's and Whooper Swan flocks recorded during the 1990–91 survey found that the number of Bewick's-only flocks did not differ from the expected frequencies, which were calculated from the proportion of sites known to receive swans of each species (Rees *et al.* 1997b; also see Chapter 4, Section 4.7). The number of flocks that included both Bewick's and Mute Swans, and also the number of mixed Whooper and Mute Swan flocks, were substantially lower than the expected values, with Mute Swans more likely to be found in single-species flocks, probably because of their wider distribution over different habitats within the UK (Rees *et al.* 1997b). The separation into single-species flocks may be partly due to seasonal variation in food choice. A study of swan flocks in the Tweed Valley, Scotland, found that Whooper Swans were more likely to feed in single-species flocks on winter cereals at the start of the winter, but that they fed with Mute Swans on Oilseed Rape from December to March inclusive (Chisholm & Spray 2002).

The extent to which Whooper Swans compete with Bewick's Swans in the breeding range has not been studied and, since Whoopers generally nest further south in the taiga zone, this has only recently been considered as a potential conservation issue. Given their relative size, however, it is likely that a Whooper Swan pair would displace Bewick's Swans in any dispute over a breeding territory, and the two species breed within only tens of kilometres of each other in the Nenetskiy State Nature Reserve of northern Russia. Monitoring programmes have shown that there has been an increase in the density of territorial Whooper Swans on the Pechora Delta just south of the Nenetskiy State Nature Reserve since the early 1980s (but not of Bewick's Swans slightly further north), despite the Pechora Delta being at the northern reaches of the Whooper Swans' range (Shchadilov *et al.* 2002). There is also evidence for a northward shift in the Whooper Swans' breeding distribution in other parts of Eurasia, with the species now breeding in the open tundra habitat previously the preserve of the Bewick's Swan (Syroechkovski 2002). Substantial increases in numbers recorded for the continental northwest European Whooper Swan population in recent years (Laubek *et al.* 1999) and predictions of climate and habitat change in northern Eurasia linked to global warming may result in further changes in the Whooper Swans' breeding distribution in future years, possibly to the detriment of their smaller relations.

CHAPTER 3

Migration and movements

Summer now seemed to have suddenly burst upon us in all its strength, the sun was scorching, the snow in many places melted so rapidly as to become almost impassable. The mud banks of the Zylma were steaming from the heat ... On the 11th [May*] we saw for the first time a pair of swans. The same day, on the half-open land between the Petchora and the Zylma, we saw some flocks of wild geese, and, near a pool of water on the ice, half a dozen Siberian herring gulls.*

Henry Seebohm (*The Birds of Siberia*)

Substantial information on the distribution and movements of an avian population can be obtained from count data. This is particularly true for relatively large birds such as the Bewick's Swan, but the monitoring of marked individuals provides even greater insight into why birds differ in their dispersal patterns. Questions such as the extent to which the same birds return to particular wintering and breeding sites, the optimal distances between migratory sites and reasons for variation in site selection can only be addressed by following the movements of individuals within a population. The long-term study of individual Bewick's Swans wintering at Slimbridge has shown that intraspecific variation in dispersal may be due to a range of factors, which can influence the birds in different ways and for different reasons. For instance, age, sex, previous experience of a site, social dominance, pairing success and breeding success may all affect not only the initial selection of a site, but also the timing of movement to and from that site over several years. These differences between individuals operate within a combination of environmental and physiological constraints, which determine the use of migratory and wintering sites by the population as a whole.

Since the Bewick's Swan is a long-distance migrant that nests in Arctic regions, its migratory and reproductive cycles must be closely synchronised. Birds have to breed, moult and develop fat deposits in just four months, while the breeding grounds are habitable. The innate and environmental cues used by the swans to control their migration therefore are of great importance for the continued survival of the species. The stimuli that trigger migration at the correct time, and how these are modified by climatic conditions are considered in some detail in this chapter. Less is known about other mechanisms regulating migration in this species, such as their orientation navigation systems, but tentative deductions can be derived from work on other closely related species, such as studies of the Icelandic Whooper Swan population.

As with the assessment of the size and distribution of Bewick's Swan populations given in Chapter 2, many questions concerning the swans' migratory patterns remain unanswered. For instance, we can only speculate about the reasons why Bewick's Swans (and, indeed, other species) undertake long-distance migrations at all, although it is thought this behaviour evolved in response to the emergence at high latitudes of protein-rich vegetation in spring, and the availability of aquatic invertebrates which promote rapid growth in the young (Murton & Kear 1973). Clearly, if the Arctic tundras are the swans' preferred summer quarters, migration becomes almost obligatory because very few animals are able to survive at these latitudes over winter. Return migrations are less evident in southern hemisphere waterbirds, probably because there is no terrestrial southern equivalent to the broad expanse of pan-Arctic tundra, with its spring bloom of vegetation and mosquitoes. The emergence of phytoplankton under the Antarctic ice-caps each spring provides food for vast numbers of krill, however, which in turn provide food for fish, marine mammals (particularly seals and whales), penguins and seabirds that migrate to Antarctic waters. Other species therefore make major journeys to Antarctica, although their movements are largely hidden within the southern oceans. Given that the Bewick's Swans undertake long-distance migrations of at least 3,000km (one way) each year, this chapter considers some of the patterns within the seasonal changes in distribution, and the possible reasons for these.

3.1 MONITORING MIGRATION: RINGING PROGRAMMES AND OTHER TECHNIQUES

In order to study the movements of individuals, it is important that individually recognisable natural or artificial markers are used. In the case of long-term studies this may require a life-long label, particularly if it is difficult to recapture individuals to replace artificial markings. Mute Swans caught on the Thames in the Middle Ages by the Dyers and Vintners Companies were marked with a cut on the upper part of the bill to denote ownership. The healed scars served to identify the marking company thereafter, for the lifetime of the birds. Although ducks have long been caught for the table in netted traps or decoys (Kear 1993), it was not until 1907 that decoys were first used to catch birds for marking with rings in early studies of bird migration (Mortensen 1950). Very few studies have the luxury of being able to identify individuals by their natural markings, so recognition of each bird by its unique black-and-yellow bill pattern is a major advantage to the Wildfowl and Wetlands Trust's Bewick's Swan study at Slimbridge (Scott 1966). This technique has enabled us to monitor the presence or absence, the associations and behaviour of all birds wintering at the site for more than 40 years (Chapters 1 and 7 give further details). Artificial markers are still needed for monitoring individual Bewick's Swans outside the Slimbridge area, however, since only trained observers (familiar with particular individuals after watching them on a near daily basis) can identify their 'known' swans by bill pattern and a combination of other natural traits such as head shape, bill shape and body size.

In the first three years of the Bewick's Swan study at Slimbridge (winters 1963–64 to 1965–66 inclusive), relatively few birds were recorded each season (24, 74 and 149 respectively, identified by bill pattern) and no serious attempt was made to mark them with leg rings. When numbers wintering at Slimbridge increased in the following three seasons, efforts were made to catch swans by raising a huge net in the path of their evening flight from the Rushy Pen (Ogilvie 1968), but in four attempts only seven swans were caught (Evans 1982). Swans caught and marked in the early years of the study therefore were mostly picked up after landing in confined spaces, from which they were unable to take off. The first 'swan-pipe' at Slimbridge, designed with a view to catching Bewick's Swans in greater numbers and on a more regular basis, was constructed in summer 1969. A corner of the lake frequented by the birds was extended into a channel and screened, and the following winter 91 new birds were caught in the pipe (Evans 1982). Initially five or six catches were made each winter, but from 1974–75 onwards just one or two were made, to minimise disturbance levels. The initial design consisted of a small lagoon, covered with netting, and although several modifications have been made over the years, the principle remains the same. Each day the member of staff responsible for feeding the birds in the Rushy Pen walks along two sides of the lake and into the pipe, distributing grain from a wheelbarrow as he or she progresses. The food attracts the swans into the pipe where, when a catch is being made, a gate closes to block the exit to the lake. For many years a 4m-wide wooden gate was used, which swung across the exit by means of pulleys (Evans 1982), but more recently a drop gate has also proved effective. After they have been caught, the swans are moved to holding pens, where they are split into smaller groups to ensure that they do not trample on each other. Each swan is then wrapped in a specially designed jacket to minimise feather damage during ringing (Figure 3.1; Evans & Kear 1975).

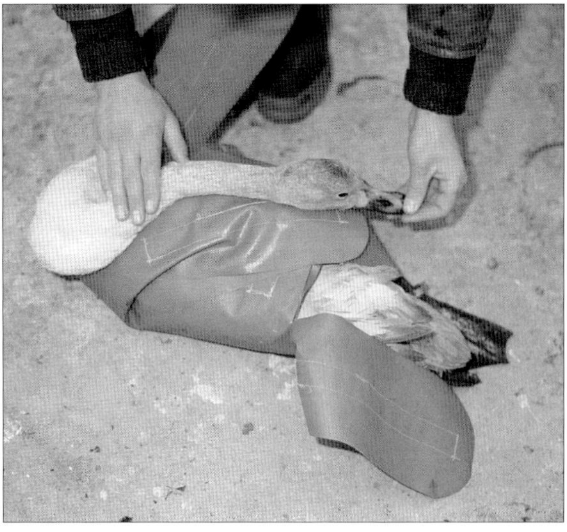

Figure 3.1 Each swan is wrapped in a specially designed jacket to minimise feather damage during ringing (Joe Blossom).

Once caught, the sex of each unringed bird is determined by cloacal examination, and the swan is then ringed for future recognition. Before winter 1967–68, the swans were fitted with metal leg rings issued by the British Trust for Ornithology (BTO), the organisation responsible for the marking of wild birds caught in the United Kingdom. Although the metal rings are durable and each is engraved with a unique code, the codes are usually too small to read except when the bird is retrapped or recovered. In the late 1960s, the Wildfowl & Wetlands Trust therefore developed plastic 35mm high leg rings for swans, engraved with alphabetical or numerical codes 15mm high, which can be read at distances of up to 300m in the field with a telescope (Ogilvie 1972a). These plastic ('darvic') leg-rings were first used on Bewick's Swans in November 1967, with the BTO metal ring being fitted to the opposite leg, and this method of marking the swans has been used ever since. For the Bewick's Swan study, yellow rings were allocated to known-age birds (i.e. those first recorded as cygnets or yearlings) and white rings to swans first seen as adults. Male birds were usually marked with the plastic ring on the right leg, and females on the left.

American scientists have long used plastic neck collars to identify Whistling Swans (Sladen 1973), and neck collars are commonly used in studies of geese. The possibility of using neck-collars for WWT's Bewick's Swan study was considered but not applied (Evans 1982) and, although collars were used in the early years of collaborative fieldwork in Russia, the practice was discontinued in the Anglo-Russian studies. Leg rings are obscured when the vegetation is high or the birds are on water, and neck-collars are easier to see, which increases the reporting frequency, but there was concern that neck-collars might have an adverse effect on the swans. This has been found for some species (e.g. Lensink 1968; Ankney 1975; Raveling 1976; Ely 1990), whereas others such as Mute Swans show no adverse effects (Spray & Bayes 1992). Additionally, neck-collars were considered inappropriate on aesthetic grounds for birds being viewed at close quarters, such as the Bewick's Swans at Slimbridge.

In addition to recording the ring numbers and the age and sex of each swan, the swan catch team also weighs the birds, measures body size (usually skull length, tarsus length and wing length, with bill length and tarsus thickness also being recorded in some years) and photographs the bill patterns, before releasing each bird on water. In the early 1970s (winters 1970–71 to 1972–73), the tail and wingtips of many birds were dyed yellow with picric acid to attract the attention of observers at other sites along the flyway and increase the resighting rate. Ornithologists and amateur birdwatchers soon acquired the habit of checking the marked birds for leg rings, and of checking all the birds in the flock for leg rings irrespective of whether or not they had yellow tail feathers. Yellow dye was again used in the late 1970s (1976–77 to 1979–80) in anticipation of a possible expedition to the breeding grounds (undertaken in 1979 by Sir Peter Scott and his family), in 1986–87 prior to the author's visit (under the Winston Churchill Travelling Fellowship) to Bewick's Swan staging sites in the former Soviet Union, and similarly in the early 1990s (1990–91 to 1991–92) before joint Anglo-Russian expeditions to the Russian Arctic. On these occasions the birds were marked to aid British participants in identifying any Slimbridge-wintering swans at their late staging or breeding sites, but tail-dyed swans were not seen in the breeding range. The yellow markings are temporary, because the birds shed their dyed feathers during the annual moult in July–August. A check was made to see if dyed birds were more likely to be shot, illegally, and there was no evidence that they received undue attention from hunters (Evans 1972, 1982).

For many years Bewick's Swans were ringed only at Slimbridge (Table 3.1), but the building of swan-pipes at Caerlaverock and Welney in 1980 extended the ringing

Table 3.1 Number of individual Bewick's Swans ringed in the UK between 1960 and 2002, and the proportion ringed at different sites in each decade. Numbers are given in parentheses.

	<1960	1960–69	1970–79	1980–89	1990–99	2000–02	Total numbers
% Slimbridge	0	100 (189)	99.8 (810)	66.1 (369)	33.8 (202)	86.5 (64)	1,643
% Welney	0	0	0	26.0 (145)	2.5 (15)	6.8 (5)	165
% Caerlaverock	0	0	0.2 (2)	6.6 (37)	0.3 (2)	0	41
% Martin Mere	0	0	0	1.3 (7)	63.3 (378)	0	385
% Other	0	0	0	0	0	6.8 (5)	5
Number ringed	0	189	812	558	597	74	2,230
% Total ringed	0	8.5	36.4	25.0	26.8	3.3	

programme to other parts of Britain. A fourth pipe at Martin Mere was brought into use in 1990. Opportunistic catches elsewhere (at Pensthorpe, Norfolk, and Little Downham, Cambridgeshire) have added to the numbers ringed in southeast England (Rees & Bowler 2002). Although several thousand Bewick's Swans winter on the Ouse Washes, catching them at Welney always proved difficult because fluctuating water levels made it difficult to operate the pipe, and in recent years the pipe has fallen into disrepair. Catching success at Caerlaverock reflects variations in the rather small numbers of Bewick's Swans wintering in the region (Table 3.1), although the pipe remains highly effective at catching Whooper Swans, for which it was originally built. Good numbers of Bewick's Swans were caught in the new Martin Mere swan-pipe in the early 1990s (Table 3.1), but more recently numbers wintering in the area have diminished, and those visiting Martin Mere are being displaced by the larger Whooper Swans, both at the grain feeds and from the pipe itself.

Dutch ringers started a regular marking programme for Bewick's Swans wintering in the Netherlands in 1985. During the 1990s, birds were also caught at migratory sites in Germany, Denmark and Estonia, and on the breeding grounds in the Russian Arctic. A total of 502 Bewick's Swans was marked by Trinus Haitjema in the Netherlands and Germany. The first birds were marked with two plastic leg rings, one above the tarsal joint on each leg, but since 1990–91 the Dutch have preferred to use yellow neck-collars. In the first winter they caught the swans by placing a mobile miniature pipe or trap over one of the waterways used by the birds, then manoeuvring them into the trap after dark, but the birds quickly became wary of this system. More recently, most birds have been caught by cannon-netting which, although it deters the swans from returning to a particular field for the rest of the winter, does not seem to affect feeding site selection in subsequent years (T. Haitjema pers. comm). Cannon-nets were also used to catch swans staging in Estonia in the early 1990s, when Jan Beekman worked with Estonian scientists in a study of the ecology and energy requirements of the birds during migration (Beekman *et al.* 1994b; Luigujõe *et al.* 1996).

Theoretically, swans are easier to catch during the summer months, because they become flightless during their annual moult. In reality, much depends on their distribution, and on their ability to avoid capture by running across the tundra (if on land) or by diving (if on water). Marking programmes in the Russian Arctic date back to the mid-1970s, when 43 birds caught in Chukotka were fitted with red neck-collars to determine their migration route along the Pacific flyway (Kondratiev 1984; Kondratiev & Kishchinski 1987). There were about 90 sightings of these birds migrating along the islands of Japan (Bird Migration Research Centre 1987; Kamiya & Ozaki 2002), but only three were recorded in Russia (Kondratiev 1984, 1991a).

Joint expeditions by Russian, British, Dutch and Danish scientists to the Russian Arctic since 1991 have resulted in several hundred birds being marked in the western part of the breeding range, most with blue neck-collars, others with leg rings only. Most of the neck-collared birds were non-breeders caught by the Dutch and Russian team working on the Korovinskaiya Gulf, north of Nar'Yan Mar, during the early to mid-1990s. In 2003 to 2005, WWT renewed its ringing effort in the region, in liaison with the staff of the Nenetskiy National Nature Reserve, and a further 350 non-breeders caught in the same area were marked with leg-rings. Breeding birds, and also the downy young, were caught mainly near Khabuicka in the Nenetskiy National Nature Reserve between 1992 and 1996.

In 1999, a ringing programme was initiated on the Lena Delta, where a Russian team fitted red neck-collars to the birds. They marked four swans in the first season and at least 16 more in summer 2000 (Solovieva 2000; Solovieva pers. comm.), with a total of 28 marked to date (Pozdnyakov 2001).

Although plastic leg rings are undoubtedly a useful means of monitoring individuals over several years, Bewick's Swans do lose their leg rings, so a failure to observe a particular ring does not necessarily indicate the death of the bird. In order to control for this fact in survival analyses, and also to estimate the ringing rate needed to maintain a marked sample of the population, the rate of ring loss was determined for 1,569 Bewick's Swans marked with 'darvic' rings in Britain up to winter 1987–88, with re-sightings included to the 1988–89 season (Rees *et al.* 1990). A further 167 swans fitted with rings made from 'vinylast' plastic shed their rings more rapidly. The percentage of 'darvic' rings known to be lost or broken after two years was 6.0% for female Bewick's Swans and 14.7% for males, with the mean minimum lifespan for the plastic rings placed on male and female swans being 3.0 and 3.5 years respectively. These figures are lower than those which would occur through ring loss alone, because damaged rings are replaced when the birds are recaptured and the ring lifespans are minimum values because most rings were still intact when the swans were last seen. Nevertheless, the figures show that male birds are significantly more likely to lose or damage their rings than females. Male swans are more frequently involved in physical combat, and it is likely that they shatter their rings during these encounters (Rees *et al.* 1990).

The larger Whooper Swans shed their rings more rapidly than the smaller Bewick's Swans and, within each species, some individuals show a greater tendency to lose or break their rings than others. It therefore seems that the birds' overall size and strength are main reasons for the ring loss. The size of the swan's leg may also be a contributory factor, but this seems less likely because the rings do not constrict the leg in any way. Ring diameter may make a ring more liable to breakage, however, and could partly account for the higher rates of ring loss by Whooper Swans, since their rings are moulded slightly larger. Of all the Bewick's Swans ringed, 176 (10.2%) lost or broke at least one ring and 30 of these (17%) lost at least one of their replacement rings. Of the 176 birds that lost their first ring, 67% were male and 33% were female. Of the 30 replacement rings lost, 80% were lost by male birds, with one male (named 'Deoch') breaking six rings in 13 years. Thus, not only do a greater proportion of males than females lose (or manage to remove) their rings, but some individuals are more likely than others to do so, and the great majority of these are males. The metal BTO ring, which is much more durable, helps to confirm a swan's identity upon recapture if the plastic ring is lost. Even metal rings are not swan-proof, however; 21 individuals (of which 16 were male) have received replacements at the time of writing, and at least three individuals ('Clairvoyant', 'The Brain' and 'Survey') returned to Slimbridge after losing both their plastic and metal rings. Again, all three were male.

The development and refinement of satellite-tracking techniques has made a major contribution to the studies of avian movements. Scientists are now able to obtain more precise information on flight time, distances between staging sites used in 'stepping stone' migrations, and the length of time spent by individuals for resting and feeding at each stage. Sensors within the transmitters and, more recently, data loggers can also record environmental and physiological data, such as air and body temperature, heart rate, wing beat and air pressure (reflecting altitude) (Beekman *et al.* 1996a). Initially the transmitters were too heavy, and their lifespan too short, for use in studies of migratory birds, but PTT (Platform Transmitter Terminal) transmitters weighing between 45g and 185g were used to study Bewick's Swan migration in both the eastern and western populations during the 1990s (Higuchi *et al.* 1990, 1991; Nowak *et al.* 1990; Beekman *et al.* 1996a, 2002). Lighter (20g and 30g) PTTs are now available, but these use smaller batteries, which currently limit the number of locations that can be recorded, and also the lifespan of the transmitter. The 100g PTT transmitters used in the 1990s had an 'active' lifespan of only 2–3 months, although the total lifespan could be prolonged by the use of a switch which delayed the start of transmission, and regulated transmission to only a few hours each day. Thus it has been possible to follow the whole migration in one direction, northbound for swans caught in winter or southbound for swans caught in the breeding range.

The development of solar-powered transmitters may enable the full return migration to be monitored, particularly if methods used to attach transmitters to the birds can be improved, and the transmitters are reduced further in size. The effect that attaching a transmitter has on a swan's behaviour or survival has yet to be determined, however, because of the small numbers involved and difficulty in observing the birds thereafter in the field. An early study tried three different methods of attaching the transmitters to the swans. This involved four birds fitted with satellite transmitters in Japan in 1990, of which two were attached to the backs of the swans by means of a harness, one was glued to the back feathers with epoxy resin adhesive and the fourth was attached with clasps to a neck-collar. There was no obvious difference between these different procedures in the length of time that the transmitters continued transmitting. The two fitted with harnesses sent location signals for 39 days and 46 days respectively; the one glued to the back feathers transmitted for 42 days; and the one attached to the neck-collar for 30 days. Although it was not certain whether the transmitters stopped transmitting because the bird had died, because the transmitter had fallen off the bird, or because the batteries were exhausted, the researchers seem to think that the last was most likely (Higuchi *et al.* 1991). Two of the swans were also found to have lost their transmitters (one glued to the back and the other fitted to the neck-collar) during the summer months.

Attachment of satellite transmitters using a Teflon harness has since proved less successful. Of eight transmitters fitted in this way in Japan (four 105g transmitters on 10 March 1994 and four 89g transmitters on 5 March 1997), six ceased transmission within two months. None of the swans were reported in subsequent years, although their having been marked with leg rings (rather than neck-collars) would have reduced the chances of re-sighting the birds on the lakes of Japan. One bird lost its transmitter near Lake Hyoko (northeast Honshu) in early April, and two others appear to have been sick (one still in Japan in June 1994 and one recaptured in late April 1997). The surviving transmitters provided data for reasonably long periods, however, with two birds spending the summer on Udly Lake, near the mouth of the Amur River. The early stages of autumn migration were noted for one of these birds. It started to move south on 24 October 1994 and was tracked along the Amur and Ussuri Rivers to near Khabarovsk, but the batteries on its transmitter finally failed on 5 November 1994 (Kamiya & Ozaki 2002).

The swans' ability to shed their transmitters was similarly noted in studies along the European flyway. Of seven Bewick's Swans fitted with transmitters by WWT and Colin Pennycuick (using an elastic harness and epoxy resin to glue the transmitter to the swan's back), only four transmitted throughout the swans' migration, and two of these are known to have fallen off the birds shortly thereafter.

In addition to receiving and processing signals from the PTTs, satellites have also been useful in assessing, through satellite imagery, the habitats and food resources available to the birds on migration. For instance, images from the LANDSAT satellite have been used to detect submerged macrophytes along the Bewick's Swans' migratory route, in order to identify potential feeding grounds for birds travelling between Arctic Russia and Western Europe (Beekman *et al.* 1994b). The Dutch team has linked the two methods of satellite tracking and satellite imagery to determine whether the combination of flight performance, the need to replenish fat reserves for onward migration and the availability of aquatic vegetation along the flyway have resulted in the birds being dependent on particular staging sites for their future survival (see Section 3.2.2).

3.2 MIGRATION ALONG THE WESTERN AND EASTERN FLYWAYS

3.2.1 Migration: northwest European population

Bewick's Swans in the northwest European population follow a relatively narrow migration route between the breeding and wintering grounds. Upon leaving the breeding grounds, the birds follow the north Russian coastline west to the White Sea, then head southwest across Karelia, via Lake Onega and Lake Ladoga to the Baltic coast (Rees 1991; Rees *et al.* 1997a; Figure 3.2). Movements from the Arctic tundras start in early September, with mass departures usually occurring in the second half of September. All birds usually have left the area by mid-October, as winter closes in (Figure 3.3). Satellite tracking of the swans' autumn migration by Dutch scientists in 1998 found that the birds followed this pattern, with all the tracked birds leaving in the third week of September (Beekman *et al.* 2002). A much later departure in autumn 2003 may be exceptional, due to unusually mild autumn weather. In that year WWT scientists visited the Nenetskiy State Nature Reserve with the staff of the reserve, Professor Colin Pennycuick and the BBC Natural History Unit, to fit Bewick's Swans and a single Whooper Swan with satellite transmitters. The aim was to identify key sites in the northern part of the swans' migration, and to determine the migration route for Whooper Swans spending the summer in this part of Russia. The birds' progress was also the focus of BBC Radio 4's 'Migration Week' in November that year. As time progressed, and the birds showed no signs of leaving the Pechora Delta into early October, we grew increasingly concerned but were reassured by reports from Russian colleagues that the autumn was the warmest for at least 20 years – and that mass migration from the delta had not yet occurred. Temperatures fell mid-month and the first two Bewick's Swans with satellite transmitters (named 'Kostya' and 'Anatoli') finally set off on 14 October, with the third ('Andrei') following on 19 October. The Whooper Swan ('Huck') remained on the southern part of the Korovinskaiya Bay, which he had frequented since mid-August, until 21 October (by which time it was –6°C), when he too started migrating

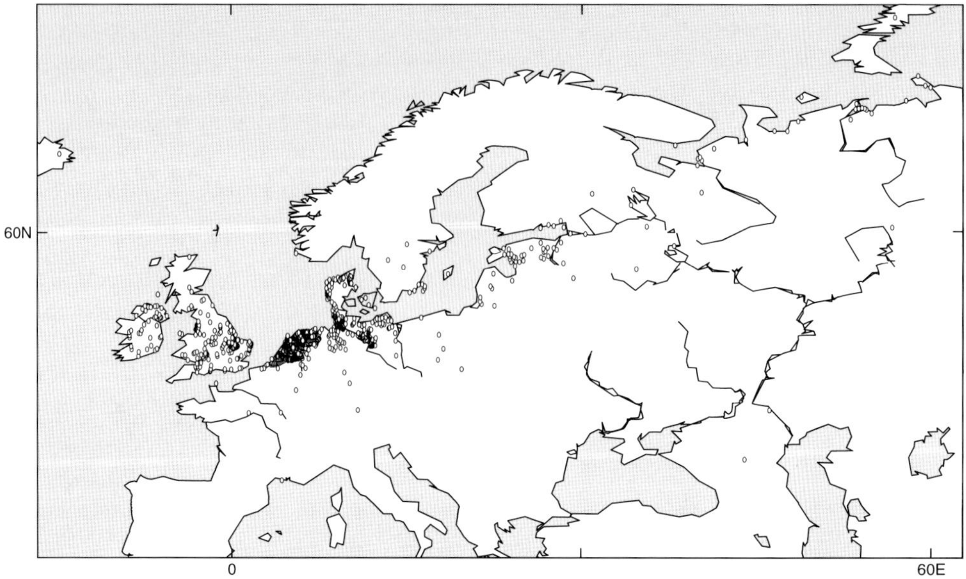

Figure 3.2 Ring re-sightings and recoveries for Bewick's Swans marked in the UK. From Rees 1991; WWT unpubl. data.

south and west, taking much the same route as that used by the Bewick's Swans a few days earlier.

Large numbers of swans are recorded in the Gulf of Finland and Estonia in early to mid-October, as the birds follow the northern or, more commonly, the southern shores of the Baltic to wintering sites in northwest Europe. Main staging sites such as Matsalu Bay, Lake Peipus and the Parnu region in Estonia are used in autumn, and more intensively in spring (Luigujõe *et al.* 1996). Parts of Denmark (such as the fjords of north and west Jutland, Laubek 1995a) and Germany (the upper and lower reaches of the Elbe) hold Bewick's Swans throughout the winter, but receive significantly more birds during the migration period. Passage through the Baltic continues into November in mild autumns (Luigujõe *et al.* 1996), with mass overnight migration through Estonia noted as late as 16 November in 2000 (A. Kuresoo, pers. comm.). The first birds usually reach their main wintering sites in the Netherlands and Britain by mid- to late October, with birds continuing to move westwards to Britain during the midwinter period. Freezing conditions and depletion of food reserves in the Netherlands can result in an influx to Britain as late as February in some years (for instance, in winter 1996–97, Waters *et al.* 1998).

The onset of the spring migration has been monitored in greater detail than the departure from the breeding grounds, due to the logistical difficulties and risks of remaining in northern regions (e.g. ensuring that scientists can be retrieved by boat or helicopter from Arctic study sites) as winter progresses. The precise timing of the spring migration is influenced by weather conditions (which probably is also true for the autumn passage), but in general birds start moving from Britain to the continental mainland in February, particularly towards the end of the month. Most birds have left Britain and Ireland by mid-March and the Netherlands by the end of March (Rees *et al.* 1997a), although on-going analyses indicate an earlier departure from the Netherlands over the last 10 years (J. van Gils, pers.

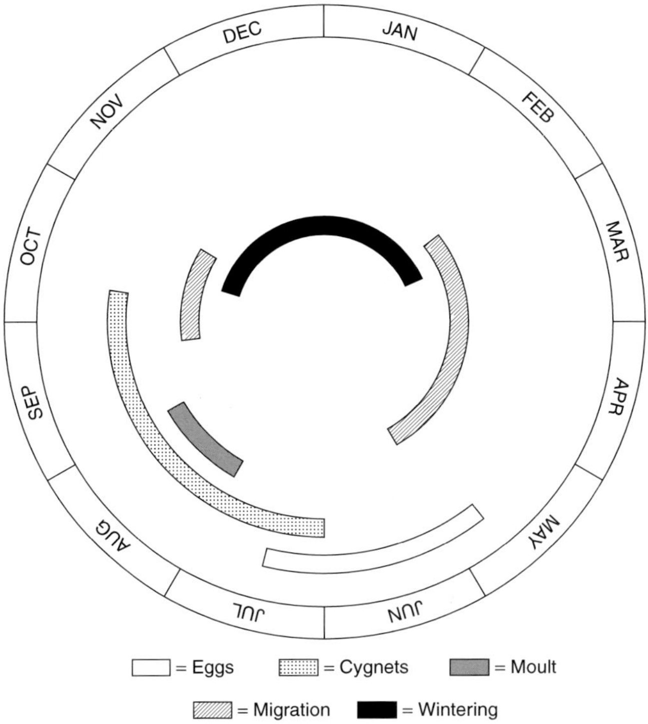

Figure 3.3 Timing of the Bewick's Swan annual migratory and reproductive cycle.

comm.). They then disperse widely across Estonian wetlands, where up to 18,000 birds (three-quarters of the population) may be observed during 2–3 weeks in April (Leito & Luigujõe 1991; Luigujõe *et al.* 1996), and although there has been a record of an early arrival near St Petersburg in the first half of April, most birds reach the Gulf of Finland in early May (Rees *et al.* 1997a; see Chapter 2).

Upon leaving the Gulf of Finland, the swans head northeast over Karelia, following – but generally not stopping at – Lake Ladoga and Lake Onega, in the shortest overland distance to the White Sea on the north coast of Russia. The Divina Delta and Divina Bay near Arkhangelsk on the White Sea are important staging areas at this time (Nolet *et al.* 2001b). Thereafter they follow the coastline east, or head across the Barents Sea to Kolguyev Island or southern Novaya Zemlya, to reach the breeding grounds by mid–late May, or possibly June in some years (Rees *et al.* 1997a). Thus, while the autumn migration takes 4–6 weeks, the spring migration takes longer, with most birds moving out of the wintering grounds by mid-March and not arriving at the nest sites until late May, 8–10 weeks later.

The swans tend to follow the same route in both autumn and spring (Figs 3.2 and 3.4), although the significance of particular migratory sites to the birds may differ between seasons, depending on food availability. For example, the Lauwersmeer in the Netherlands is an important autumn staging area for the birds. There, the availability of Fennel Pondweed in the lake and waste sugar beet on the surrounding fields has attracted up to 7,000 Bewick's Swans in 1996 and 6,461 in 1988, with average peak counts (1985–2001) of 1,600 birds (Koffijberg *et al.* 1997; SOVON 1998; Appendix 2). Similarly, the emergence of aquatic vegetation (*Zostera* and *Chara*) in the shallow waters of the White Sea in April–May provides a valuable food supply for up to 18,000 birds in the spring (Nolet & Drent 1998; Nolet *et al.* 2001b).

Reports of ringed birds are less frequent in autumn than in spring (Figure 3.4). The fewer autumn sightings is probably because the birds do not congregate in such large numbers at their migratory sites in autumn, but move directly to their wintering grounds, thus reducing ring-reading opportunities en route (Evans 1982). Total numbers seen staging in Estonia are two or three times higher in spring than in autumn (Luigujõe *et al.* 1996), and counts of Bewick's Swans at sites in Germany in spring are more than double those in autumn (Schubert 1974; Meier 1974; Dahms & Grosskopf 1978). Autumn migration through Germany tends to be along the coast (e.g. 433 Bewick's Swans seen flying west in the Wadden Sea between Norddeich and Bensersiel on 18 October 1997), whereas in spring the swans head inland via the lowlands of the large rivers (e.g. Ems, Weser, Elbe) and staging sites in Schleswig-Holstein (A. Degen, pers. comm.). In Denmark peak numbers similarly occur in late March and early April (Joensen 1974). More detailed monitoring has shown that the migration period in Estonia is longer in autumn (running from September to early November) than in spring, but occurs in two waves thought to reflect non-breeders arriving before family groups, rather than individual birds staying longer at the site. Information on the proportion of cygnets in the Estonian flocks is needed to confirm this hypothesis (Luigujõe *et al.* 1996; see also Section 3.3). The reason for the autumn migration being more rapid than the spring migration may be at least partly due to the birds being held up at spring staging sites upon meeting freezing conditions during their journey north (Rees & Bowler 1991). This may be accentuated by the tendency for migratory birds to feed more intensively in spring, possibly with a view to developing nutrient reserves in preparation for the breeding season.

Satellite-tracking of Bewick's Swans on autumn migration, undertaken in 1998 (Beekman *et al.* 2002) and 2003 (WWT, unpubl. data) reinforced this view of a more rapid autumn passage. Five birds fitted with transmitters in the Pechora Delta region of Arctic Russia during the 1998 summer flew almost non-stop from the breeding grounds to the Baltic countries, with 1,800–2,300km flights that included very short stops probably only to drink. Three swans stopped only briefly (<3 days) on the White Sea in autumn, which is an important staging area in the spring, and three (including two of the White Sea birds) also stopped briefly on the Gulf of Finland near St Petersburg. One bird traversed Russia in a single flight, alighting on Lake Peipus on the Russia/Estonia border three days after leaving the Pechora Delta (Beekman *et al.* 2002). The three swans tracked in autumn 2003 similarly flew directly from the Pechora Delta to the Baltic region, taking approximately two days to complete the journey. All three used Lake Peipus, with a smaller lake near Narva also used by one of the birds (WWT, unpubl. data, Fig. 3.5).

The swans remained in the Baltic countries to rest and feed in both years before moving onwards to their wintering sites. The satellite tracks in 1998 indicated that, in addition to staging at Lake Peipus and on the west coast of Estonia, the swans also used several sites in Lithuania, both the coastal waters of Nemunas Delta and Kursiu Lagoon, and inland lakes and fishponds such as the Raseiniai fishponds, Parseliai ponds near Kazlu Ruda and the Zuvintas Lakes Nature Reserve (30km SE of Kazlu Ruda). In 2003 the birds spent two to three weeks on Lake Peipus, then headed more directly towards the wintering grounds. It is therefore not yet clear whether the swans regularly disperse into small groups in the Baltic region in autumn or, indeed, if some important autumn staging sites have yet to be discovered. Certainly, there is increasing evidence that the swans stop for several hours on the

(a) Autumn

(b) Spring

Figure 3.4 Re-sightings/recoveries of Bewick's Swans during movements to and from Slimbridge. (a) = autumn migration; (b) = spring migration. From Evans 1982, WWT unpubl. data.

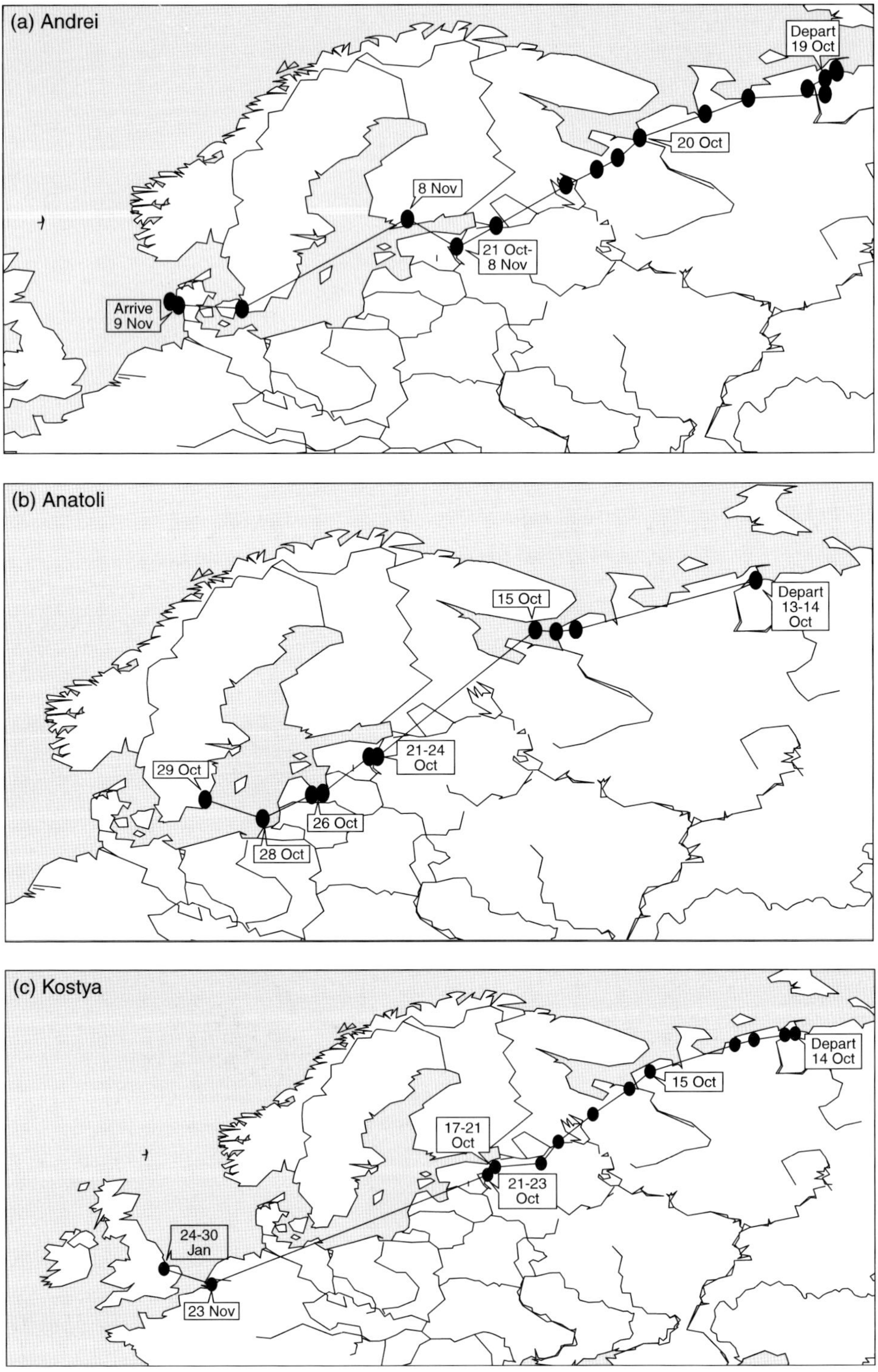

Figure 3.5 Satellite-tracks for three Bewick's Swans from the northwesr European population on autumn migration. WWT unpubl. data.

Malozemelskaya tundra, between the Indiga River and Cheshkaya Bay, although this is probably for resting rather than for feeding.

3.2.2 Frequency of staging: northwest European population

Information on the spacing of migratory sites, together with an understanding of the biological reasons why the birds use these sites, is important for ensuring that key areas are protected. Habitat loss may be a major problem for Bewick's Swans, which migrate on a narrow front, since the loss of a single wetland could result in the birds being unable to obtain the food and therefore the energy necessary for them to complete the onward journey to the breeding or wintering grounds. Transcontinental migrants may be further limited in their choice of staging posts if they are unable to utilise the forested taiga zones. Recoveries of ringed birds (included in Fig. 3.2) suggest that the birds use a number of staging areas but, since several were shot, recovery locations may not reflect the importance of the sites to the species (Rees & Bowler 2002). Satellite-tracking studies initiated in the early 1990s therefore provided valuable information on the maximum distances flown by the swans before stopping for refuelling, and the amount of time spent feeding at the staging sites.

The first satellite tracking of Bewick's Swans in the northwest European population was of 13 male Bewick's Swans fitted with satellite transmitters at staging sites in the Netherlands, Germany and Estonia between 1990 and 1993, and a further 10 birds fitted with transmitters in Denmark in 1996 (Nowak *et al.* 1990; Beekman *et al.* 1996a; Beekman & Laubek 1997). Preliminary results showed that most individuals took routes across the Baltic Sea, heading directly from northern Germany or southern Sweden to Lithuania and Estonia and taking about a day to cross the sea, in one case at an estimated speed of 70km/h (Nowak *et al.* 1990). Sites on the Elbe (in northern Germany), in Gotland (Sweden), western Estonia, Finnish Bay (near St Petersburg), in the vicinity of Lake Onega (near Petrozavodsk) and on the delta of the Sevemaya Dvina River (near Arkhangelsk) were used by several individuals, confirming that these are regular staging sites for the birds (Nowak *et al.* 1990; Beekman *et al.* 1996a). Upon leaving Estonia, three birds with transmitters were tracked to St Petersburg before their transmitters failed, and two onwards to Arkhangelsk. One individual male (902U), caught in Estonia in 1993, spent 22 days near Lake Onega and 16 days on the White Sea before reaching the Velta River (northern Russia) in mid-June. He returned to the Netherlands, having lost his transmitter, in the 1993–94 winter (Beekman *et al.* 1994b, 1996a).

Of two birds fitted with satellite transmitters by the WWT at Slimbridge in the 1995–96 winter, one ('Abelhard') was tracked to Kolguyev Island in the Barents Sea (presumably his regular breeding site), staging in Estonia, the White Sea region, and Cheshkaya Bay en route, and arriving in mid-June in what was thought to be a late breeding season. He and his mate returned to Slimbridge the following winter, without any young, 1996 having been a poor breeding season for the population as a whole. The second bird ('Pedro') sadly was found dead in Sweden during the spring migration. Although the cause of death was not recorded, he had been found to have a lead pellet in his body tissues in January 1995, rising to five by January 1996, suggesting that illegal hunting may have been a contributing factor.

Overall, the satellite-tracking studies have shown that the swans use just two or three stopover sites during the north-bound journey of 3,000–3,500km from the wintering range to the breeding grounds. This agrees with initial results of studies of the energy demands of migration, which estimated that the swans can fly some 1,450km before needing to refuel by feeding (Beekman *et al.* 1996a). These results also confirm the pattern indicated by ring re-sightings and recoveries, which suggested that clusters of recoveries in Estonia, the White Sea region and the Pechora Delta showed that these were vital staging and breeding areas for the birds (Rees 1991; Rees *et al.* 1997a; Figure 3.2). The results further serve to emphasise that individuals use most or all of the staging sites in a single season. The more recent satellite tracking of swans on autumn migration, which found that individuals may fly 1,800–2,300km during the southbound journey without prolonged feeding in the White Sea region (Beekman *et al.* 2002; WWT unpubl. data), indicates that flight performance calculations and the energy required by Bewick's Swans on migration may need to be recalculated, although wind assistance during migration could be an important contributory factor.

The observation that Bewick's Swans feed mainly on aquatic vegetation at key staging areas, notably on the Lauwersmeer (where they eat *Potamogeton*) in autumn and Matsalu Bay (*Potamogeton*) and the White Sea (*Zostera* and *Chara*) in spring, has led Dutch scientists to suggest that the availability of submerged macrophytes (especially the *Potamogeton* species) along the migratory route may limit the size and future viability of the northwest European Bewick's Swan population (Beekman *et al.* 1996b). In particular, the swans' preference for the rhizomes and tubers of Fennel Pondweed, which are particularly rich in carbohydrates, upon arrival in the Netherlands in autumn, suggested that these might be a crucial part of the swans' diet throughout the autumn migration (Beekman *et al.* 1991; Beekman *et al.* 1996b).

The Dutch team therefore used LANDSAT satellite images to detect submerged macrophytes along the flyway for the western Bewick's Swan population, with a view to determining potential feeding areas for the species during migration, and to make preliminary assessments of the quantity of food available for the birds. Images were taken of nine areas with waterbodies potentially able to sustain aquatic vegetation, along or close to the main migratory route. These included lagoons and shallow bays along the southeast coast of the Baltic Sea, Lakes Ladoga and Onega in Karelia, the White Sea area and the Pechora Delta.

The images confirmed that four areas known to hold large numbers of Bewick's Swans (Korovinskaiya Gulf in the Pechora Delta, Russia; Matsalu and Haapsalu Bays and the Vaike Vainen Strait in Estonia; Coulandic Haff/Neumunas Delta on the Kaliningrad/Lithuanian border; and Lake Veluwemeer in the Netherlands) contained good quantities of aquatic vegetation (Beekman *et al.* 1996b). There was no sign of aquatic vegetation at Vistula Bay in the Gulf of Gdansk (Poland), at Finnish Bay near St Petersburg, and at Lake Ladoga in Karelia (Russia), but Bewick's Swans are thought to feed only at the second of these sites.

An estimate of the area of submerged macrophytes along the flyway suggested that the northwest European population may be using just a few hundred km^2 for feeding (Beekman *et al.* 1996b), but there remains the possibility that terrestrial vegetation during the journey north is also important in providing the swans with the food necessary to enable them to complete their migration. Certainly the swans frequent stubble fields in Estonia in spring, feeding on spilt grain (left from the harvest) and shooting cereal crops, and also spend shorter periods on damp pasture (Rees & Bowler 1991; Luigujõe *et al.* 1996). Only the aquatic vegetation in Matsalu Bay, however, holds large numbers of Bewick's Swans for the whole staging period in Estonia in the spring (Luigujõe *et al.* 1996). Historic information on the distribution of wetlands and aquatic vegetation would be needed, together with early

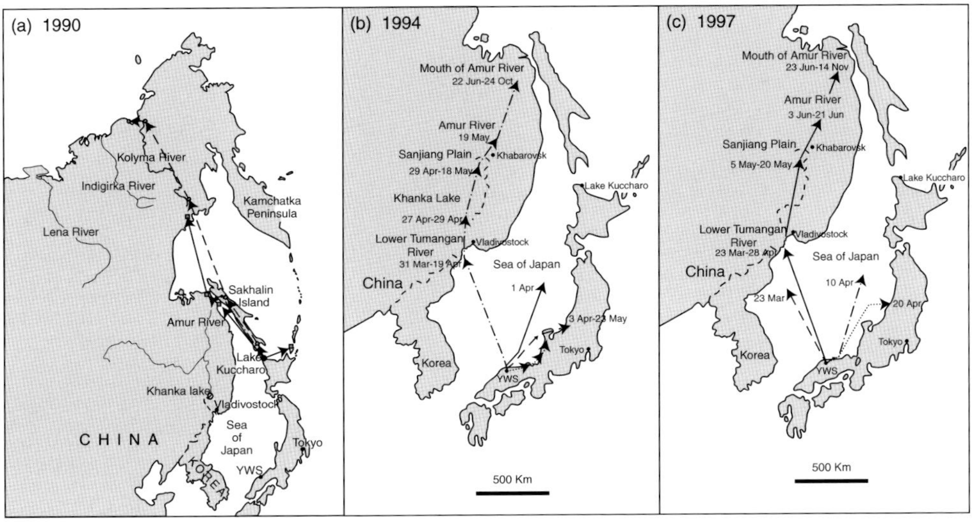

Figure 3.6 Satellite tracks of eastern population Bewick's Swans on spring migration in (a) 1990, (b) 1994, and (c) 1997. YWS = Lake Nakaumi, Yonango Waterbirds Sanctuary. From Higuchi *et al.* 1991; Kamiya & Ozaki 2002.

information on swan numbers and distribution, to determine whether eutrophication and the drainage of traditional wetland habitats have restricted the Bewick's Swans to a handful of migratory sites in recent decades, or whether they have always flown long distances between staging sites. Unfortunately, it seems likely that we will never know.

3.2.3 Migration and frequency of staging: eastern population

Just as our knowledge of the trends in numbers and distribution of swans in the eastern population is not as comprehensive as for the northwest European population (Chapter 2), less is known about the migration routes along the eastern flyway. Four birds marked with satellite transmitters at Lake Kuccharo, Hokkaido, in northern Japan on 10 April 1990 gave some preliminary information about the staging sites used during the spring migration (Higuchi *et al.* 1991). A second satellite-tracking study followed four birds on spring migration from Lake Nakaumi in southern Japan in 1994, and a further four birds from the same site in 1997 (Kamiya & Ozaki 2002).

All four birds monitored in 1990 followed a similar migratory route, flying north along the Russian island of Sakhalin, then staging in northern Sakhalin (two birds staged there for four days and nine days respectively in early to mid-May) or at the mouth of the Amur River (two birds staged there for 13 days and 16 days respectively between mid-April and mid- to late May) (Higuchi *et al.* 1991; Figure 3.6). One bird (F1287), recorded at southern Sakhalin on 26 April and northern Sakhalin on 1 May (where she remained until 9 May), arrived at the mouth of the Kolyma River (68.6N, 161.3E) within the breeding range on 17 May, covering the 3,083km migration from Lake Kuccharo in 37 days. A second swan (M1289) was recorded on the south coast of continental Russia west of Magadan when the transmitter battery failed, having crossed the Russian coast on a similar trajectory to swan F1287. The records from swans F1287 and M1289 suggest that, like the western population, Bewick's Swans fly long distances between staging sites along the eastern flyway, although it should be remembered that the sample size is very small.

Since swan M1288 spent nearly a month, and M1289 two weeks, around the mouth of the Amur River, it seems that the many lakes and marshes there form an important resting and feeding area for eastern Bewick's Swans on migration (Higuchi *et al.* 1991). Happily, swan F1287 (tracked to the Kolyma River) returned with one cygnet to Japan, where she was identified by her leg ring, and swan M1288 was also seen in Japan the following winter.

The eight swans tracked from southern Japan in 1994 and 1997 showed a slightly different pattern. Three crossed the Sea of Japan, with two of these birds reaching the Russian mainland near Vladivostock, then travelling north via Khanka Lake and the Sanjiang Plain in China to the mouth of the Amur River in northern Russia. The transmitter on the third bird failed whilst it was over the sea. The other five birds took a different route, and were heading north along the Japanese islands when their transmitters failed (Figure 3.6). These observations suggest that Bewick's Swans wintering in southern Japan follow at least two migration routes in spring, one along Japanese islands (where they join the route along Sakhalin taken by birds wintering in the northern part of the country), and the other directly across the Sea of Japan (Kamiya & Ozaki 2002). For both routes, however, the mouth of the Amur River appears to be an important spring staging site.

3.2.4 Turnover and duration of staging

The extent to which flocks of birds at migratory or wintering sites consist of the same individuals, or whether birds are constantly moving out and being replaced by new birds moving in (the 'turnover' of individuals) is an important issue for site protection. A high rate of turnover may mean that most or the whole of the population is dependent on several sites during the year, even if the maximum counts at some of these sites are relatively low. Turnover is difficult to estimate when only a small proportion of a population is marked and ring-reading is patchy, and ring sightings are almost bound to underestimate the duration of staging and the number of birds using a site during migration. Even at Slimbridge, where adults and yearlings in the flock can be recognised instantaneously by their bill pattern, a single trained observer can identify only *c.*400–500 birds during the rather short winter daylight hours. The maximum number recorded in a single day was 420 in January 1979, when several hundred swans arrived and up to 610 were counted at roost in the Rushy Pen. In most winters, however, the total number of swans present does not exceed 400, and the number of swans identified during the winter is a fair representation of the numbers of individuals visiting the site. Excluding 1963–64, the first year of the study when not all the swans on the Severn Estuary visited the Rushy Pen, the number of individuals identified throughout the season ranged from 103% to 185% of the maximum count of the winter (Figure 3.7). This indicates some annual variation in the turnover rates for the site. Although a high proportion of birds depart after just one day, particularly new birds not familiar with the distribution of food at Slimbridge (Evans 1980; Rees 1989), those that do not leave at once usually stay for most of the winter. A frequency distribution of the number of days that individual swans spend at Slimbridge each winter shows that the distribution peaks at 51–60 days for all birds, but at 71–80 days for swans that spend at least five winters at the site (Figure 3.8 from Rees 1989).

At migratory sites, where turnover rates may well be higher, due to the rapid movement of birds through the area, first and last sightings of individually marked birds still give some indication of whether the same birds remain for several days or weeks before moving on. Behavioural patterns also help to indicate whether there have been any newcomers, since swans spend more time sleeping immediately after arrival (Rees & Bowler 1991) and the

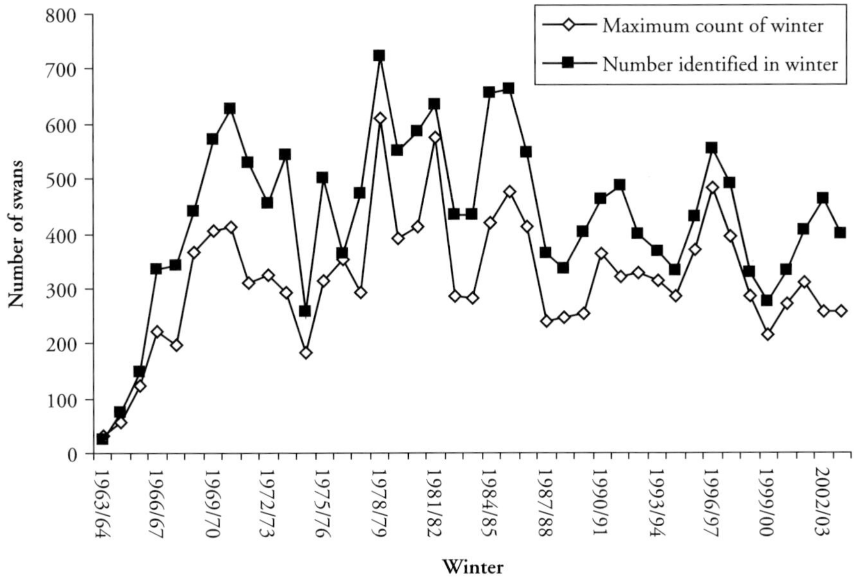

Figure 3.7 Maximum numbers of birds counted at Slimbridge each winter, and numbers identified during the winter season.

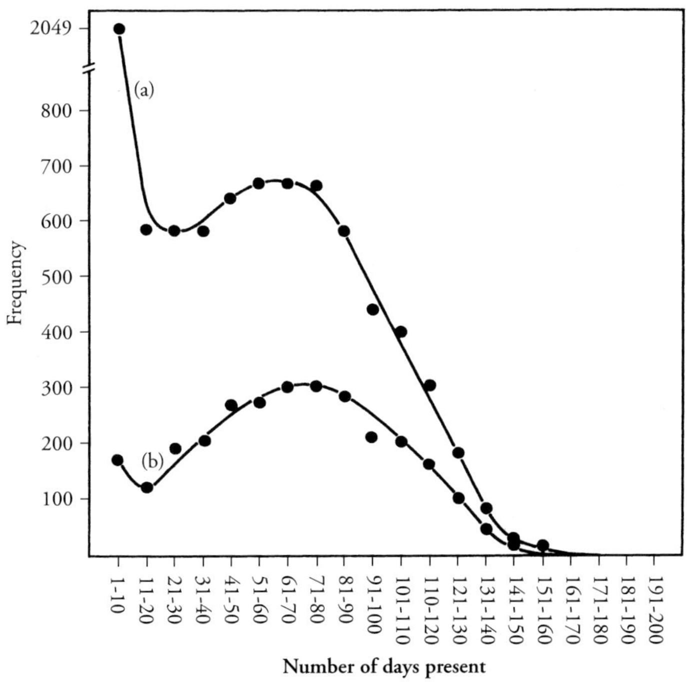

Figure 3.8 Frequency distribution of the number of days that individual Bewick's Swans spent at Slimbridge each winter. (a) All swans included. (b) Swans with 5+ years' data included. From Rees 1989.

frequency of aggressive encounters is also higher, as the swans attempt to gain a dominance position within the herd (Scott 1978; Chapter 6). Detailed monitoring of Bewick's Swan migration through Estonia in spring 1993 found that a high proportion (60%) of birds sighted were recorded only once, with the average staging period for all birds identified that season being 3.3 days (maximum duration 12 days). Local dispersal of birds made the following of individuals more difficult, and several of the birds reported more than once were seen in different parts of Estonia, generally heading in a northeast direction (Beekman *et al.* 1994b). Further monitoring suggested that individuals spent longer periods in Estonia in autumn, averaging 14.7 days ($n = 14$, SD +/- 5.3, range 9–31 days) than in spring — 12.3 days ($n = 39$, SD +/- 4.2, range 4–25 days) for the years 1992–1994 inclusive (Luigujõe *et al.* 1996). Information gained from observing birds marked with leg-rings (less so with neck collars) is likely to underestimate the time spent at a site, since the rings may be missed on the initial arrival and final departure dates. Nevertheless, staging in the Baltic countries (particularly Estonia) is clearly important to the birds for resting and feeding during both autumn and spring migration.

More precise information on the timing of individual movements has been obtained from the satellite tracking studies, which have shown that the swans fly directly from the breeding grounds to the Baltic, covering some 1,800km in as little as 2–3 days (Section 3.2.1). Data from the swans tracked in 1998 indicate that the birds then spend 2–6 weeks (average 32 days in 1998) at staging sites in Estonia and Lithuania before moving to the wintering grounds (Beekman *et al.* 2002). Autumn 1998 is thought to have been an unusual season, however, due to high water levels and a lack of vegetation in the lakes of western Russia and eastern Estonia (especially Lake Peipus), which resulted in swans congregating in large numbers on the west coast of Estonia. Moreover, once in the Baltic countries, strong westerly gales appear to have delayed the onward flight to the wintering grounds in western Europe, resulting in the long stopover period recorded (Beekman *et al.* 2002). The three swans tracked in 2003 spent less time in Estonia and Lithuania (7–14 days), although the precise timing of 'Kostya's' onward flight is not known because he lost his transmitter at Lake Peipus (WWT unpubl. data).

The journey from the Baltic staging sites to the wintering grounds can be quite rapid. On average, marked individuals are seen at the wintering sites 7.9 days (+/- 3.2 days; $n = 17$) after they are last sighted (by ring or neck-collar code) in Estonia (Luigujõe *et al.* 1996). More precise data for the five swans satellite-tracked in autumn 1998, found that they took 3–4 days to reach Germany and 4–6 days to reach the Netherlands. However, one bird made a more leisurely journey from Lithuania via southern Sweden (where it spent 29 days) and Denmark (5 days) to the Netherlands. All five birds reached the Netherlands between early November and early December (Beekman *et al.* 2002). The satellite tracking therefore confirms not only the importance of the autumn staging posts, but also indicates that the time spent there by individual swans can vary considerably even within one autumn season. For instance, if head winds set in for several days before the birds have left the Baltic region (Beekman *et al.* 2002), this may affect the time spent at a staging post.

The apparently shorter spring staging period observed by Estonian scientists in the early 1990s (Luigujõe *et al.* 1996) is perhaps even more likely to be subject to annual variation in weather conditions, for the swans may encounter freezing conditions during their journey north in late springs (Rees & Bowler 1991). Certainly, re-sightings of swans marked with neck collars have led Estonian scientists to propose that spring migration from western Europe (Germany and the Netherlands) to Estonia varies depending on the prevailing weather conditions, with birds flying direct from the wintering grounds to Estonia during mild springs (first seen in Estonia on average 3.4 days after last being recorded at their wintering sites), whereas movement through Europe may be delayed by adverse weather and poor feeding opportunities in colder years (Luigujõe *et al.* 1996). Satellite tracking indicates that swans may spend up to a month at one or more staging sites in Estonia in the spring, with one swan (fitted with a transmitter in Denmark in April 1996) recorded on the Vaike Vainen Strait, Estonia, for 31 days from 16 April 1996 to 16 May 1996 (Beekman & Laubek 1997; Beekman *et al.* 2002). With counts of up to 18,000 birds in the region in the spring during the early to mid-1990s (two-thirds of the total western population), it seems that most of the population spends at least a few days in the area when heading north. The Baltic countries are clearly an essential link in the chain of widely separated stopover sites between the breeding grounds and the wintering range.

3.2.5 Migratory routes used by swans from different wintering sites

Despite the well-defined migration route between the wintering and breeding grounds, at least for the western population, there is some evidence that birds wintering in different parts of Britain may have different migratory patterns. A review of some 22,000 sightings for Bewick's Swans caught in different parts of Britain between 1966 and 1994 found that those ringed in northwest England (at Martin Mere) and southwest Scotland (at Caerlaverock) appear to have a more northerly migration route through Europe than those marked in southern Britain (Figure 3.9; Rees & Bacon 1996; Rees & Bowler 2002). For instance, the Martin Mere-ringed birds were significantly more likely to be reported in Denmark, Germany and Northern Ireland than those marked in the southwest at Slimbridge, which were seen more frequently in the Netherlands and other parts of Britain (Table 3.2; Rees & Bacon 1996). Birds ringed at Welney, in East Anglia, had a more southerly and easterly distribution, being re-sighted mainly in the Netherlands and Germany. Although relatively few Bewick's Swans have been ringed at Welney, none of these has been reported in Ireland (Table 3.2).

Most Bewick's Swans in Britain winter on the Ouse Washes, so it is likely that most birds both enter and leave Britain via East Anglia, with an unknown proportion of swans in more northerly sites in Britain and Ireland flying directly from and to the continental mainland

Table 3.2 Number of British-ringed Bewick's Swans identified in the countries of northwest Europe (excluding ring site) over time. From Rees & Bacon 1996.

Ring site	No. Swans identified	% Individuals reported in each country							
		GB	NL	EIR	IRE	BRD	DDR	DK	Other
Slimbridge									
(seen pre-1975)	267	50.6	15.0	7.9	0.0	25.5	0.0	0.7	0.4
(seen 1975–84)	1,529	40.5	22.1	1.1	0.1	24.5	8.0	2.9	0.7
(seen 1985–94)	1,123	37.0	33.7	1.6	0.9	12.8	7.4	3.1	3.5
Martin Mere	516	21.5	30.8	1.7	2.7	17.1	11.4	11.6	3.1
Caerlaverock	88	44.3	19.3	0.0	1.1	21.6	10.2	3.4	0.0
Welney	170	19.4	30.6	0.0	0.0	21.8	14.7	9.4	4.2

Figure 3.9 Frequency of re-sightings of Bewick's Swans marked at (a) Slimbridge, (b) Martin Mere, (c) Caerlaverock, and (d) Welney. Large circles indicate > 10 swans identified at the site; small circles ≤ 10 sightings. From Rees & Bacon 1996; WWT unpubl. data.

(Rees & Bowler 2002). Interestingly, a high proportion (44.3%) of sightings of swans that had been ringed at Caerlaverock were from other British wintering sites. Some 70% of these cases were due to birds switching to other wintering sites in Britain (notably the Ouse Washes, although four birds transferred to Martin Mere), with the other sightings being from sites in the area around Caerlaverock (at Islesteps and Powhillon). There was no evidence that individuals staging on the Ouse Washes and in northern England were en route to Caerlaverock, yet one of the swans ringed at Caerlaverock was subsequently seen in Northern Ireland.

3.2.6 Variation in migration routes over time

In the early part of the 20th century it was thought that swans had a more northerly migration route to Britain, and presumably onward to Ireland. Although rare visitors to Ayrshire, they could be seen regularly in Renfrewshire in the 1930s (albeit in smaller numbers than Whooper Swans), and wintered in 'substantial numbers' on freshwater lochs in Bute (McWilliam 1936). There are also reports of a regular autumn and spring passage

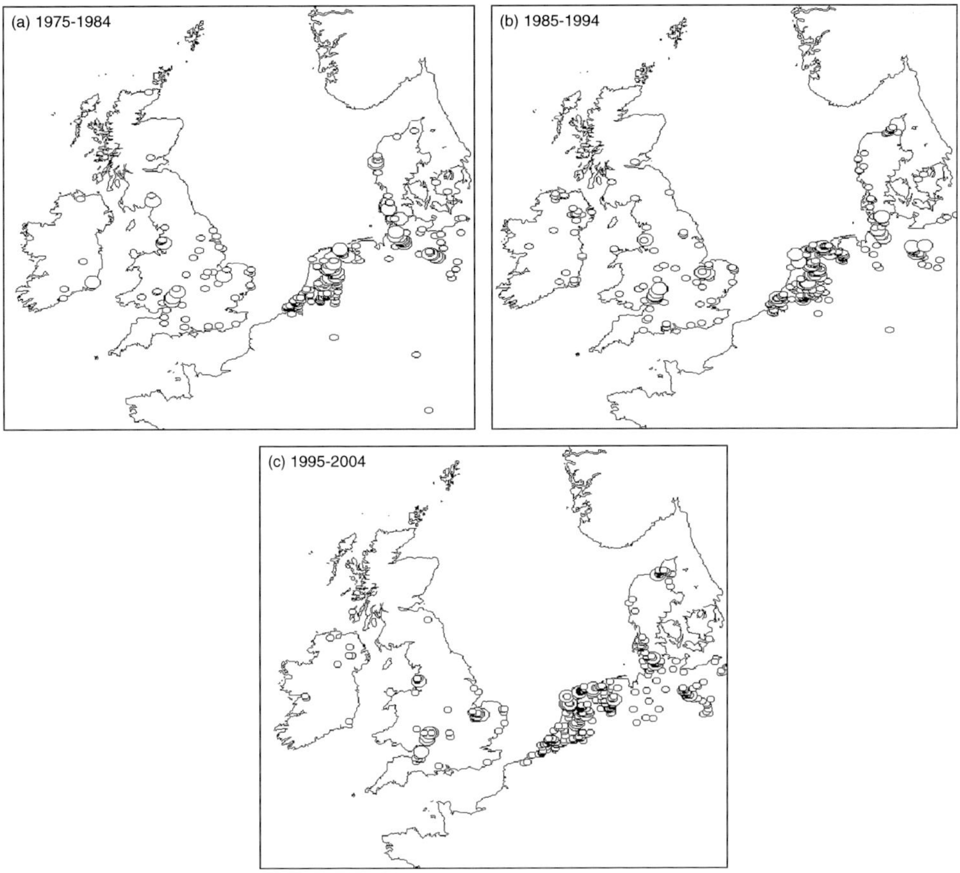

Figure 3.10 Frequency of re-sightings per site of Bewick's Swans marked at Slimbridge: (a) for sightings from 1975-1984, (b) for sightings from 1985-1994, and (c) for sightings from 1995-2004. Large circles indicate > 10 swans identified at the site; small circles ⩽ 10 sightings. From Rees & Bacon 1996; WWT unpubl. data.

through the Shetland Islands, with main wintering sites located in the Inner and Outer Hebrides, particularly on Tiree and North Uist (Baxter & Rintoul 1953; Ogilvie 1969; further details in Chapter 2). It therefore seems that migration to Britain followed a more northerly route in earlier times, and that changes in migration and winter distribution have occurred in the last 100 years (Ogilvie 1969). Certainly, selection of staging and wintering sites may be influenced by changes in environmental conditions along the flyway, such as creation of polders in the Netherlands over several centuries, including the 20th century. The eutrophication of wetlands in the Netherlands in the early 1960s is thought to have been a reason for Bewick's Swans moving on to Britain at this time.

Ring re-sightings since the mid-1970s give little concrete evidence for a major change in migration routes in recent decades (Fig. 3.10 a,b,c; Rees & Bacon 1996). On a broad scale, the birds' use of sites in the Netherlands appears to have changed little during the late 20th century (Table 3.2), although the colonisation of new sites following the increase in population size suggests a shift in distribution within the country during the 1990s (Chapter 2).

More detailed analyses are now underway to investigate this further, and to determine how changes in land use influence the movement of Bewick's Swans to and from the Netherlands (J. van Gils, pers. comm.). The increase in the number of swans identified in Northern Ireland in the late 1980s and early 1990s (Fig 3.10b) probably reflects an increase in the efforts made to read rings in the province (Rees & Bacon 1996), since numbers counted in Ireland have declined during the 1990s (Colhoun 2000; Cranswick *et al.* 1999). Similarly, the post-1985 increase in reports from 'other' sites (Table 3.2) reflects an increase in monitoring activity along the flyway, particularly in Estonia where the birds use a wide range of wetlands, although most were recorded near Matsalu Bay in the western part of the country.

Generally speaking, Bewick's Swans use the same wintering, migratory and breeding sites each year (see Section 3.4.1), and adaptation to new conditions may take several generations. The most common reason for a shift in distribution is habitat change brought about by man. Of these, a major reason for swans changing their use of a wintering site has been ascribed to changes in agriculture, notably land drainage and the increase in arable cultivation (Kear 1963). The swans traditionally fed on aquatic and marshland plants and on flooded pasture, but have increasingly moved onto crops over the last 30 years (Rees *et al.* 1997a). Human intervention through active conservation measures has also influenced site selection by the birds. For instance, the development of reserves by the WWT has resulted in a substantial increase in the number of migratory swans wintering at these sites, indicating that they may be attracting birds from other areas (Rees & Bowler 1996). This has been most evident at Martin Mere and Welney, where the location of a secure roost in areas of prime arable land has attracted not only Bewick's Swans but also large numbers of Whooper Swans in winter (Rees & Bowler 1996; Spray & Chisholm in press). Future shifts in distribution could be expected from major environmental changes due to global warming. Since the swans breed along a very narrow band of High Arctic tundra, there is a serious risk that they may not be able to adapt to any habitat loss at high latitudes, which are the areas thought most likely to be affected by global climate change (Chapter 8).

3.3 EFFECTS OF EXPERIENCE AND BREEDING STATUS ON ARRIVAL AND DEPARTURE PATTERNS

Information from various sources indicates that breeding status affects the timing of migration. Non-breeders generally leave the breeding range earlier than failed breeders or family parties (Mineyev 1991), with cygnet development perhaps influencing the departure patterns of pairs with young. The birds pass through Estonia in two waves in autumn, but not in spring, and this again is thought to be due to non-breeders arriving ahead of the birds with young in autumn (Luigujõe *et al.* 1996). Information on the percentage of cygnets recorded in each wave is needed to confirm this hypothesis, but parallel data available for Whooper Swans similarly suggest that non-breeders migrate before the family groups (Spray & Chisholm in press).

More precise information on the effects of breeding status on movements of Bewick's Swans to and from a wintering site has been reported by Evans (1980), who also assessed the effects of previous experience of the site on the swans' arrival and departure patterns. The detailed study of the swans at Slimbridge meant that she was able to classify all birds seen in a winter as 'new' (with no known experience of Slimbridge) or 'experienced' (known to have been present for at least one previous winter). Since members of pairs or family

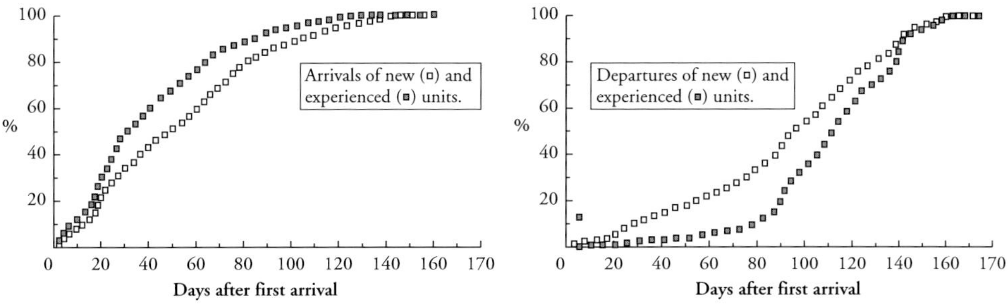

Figure 3.11 Arrival and departure patterns of 'new' and 'experienced' units (yearlings, singles, pairs and families) recorded at Slimbridge. From Evans 1980.

parties travel together on migration, pairs and families with only one experienced bird were considered to be 'experienced' for the analyses. Evans found that experienced swans generally arrived earlier, stayed longer and were absent for fewer days during the winter than birds new to the site (Figure 3.11; Evans 1980). Moreover, pairing or breeding success affects the arrival patterns, with different results being obtained depending on whether or not the swans had visited Slimbridge on previous occasions. Of the swans new to the site, yearlings were the earliest arrivals, followed by single birds, pairs and families (Figure 3.11; Evans 1980). This partially reinforces the view that non-breeding swans migrate ahead of the successful and possibly failed breeders, although the prompt arrival of experienced birds in all categories suggests that experience may be the overriding factor in a high proportion of cases.

Evans' study found that breeding status did not appear to affect the arrival patterns for experienced swans, and a longer run of data similarly showed that only the relatively late arrival of single swans was statistically significant (Rees 1988). This could be due to an individual initially being taken to a different wintering site by a new mate or associate and the pair subsequently separating, to unattached birds spending more time in exploring alternative wintering grounds in the autumn, or to the greater advantages accruing to families in returning early to a known site to achieve dominance in the social hierarchy. There is some evidence to support all three of these hypotheses, with the temporary loss of a partner affecting the timing of migratory movement (see Section 3.5.4), single swans being reported at more sites in a season than pairs and families (Rees & Bacon 1996; Section 3.4.3) and a positive but non-significant correlation between early arrival and dominance rank recorded in the early 1980s (Rees 1988).

Departure patterns also vary with pairing and breeding status, and these are modified according to previous experience of the site. Single swans generally leave first, followed by pairs, yearlings and family groups (Rees 1988). Further analysis by Bowler (1996) found that families feed closest to the roost and single swans furthest from the roost, which suggests that families (which are dominant in the social hierarchy) displace singles from the best feeding areas, and potentially from the site altogether. The relatively late departure of yearlings is probably due to the tendency to associate with their parents when at the same wintering site. The observed late departure of families differs from earlier results obtained by Evans (1980), who found that they were the first to leave, and this reflects a genuine change in the recorded departure of families since 1975–76, when Evans completed her study. The reason for this is not certain, but seems to be due to families utilising a range of feeding sites in the Gloucestershire area (particularly Walmore Common on the opposite side of the River Severn from Slimbridge), where monitoring was less intensive (and birds therefore may have been missed) in the early years of the study.

3.4 MIGRATORY TRADITION

3.4.1 Site fidelity

Wildfowl show a high level of site fidelity in both the breeding and wintering ranges. Moreover, females breeding for the first time generally return to nest in the region in which they hatched, with males dispersing more widely in search of a mate. This sex bias in dispersal has been recorded for a range of species, such as Mute Swans (Coleman and Minton 1979), Lesser Snow Geese *Chen caerulescens* (Cooke *et al.* 1975), Long-tailed Ducks *Clangula hyemalis* (Alison 1977) and Canada Geese (Lessells 1985), with the reason for females returning to natal areas to breed being attributed to their knowledge of the availability and distribution of food in the area (Greenwood 1980). Clearly, both sexes would benefit from previous experience of a site for this reason, but it is considered particularly important for females, which suffer greater weight loss during egg -laying and incubation (Lessells 1985). Differential dispersal of the two sexes is also considered important to avoid inbreeding (Bateson 1983). Any disadvantages to the male of moving to a new part of the breeding range are likely to be felt only in the short term, however, since he will benefit from experience gained of the new site when the pair return there in subsequent years.

Since it takes several years for Bewick's Swans to join the breeding population (Chapter 7), studies of the birds nesting in Arctic Russia are too recent to show whether female cygnets are more likely than males to return as adults to breed in the area where they hatched. So far, only one of 153 birds ringed as cygnets have been observed nesting at the same Arctic study site, and that individual happened to be male! Once a pair has bred on a territory, however, it is likely to return to the same territory over several years, often using the same nest mound (see also Chapter 5). The advantages of site fidelity in the breeding range are becoming clearer. In Arctic breeding species, the shortness of the summer means that the breeding season must follow a tight schedule, leaving little opportunity for exploration of alternative nesting areas. Pairs familiar with the breeding area nest earlier in spring and produce larger clutches than birds new to the site, and this is particularly pronounced in late springs when the snow is slow to thaw (Chapter 5).

The fidelity of Bewick's Swans to their wintering sites is very well documented. It was first noted in the early 1960s, when 16 of the 24 Bewick's Swans recorded at Slimbridge in the 1963–64 winter returned for the 1964–65 season (Scott 1966), forming the basis for the long-term study of the Slimbridge wintering flock. The years 1964–65 to 1969–70 saw a steady increase in numbers, and during this period some 35.1% of adults and yearlings recorded in a season had been seen at the site in previous years. Thereafter the proportion of birds with previous experience of the site ranged from 35.7%–62.7% (Fig 3.12). More 'new' birds were seen in cold winters, when freezing conditions on the continent displaced swans from their usual wintering sites in the Netherlands and Germany. Thus there was a positive correlation between the numbers identified and the proportion of 'experienced' swans recorded each winter during the build-up of the Slimbridge flock (r_s = 0.909, n = 8, $p<0.005$ for the years 1964–65 to 1970–71 inclusive), but a negative correlation between these two variables thereafter, when movement to Britain was due to birds being displaced

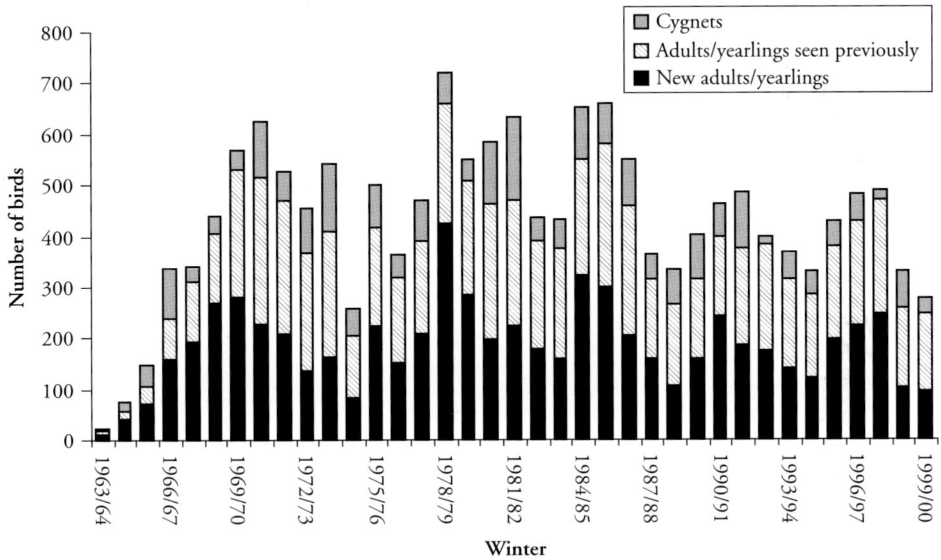

Figure 3.12 Number of Bewick's Swans recorded at Slimbridge that had been seen at the site in previous years. From Rees 1988 and WWT unpubl. data.

from traditional wintering sites further east (r_s= -0.575, n = 29, p<0.002). Certainly, of the 165 birds ringed at Slimbridge for the first time during the cold 1978–79 winter, 85 (51.5%) wintered at Slimbridge only in that season and 49 (57.6%) of these were reported in later years in the Netherlands, Germany or more easterly sites in Britain. This is lower than return rates recorded for 1,396 birds ringed at Slimbridge in other years up to the 1998–99 winter, when 522 (37.4%) ringed birds visited Slimbridge for only one season, and 787 (56.4%) were seen at Slimbridge for at least one winter after ringing. The advantages of site fidelity have generally been attributed to the birds' acquired knowledge of the area, but in winter this may be more important in males, because males generally take the lead in establishing the dominance rank of the pair or family in the hierarchy and thus, if successful, in obtaining the best feeding and roosting sites.

The question of whether swans show a similar tendency to use the same sites on migration was addressed by considering 3,153 sightings of 580 individuals recorded at migratory sites in more than one season (Rees & Bacon 1996). Only 4.3% of birds used the same migratory site in every year in which they were sighted, but 38.1% returned to the same site for at least a second season. Similarly, in 845 (26.8%) of sightings, a swan identified at a site had been recorded there previously, but although sightings were more frequent in spring than in autumn, differences in site loyalty did not differ between the seasons (Rees & Bacon 1996). Since some of the sites were as little as 2km apart, these results indicate a higher level of migratory site fidelity than is immediately apparent. Grouping clusters of sites into a single region should give further evidence of site fidelity during migration in future analyses. In particular, improved site definition would help to show whether the comparatively low levels of fidelity to the staging areas, in comparison with the Slimbridge wintering site, do occur. Meanwhile, some 2,550 sightings of 421 swans seen during both autumn and spring migration for at least one year were investigated to determine whether swans use the same site on arrival in the autumn as they do upon leaving the wintering range. Only 333 (13.1%) of all sightings were of swans using the same migratory sites both in autumn and in spring, even when information from different years was included, but 94 (22.4%) individual birds used the same site in autumn and in spring on at least one occasion (Rees & Bacon 1996). Thus it seems that although individual swans do tend to return to particular migratory sites, they do not consistently use the same sites in autumn as in spring, and their use of migratory sites may also vary between years.

Site fidelity has the benefit of enabling paired birds that have become separated during migration to meet up again at their wintering site. During the 40 years of studying Bewick's Swans at Slimbridge, 59 such separations and reunions have been noted (Table 3.3). Among these, one pair ('Mistletoe' and 'Ivy') spent two consecutive winters apart, 1971–72 and 1972–73, and another ('Bentall' and 'Benita') lost each other twice — in the winters of 1986–87 and 1990–91 — before being reunited. In most cases just one member of the pair was seen, but then returned the following winter with its mate. With other pairs, such as 'Amazon' and 'Lofty', both birds visited Slimbridge but at different times during the winter, apparently just missing each other, before returning together the following year. One bird ('Ron') appeared to go back to look for his mate on losing her. He arrived at Slimbridge without 'Ami' on 29 October 1982, then was seen at Welney on 2 and 3 November before returning to Slimbridge on 27 November for the rest of the winter. Meanwhile, 'Ami' was at the Lauwersmeer in autumn 1982 and wintered on the Noordoostpolder in the Netherlands, but the pair returned together the following season.

3.4.2 Conflict within pairs regarding winter-site selection

Although individual Bewick's Swans show a high level of site fidelity, with birds returning to traditional wintering sites over several years, they also form strong pair bonds, with paired birds migrating together and remaining together throughout the year. The birds usually pair for life; on only three occasions have members of a breeding pair been recorded with a new mate in the lifetime of the original partner (Chapter 7), but 56% of males and 63% of females losing a mate re-pair within a year (Scott 1988) and the mean time taken to re-pair is 2.6 years (Rees *et al.* 1996). Moreover, although single birds may form temporary attachments during the winter months, definite pair formation still has not been recorded in the wintering range (i.e. with both members of a pair being recorded as single swans at Slimbridge in the previous winter), so pair formation appears to occur mainly at either the migratory sites or in the breeding range. A conflict of allegiance is likely to arise, therefore, for newly formed pairs on their autumn migration, with attachment to their separate traditional wintering sites conflicting with attachment to their new mate.

Sexual bias in dispersal to the breeding range has been described for several avian populations (see Section 3.4.1), but sexual differences in site fidelity in winter has received comparatively little attention. A study of the winter dispersal of ringed birds therefore was undertaken, and the return rates of newly mated birds described, to ascertain whether one sex predominated in determining the wintering area for the pair. Overall, there is no significant difference in the ratio of males to females at Slimbridge (50.8% of 1,580 birds ringed and sex-checked at the site by 1999–2000 were male). When the total number of swans sighted elsewhere was compared with the number re-sighted only at Slimbridge, however, birds reported at other wintering grounds in years when they were absent from Slimbridge were significantly more likely to be female (Figure 3.13; Rees 1987). Moreover, return rates recorded for paired Bewick's Swans that had acquired or changed a mate showed that significantly more males than females returned to Slimbridge with a new

Table 3.3 Previous experience of the study area recorded for members of pairs of swans known to have been separated for a season.

Names of paired birds	Winter missed	Sex of bird that returned	No. prev. years at site for swan that returned	No. prev. years at the site for missing mate
Offset/Pouff	1968/69	M	1	1
Leopard/Puma	1969/70	M	1	1
Stars/Stripes	1969/70	M	4	4
Groucho/Chico	1970/71	F	4	2
Panto/Mime	1970/71	M	1	1
Muzzy/Fuzzy	1970/71	F	2	2
Order/Law	1970/71	F	1	1
Alexander/Angus	1971/72	F	4	2
Ivy/Mistletoe	1971/72 and 1972/73	F	5	3
Safe/Sound	1971/72	M	1	1
Crane/Forklift	1971/72	M	2	2
Dingaan/DooDah	1971/72	M	3	1
Estralita/Dougie	1971/72	F	3	1
Ashes/Dust	1972/73	F	1	1
Negretti/Zambra	1972/73	F	3	3
Work/Play	1972/73	M	2	2
Raquello/Laura	1973/74	M	4	1
It/Itma	1975/76	M	5	2
Ho/Yo	1976/77	F	3	3
Fever/Scarlet	1976/77	M	2	1
Wong/Steppes	1977/78	M	3	1
Jamaica/Antigua	1977/78	M	1	1
Bow/Sling	1977/78	M	2	1
Jubilee/Coronation	1977/78	F	1	1
Waves/Foam	1977/78	M	1	1
Amazon/Lofty	1978/79	M	1	1
Purser/Green	1979/80	M	4	1
The Brain/Lowbrow	1979/80	M	7	7
Strelitzia/Trellis	1981/82	F	4	1
Grimes/Griselda	1981/82	M	6	1
Gov.Rockef./Cheery	1981/82	M	9	3
Flush/Blush	1981/82	M	5	1
Crocus/Snowdrop	1982/83	M	6	2
Ron/Ami	1982/83	M	6	3
Fungirl/Hargreaves	1982/83	F	3	6
Islander/Gurkha	1982/83	M	5	3
Ku/Lummy	1983/84	F	16	7
Deoch/Dorothy	1983/84	M	12	3
Benjamin/Bentine	1984/85	M	1	1
Valois/France	1985/86	M	1	1
Safeway/Tesco	1986/87	M	6	6
Bentall/Benita	1986/87	M	5	1
Prattle/Gossip	1986/87	F	9	6
Seedling/Madonna	1987/88	M	2	2
Fort/Hill	1987/88	F	3	3
Casino/Punter	1989/90	F	18	1
Bentall/Benita	1989/90	M	7	2
Leon/Lioness	1990/91	M	3	3
Splash/Iceing	1991/92	M	1	2

Table 3.3 continued

Names of paired birds	Winter missed	Sex of bird that returned	No. prev. years at site for swan that returned	No. prev. years at the site for missing mate
Windfall/Windswept	1992/93	M	10	2
Shark/Silvertip	1992/93	M	10	4
Freddie/Frodo	1994/95	M	11	1
Rochelle/Rush II	1994/95	M	5	1
Nik/Nak	1994/95	M	1	1
Strelba/Melba	1997/98	F	12	2
Jacob/Sofia	1999/00	M	1	1
Dolman/Angelis	2000/01	M	9	5
Magor/Penarth	2002/03	F	3	1

partner (Table 3.4; Rees 1987). Thus, in the case of birds with new mates, the males are more likely to return to their traditional wintering grounds, while females may be readier to change their allegiance to accommodate a new mate. Male swans therefore appear dominant in the pair's selection of a wintering site (Rees 1987).

The question of whether males generally lead the movements of a pair throughout the winter was addressed by studying the swans' pre-flight behaviour (Rees 1987). The pre-flight display involves a pronounced pumping of the head and neck, accompanied by a call, apparently used as a contact signal to ensure that the departure of members of a pair is coordinated. Previous experience of the wintering site (in this case, Slimbridge) was taken into account since anxiety, due to unfamiliar and somewhat artificial surroundings, was considered likely to make newer birds more prone to flight. Two methods were used to measure the pre-flight signalling activity: the frequency of head-bobbing movements for each bird (recorded from the start of the display until take-off) and the number of times that each sex initiated actual flight by being the first to open their wings.

The relationship between the frequency of head-bobbing and flight initiation proved strong, with 68% of 60 males and 87% of 52 females displaying more frequently than their mate, and ultimately leading the pair's departure from the lake. The individual that invested most effort in display therefore seemed most likely to risk separation from the mate by being the first to fly. The effect that previous experience of a site has upon flight initiation proved to be a significantly important factor. In pairs consisting of one new and one experienced bird the new swan (of either sex) usually led flight in local movements throughout the season (Rees 1987). Also, discrepancies in the head-bobbing frequency recorded for

Table 3.4 Return rates of paired Bewick's Swans upon acquiring or changing a mate. Two-tailed chi-squared tests: * = $p < 0.05$; ** $p < 0.02$, NS = not significant. From Rees 1987.

	Males		Females		p
No. swans that returned	385		327		
% paired	77.1%	(297)	77.4%	(253)	NS
% 1st recorded as single that returned with a mate	31.6%	(94)	26.1%	(66)	*
% 1st recorded as paired returning with new mate	15.6%	(45)	9.9%	(25)	**
% 1st recorded as single returning with at least 2 (successive) mates	9.4%	(28)	9.9%	(25)	NS
Overall % paired birds returning with new/changed mate	56.2%	(167)	45.5%	(115)	**

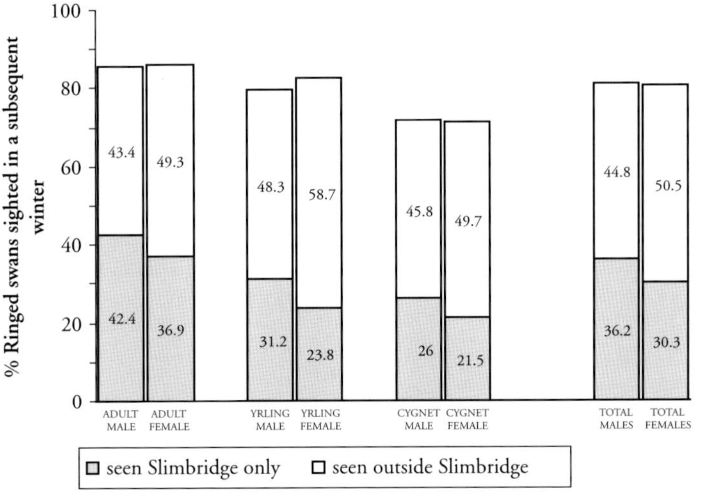

Figure 3.13 Sex differences in Bewick's Swan site fidelity. Open columns indicate swans observed wintering outside the Slimbridge area; closed columns indicate swans observed wintering at Slimbridge only. From Rees 1987.

established pairs (mean difference +/- 0.03 bobs per second, SE +/- 0.03) were significantly lower than for new pairs (mean difference +/-0.32 bobs per second, SE +/- 0.06), indicating that where paired birds had spent more time together, less time was needed in pre-flight display, presumably because their movements were better coordinated (Rees 1987). Nevertheless, the pre-flight study confirmed that males lead the movement of the pair in the autumn, but that flight initiation by the female becomes increasingly common as the winter progressed (Figure 3.14; Rees 1987). A shift in responsibility for leading movement therefore appears to occur as spring approaches, supporting the general trend in waterfowl for females to dominate in movements to the breeding grounds.

3.4.3 Exploratory dispersal by sub-adult and single birds

Young birds often migrate earlier and range further than adult birds. In the breeding range, this has been attributed to advantages to be gained by birds not yet of breeding age in exploration and habitat assessment prior to establishing their own territories (Baker 1978). In the case of the Bewick's Swans, where cygnets remain with their parents throughout the first winter, such exploratory dispersal is more evident in yearlings and single birds. Two of only three individuals known to have switched flyways, having been ringed in Britain and recovered towards the Caspian wintering grounds, were young birds; one was ringed as a cygnet and the other as a yearling (named 'Peanuts' and 'Tropical' respectively; Rees 1991; WWT unpubl. data). The third ('Lincoln') also may have been a young swan (perhaps in her third winter) since she was ringed at Welney as a single female. A comparison of the arrival patterns for swans with no previous experience of Slimbridge showed that yearlings new to the site generally arrived first, followed by singles, pairs and finally family parties (Figure 3.11; Evans 1980). Bearing in mind that the bulk of the population winters further east, on the Ouse Washes or in the Netherlands, many yearlings (and perhaps two-year-olds) thus apparently overshoot their previous wintering grounds for more westerly sites. The late arrival of new paired birds and families could be because, being older, they probably already have a traditional wintering site, which may have become less favourable; hence

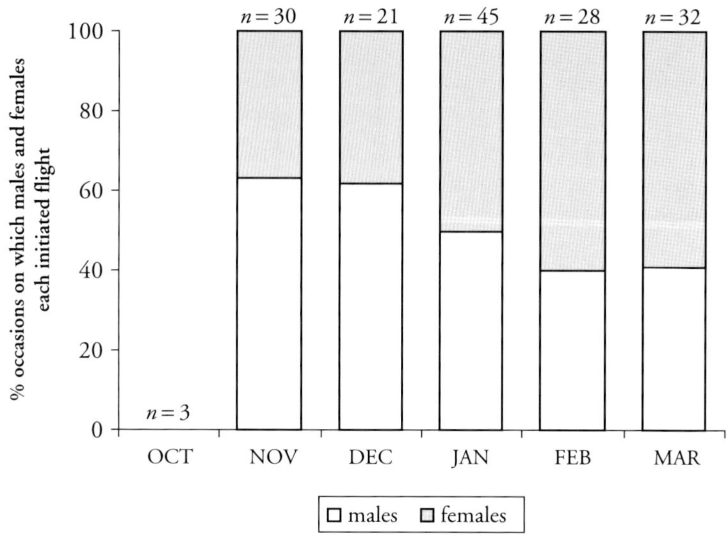

Figure 3.14 Changes in the number of occasions on which paired males and females initiate flight at different stages of the winter; *n* = number of pairs recorded. From Rees 1987.

the delayed onward movement to Slimbridge. Having arrived at Slimbridge, new yearlings and then new singles were more likely to remain there for the rest of the winter, with fewer one-day attendances recorded for these than for pairs (with or without cygnets) new to the site (Evans 1980). The prompt arrival of individuals that have wintered at Slimbridge in previous years does, however, indicate that pairs and families move directly to familiar sites on arrival in the wintering range.

Further consideration of the number of sites used by individual birds, between arrival in the wintering range in October to departure for more northerly latitudes in late March, showed that swans were recorded at more sites in years when they were single or yearlings (mean = 1.30 sites, SD +/-0.73, $n = 2553$) than when paired (mean = 1.21 +/- 0.60, $n = 2344$) or accompanied by cygnets (1.18, +/-0.51, $n = 1506$; Rees & Bacon 1996). The effect of pairing and breeding success on the number of sites used in a winter was significant after controlling for observer effort and annual variation (Rees & Bacon 1996). So the level of movement between sites is reduced for swans accompanied by a mate, and even more restricted if pairs are accompanied by their offspring. The interaction of year and social class (i.e. whether a bird was single, paired or with a family) was not significant, indicating that the annual variation in the number of sites visited (probably linked to weather conditions and food availability) does not affect single birds more than pairs or family parties. Since birds without young do not have to allow for the food requirements of their offspring, they may use the opportunity to assess food availability across a wider part of the wintering range. Single birds also tend to be subordinate in the social hierarchy and are thus displaced from optimal feeding areas, which may explain the larger number of sites visited during the winter. This does not necessarily explain the more rapid movement to Slimbridge of yearlings and single birds new to the site (described by Evans 1980), since an early arrival implies overshooting more familiar territory rather than testing a variety of alternative wintering areas. However, a longer run of departure data showed that swans left significantly earlier in years when they were single, and rather later when paired, and that families were amongst the last to start their spring migration (Rees 1988). The early departure of single swans could be due to exploratory dispersal, to displacement by the more dominant pairs and families wintering at the site, or to both of these factors. Interestingly, single adults (of both sexes) also proved less likely to return in subsequent years than paired birds, which in turn were less likely to return than pairs recorded with young (Bowler 1996).

3.5 THE TIMING OF MIGRATION

3.5.1 Importance of day-length in controlling migration

Like many long-distance migrants breeding at high latitudes, the timing of the Bewick's Swans' migratory and reproductive cycles must be closely synchronised, because the birds have just 3–4 months in which to breed, moult, develop fat reserves and head south before the winter closes in. Since day length is a highly reliable indicator of season outside the equatorial regions, it has been adopted by many species as the stimulus which triggers the onset of the migratory season (Lofts & Murton 1968; Murton & Westwood 1977), with studies of passerines showing that innate circannual rhythms of migratory restlessness can be modified by artificially changing day length (Berthold 1996 gives a review).

The floodlighting of the Bewick's Swans' roost in the Rushy Pen at Slimbridge provided a unique opportunity for analysing the swans' departure patterns under different light regimes (Rees 1982). The floodlights were installed to provide an evening spectacle during a visit to Slimbridge by HRH Prince Philip, Duke of Edinburgh (Lady Scott pers. comm.). They were retained to delight other visitors because the artificial illumination did not initially seem to affect the swans' behaviour or welfare. In the early years of the study, the swans flew to the Severn Estuary to roost so, although the pen was lit until 22:00 hours, the swans mostly received a natural day length. Then, in the early 1970s, the remains of 12 dead swans were found in the surrounding fields (all at least partly eaten by foxes), and the swans were encouraged to roost within the fox-proof perimeter fence by distributing grain under the lights. Earlier departure in subsequent years was attributed initially to mild winters during the mid-1970s rather than to the artificially prolonged day length giving the birds a false indication that spring had arrived. During the 1977–78 winter, however, the onset of spring migration was exceptionally early; all the swans had left Slimbridge by 6 February, and 18 of these were identified in western Germany by 9 February. When the weather deteriorated later in the month, 14 of the swans identified in Germany returned to Slimbridge and one to Welney (Evans 1982), indicating that the birds had left prematurely. Furthermore, swans in Ireland might be expected to leave the wintering grounds before or at least as early as those wintering at Slimbridge, yet in 1977–79 counts at the Wexford Wildfowl Reserve in southeast Ireland peaked at the time when all the birds had first left Slimbridge (Figure 3.15; Evans 1982).

In view of this exceptional 'reverse migration', the floodlights were dimmed at 20:00 hours in the winters of 1979–80 to 1981–82 to see whether the reduced photoperiod delayed the swans' departure patterns. Multivariate analyses were undertaken to control for annual variation in weather conditions, food supply (i.e., the distribution of grain) and frequency of disturbance (mainly swan catches); the results confirmed that swans floodlit from dusk until 22:00 hours departed on spring migration significantly earlier than those illuminated until 20:00 hours. These in turn left before those receiving a natural day length (Figure 3.16; Rees 1982). Moreover, a comparison of the Slimbridge departure pat-

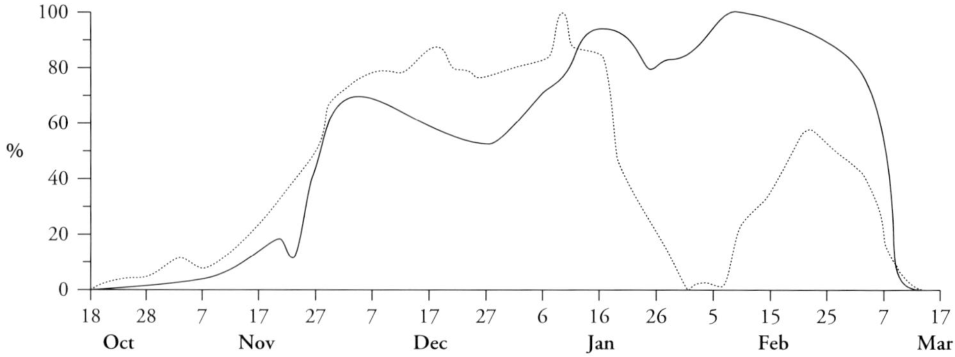

Figure 3.15 Daily counts of Bewick's Swans at Slimbridge (dotted line) and at the Wexford Wildfowl Reserve (solid line) in 1977-78, expressed as percentages of the maximum count at each site. From Evans 1982.

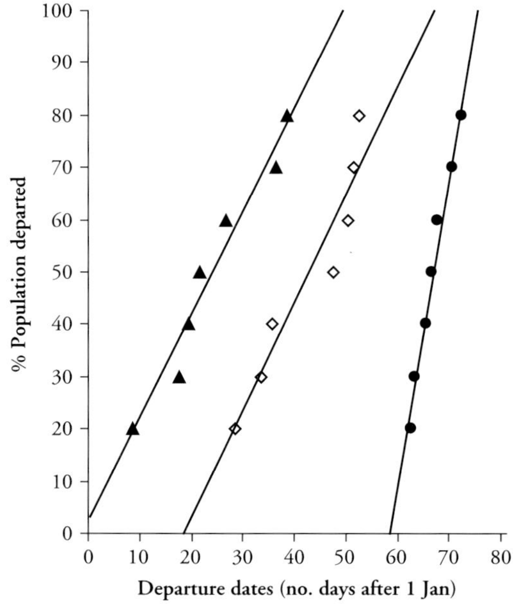

Figure 3.16 The relationship between the percentage of the swan population departed from Slimbridge and the number of days after 1 January for each of the light regimes. ▲ = Illumination until 22:00 hrs. ◇ = Illumination until 20:00 hrs. ● = Natural day-length. From Rees 1982.

terns with those recorded at Welney and Wexford showed that there was no significant difference in the timing of migration for birds wintering at Welney and Wexford, but that departure from Slimbridge was significantly earlier than from the other two sites when the floodlights were on until 22:00 hours. As a result of this study the floodlights have been dimmed at 19:00 hours since the early 1980s, and the swans now leave Slimbridge in late February to early March, which coincides with the timing of departure from other wintering sites in Britain and Ireland. Thus, although day length clearly regulates the onset of Bewick's Swan migration, at least in spring, it seems that the period of evening

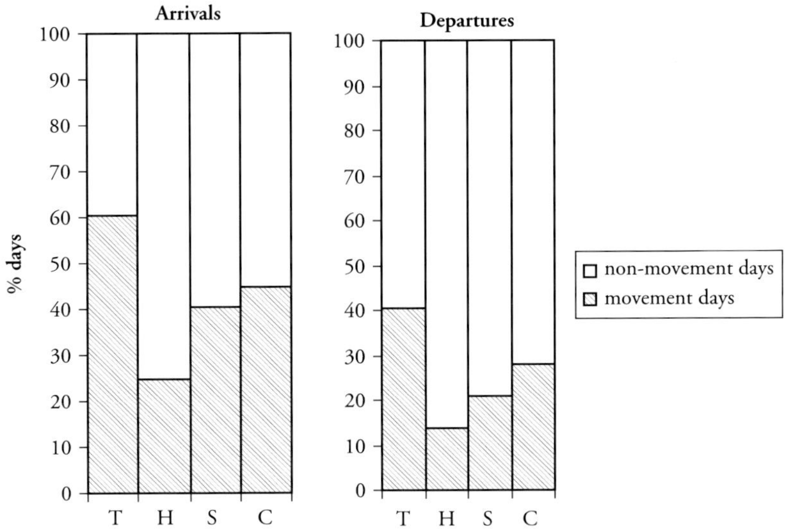

Figure 3.17 Proportions (%) of movement and non-movement days in each wind condition. T = tail, H = head, S = side, C = calm. From Evans 1979b.

floodlighting at Slimbridge is now short enough that it does not impinge on the swans' departure patterns.

3.5.2 The effects of weather on migration patterns

Ornithologists have long been aware of the importance of short-term variation in weather conditions in inhibiting or stimulating bird migration (e.g. Nisbet & Drury 1968). In the Slimbridge Bewick's Swan study, Evans (1979b) initially showed that, for both arrivals and departure, wind direction is an important factor with tail winds being preferred for migration, followed by calm and side winds, and with head winds generally being avoided. She classified major movement periods according to whether 5% or more of the total number of swan units (i.e. families, pairs or singles) recorded at Slimbridge during the winter arrived or departed on that day. The proportion (%) of movement days associated with each wind direction (tail wind, head wind , side wind or calm) clearly indicated that the birds were more likely to arrive on north to east winds (tail winds from northwest Europe) in autumn and, conversely, to leave on south to west winds (tail winds for the journey to Arctic Russia) in spring (Figure 3.17; Evans 1979b). Other climatic variables, such as wind speed, were not significant.

When the effects of day length, weather conditions (including wind direction), food supply and disturbance levels on the timing of the swans' departure from Slimbridge were all taken into account, day length proved the most important factor. It became progressively less important in inducing migration later in the season, however, whereas the relative significance of wind direction (particularly north to east head winds) in regulating departure dates increases (Table 3.5; Rees 1982). This finding reinforces the classical view that photoperiod is one of the main factors regulating physiological and behavioural cycles (including migration) but that, having determined the broad migratory season, more immediate variables such as climatic conditions trigger or inhibit movement on a day-to-day basis. Thus, once the swans have received sufficient photostimulation to signal the onset of

Table 3.5 The influence of day length and wind direction on Bewick's Swan departure dates. * = significant at the 5% level; ** = significant at the 1% level. Degrees of freedom for the Final F tests = 1,10. From Rees 1982.

% of population departed	Variable (as added to regression analysis)	% variation in departure data explained by each variable (cumulative)	Sequential F value	Final F value
20%	Day length	71.8	35.69 **	40.01 **
	% calm	77.0	2.95	2.48
	% side	80.9	0.93	0.96
	% head+tail	84.7	2.47	2.47
40%	Day length	64.4	25.30 **	27.57 **
	% side	68.3	1.60	1.06
	% calm	71.6	1.47	2.22
	% tail+head	77.0	1.50	1.50
60%	Day length	35.5	7.36 *	9.93 *
	% head	47.0	3.08	4.23
	% tail	53.3	1.60	2.42
	% calm+side	59.0	0.48	0.48
70%	Day length	28.3	5.52 *	15.14 **
	% head	48.2	5.00 *	8.61 *
	% tail	61.7	4.22	4.94
	% calm+side	67.4	0.10	0.10
80%	Day length	19.7	3.43	13.21 **
	% calm	44.9	5.95 *	4.74
	% head	52.6	1.96	5.58 *
	% tail+side	66.7	0.01	0.01

the migratory season, the final departure date is determined by weather conditions, particularly wind direction (Evans 1979b; Rees 1982).

The satellite tracking of Whooper Swans between Britain and Iceland has provided further insight into decisions made by birds migrating under adverse weather conditions; two swans that encountered strong head or side winds during their migration landed on the sea for up to 31 hours before resuming their flight towards Iceland (Pennycuick *et al.* 1996). Similarly, Bewick's Swans observed staging at Matsalu Bay, Estonia, in spring 1988, encountered blizzard conditions shortly after arrival and spent several days sleeping on the ice until the feeding areas thawed (Rees & Bowler 1991). Thus, whilst the swans use Estonian wetlands (and other staging areas) mainly to rest and accumulate body reserves, they may also use them as safe havens if they are deterred from onward migration by adverse weather encountered during the journey north, particularly if spring arrives late in the higher latitudes.

3.5.3 Effect of body condition on migration patterns

Since birds have to carry enough body fat to provide the energy needed for migration, a further limitation on the timing of migration might be their body condition before departure. Surprisingly, there is no evidence to show that the extent of body reserves affects the birds' departure patterns; the provision of additional food for Bewick's Swans in winter does not seem to trigger early migration (Rees 1982). Bewick's Swan weights generally increase until the second half of December, then plateau (Evans & Kear 1978). Studies of other waterfowl show that the weights of similar migrant species peak in early to midwinter (Matthews & Campbell 1969; Owen & Cook 1977; Owen 1981). A study of both the weight and body condition of Bewick's Swans wintering at Slimbridge confirmed that the swans gain weight rapidly from their arrival, reaching a maximum in late December to early January. There is then a midwinter decline, but body condition improves again prior to migration at the end of March (Bowler 1994). Since most weight is gained between mid-October and mid-November, it is likely that variables other than body condition trigger departure, although food shortage may inhibit departure in exceptional circumstances. In the season when the Bewick's Swans left Slimbridge prematurely in January, then returned to the United Kingdom in February because of frozen conditions on the European mainland, researchers estimated that the birds would need 24 days to replenish their fuel reserves (Evans 1982). Fifteen to 26 days later, in early March, the same individuals duly departed for a second time on their spring migration (Chapter 4).

3.5.4 Consistency in the migratory patterns of individual birds

Although variation in day length and weather conditions clearly are important in determining the timing of Bewick's Swan migration, the swans do not all leave a wintering site on the same day. Thus it seems that other variables must influence the timing of movements for individual birds, and several of these factors have already been discussed. For instance, swans with previous experience of the site tend to arrive earlier in autumn, with age, paired status and breeding success also influencing the timing of migration for individual birds (Evans 1980; Rees & Bacon 1996; see Sections 3.3 and 3.4.3). The temporary loss of a partner also affects the timing of migratory movement. Nineteen birds that temporarily lost their mate, for a total of 31 swan-winters, were also seen as single swans before or after pairing. In 22 of these 31 swan-winters, the swans arrived later when apart from their mates than in years when they were recorded as single swans ($z = 2.16$, $p = 0.03$, binomial test; Rees 1988), suggesting that the delayed arrival might in some way be linked to the temporary separation, such as waiting for the mate at a previous site. Additionally, 19 out of 24 male swans seen for at least three seasons with different mates (excluding winters in which they were recorded with a family) arrived at the same stage each season, relative to the rest of the Slimbridge-wintering herd, whereas nearly half the females (7 out of 16) arrived at a different stage of the winter with a new mate (Rees 1987). This tendency for mate change to have a significant effect on the timing of autumn migration for female swans, but not for males, reinforces the view (in Section 3.4.2) that males tend to lead the movements of the pair to the wintering site.

This chapter has shown that many factors influence the timing of migration, at both the broad population level (e.g. day length and weather conditions) and for individuals within the population (e.g. variation in age, pairing/breeding success and previous experience of the site). Nevertheless, after controlling for these factors, the timing of migration for individual Bewick's Swans tends to be consistent over the years, with some swans generally tending to arrive early in autumn, in comparison with the rest of the population, or to leave late in spring (Rees 1989). An analysis of the arrival dates obtained for 67 swans wintering at Slimbridge, with 10 or more years' information per bird ($n = 829$ bird-years), showed that there was significant consistency in the arrival data obtained for individual swans ($F=3.16$, $p<0.001$; Rees 1989). Similarly, the variation in departure dates between

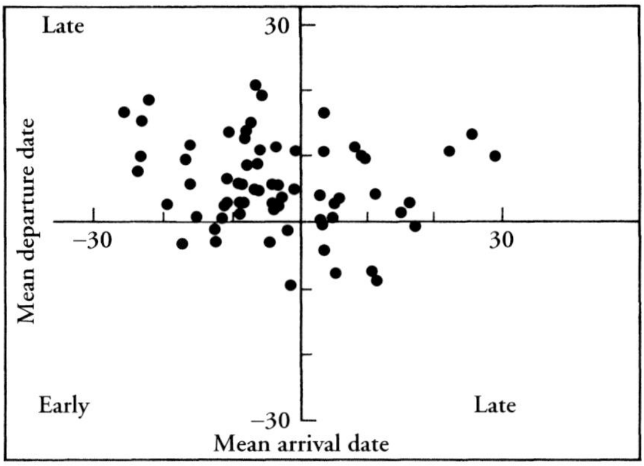

Figure 3.18 Correlation between the mean arrival and mean departure dates obtained for 67 individual Bewick's Swans wintering at Slimbridge. Arrival and departure data are expressed as the number of days' difference between the individual mean and the population mean. From Rees 1989.

swans is significantly greater than the departure dates recorded for an individual bird (F=1.72, p=0.005; Rees 1989).

Since day length appears to be a key factor in regulating the timing of migration, such consistency in individual migratory patterns may perhaps be due to differences in the individuals' response thresholds to changes in day length. If so, this mechanism appears to be quite sophisticated, since consideration of the mean arrival and departure dates obtained for the same 67 swans found a significant inverse correlation between the two sets of data: swans that tended to arrive early were more likely to leave late and vice versa (Figure 3.18; Rees 1989). There are several possible explanations for this inverse relationship between arrivals and departures. For instance, if the onset of autumn migration is determined by the termination of a photo-refractory period following the breeding season, swans with more northerly nests may have a lower threshold to photostimulation, which would result in them being the first to respond to increasing day length in spring. Alternatively, the onset of autumn and spring migration may be governed by the same control mechanism, so that swans sensitive to the initial decrease in day length would be the first to leave Russia, and so arrive correspondingly early in the wintering range, remaining there until the days were relatively long in spring. It remains to be seen whether consistency in the timing of migration is related to the location of individuals in the breeding range, and thus the time needed to complete the migratory flight, or whether every bird has its own innate response threshold to increasing day length in the spring.

To determine whether parents and their full-grown offspring tend to have similar migratory patterns, swans that had been observed for at least five winters at Slimbridge were considered. The mean arrival date of each parent differed from that of its independent offspring by less than 10 days on 61.3% of occasions (n = 31 cases where parents and their full-grown offspring both wintered at Slimbridge for at least five years), whereas the differences obtained for unrelated birds were less than 10 days on 46.2% occasions and exceeded 35 days on 21.5% of occasions (n = 39 random pairings of birds known to have bred and birds of known parentage; Rees 1989; Figure 3.19). Thus, the full-grown independent offspring tend to have arrival dates similar to those of their parents, indicating that the offspring acquire, either genetically or by learning, their parents' migratory thresholds. There was no diff-

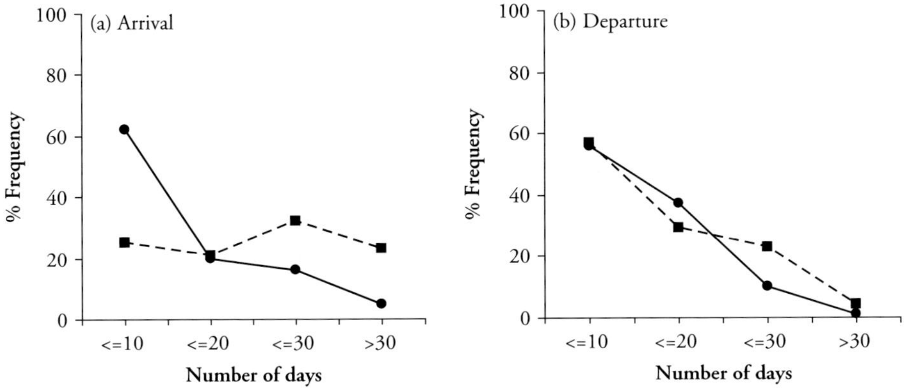

Figure 3.19 Frequency distribution of the number of days between (a) mean arrival dates and (b) mean departure dates obtained for parents and independent offspring; ——— = related parents and offspring; - - - - - = unrelated parents and offspring. From Rees 1989.

erence in departure patterns for related and unrelated birds (Figure 3.19), but the variation in departure dates for all swans wintering at Slimbridge was less than the variation in the arrival data, which may partly explain the non-significant result. Further investigation also showed there was still no evidence for a similarity in mean departure dates in the case of male parent–offspring comparisons, but that female parents and offspring did depart at a significantly similar stage each season, which may be due to females tending to lead movements of the pair in spring (as described in Section 3.4.2).

3.5.5 Effects of dominance rank on migratory patterns

Differences in the competitive ability of individuals, pairs and family units may also influence the dispersion of waterfowl populations and the carrying capacity of their habitat. Intraspecific competition for food is discussed elsewhere (in Chapter 4) but is also mentioned here since the presence of competitors may result in subdominants moving to less crowded areas. In gregarious species, such as Brent Geese, where interference between individuals is low, the birds tend to aggregate and deplete the food supply at the best site, then disperse en masse (Charman 1979; Sutherland & Parker 1985). Detailed studies of Bewick's Swan feeding strategies at Slimbridge showed that, although the birds also congregate in wintering flocks, and dominant male parents spend less time feeding and more time being vigilant than other males, it is the subordinate single males that are less likely to return to the site in subsequent years. Similar results were also obtained for female swans, but the difference in return rates for birds with and without cygnets was less marked (Bowler 1996; Figure 3.20). Dominance rank does not have a significant effect, *per se*, in determining return rates, probably because it is highly correlated with social class (i.e. whether a swan was recorded as paired with young, paired without young or single), but social class did prove significant (Bowler 1996).

Whether social dominance is associated with the timing of migratory movement has been assessed; in particular, whether early arrival enables the birds to establish a relatively high position in the hierarchy that they are able to retain as more birds reach the site. An ability to retain their rank might be achieved either through their improved knowledge of the area, or through having regained body condition promptly following autumn mi-

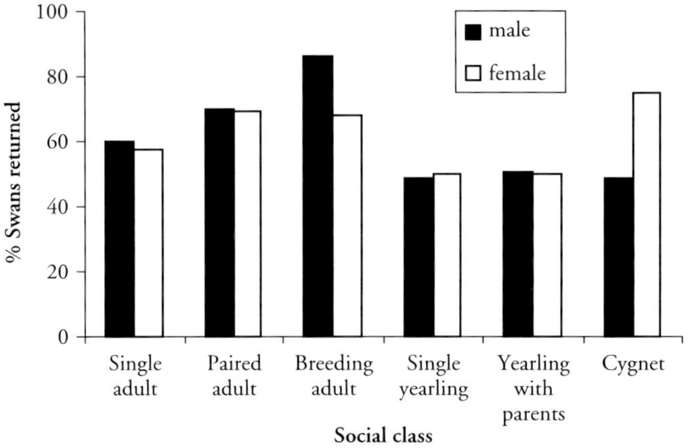

Figure 3.20 Proportion of male and female Bewick's Swans from each social class which returned to Slimbridge the following winter (n = 445 individuals monitored from 1989-90 to 1992-93). From Bowler 1996.

gration. Initial results found that, although dominant birds or social groups generally tended to arrive early and to leave late, there was no significant difference in the timing of migration between swans of different dominance ranks within each of the social classes (Figure 3.21; Rees 1988). Although further investigation is warranted, this may be at least partially due to feeding opportunities for the birds in more easterly parts of the wintering range (Denmark, Germany and the Netherlands) enabling the birds to refuel at these sites. Meanwhile, the swans' social behaviour, including the displays used in the aggressive encounters which determine the dominance ranking within the wintering flock, is described more fully in Chapter 6.

3.6 OTHER ASPECTS OF MIGRATION

3.6.1 Physical limitations

Most of the work to date on the mechanics of swan flight has been undertaken for Whooper Swans migrating between Iceland and Britain or Ireland (Pennycuick *et al.* 1996, 1999). However, the satellite tracking of Bewick's Swans along the western flyway has described their speeds (Beekman *et al.* 2002; WWT unpubl. data), and satellite transmitters have also been used to study the timing of Whistling Swan migration in North America (Petrie & Wilcox 2003). It was initially thought that swans fly at very high altitudes during migration. An exceptional flock of about 30 swans, presumed to be Whoopers, seen by a pilot at 8,200m off the Outer Hebrides, was tracked by radar until they descended over the coast of Northern Ireland (Stewart 1978). More recent studies, however, have shown that Whooper Swans usually migrate at between 100 and 400m (maximum height being 1,856m) when crossing the ocean, and that when crossing land none flew any higher than necessary to clear the terrain (Pennycuick *et al.* 1996, 1999). This agrees with estimates for another swan species, the Mute Swan, which suggest that these birds have sufficient muscle power only for a marginal rate of climb from sea level when carrying the fat needed to fuel long-distance migration (Hedenstrom & Alerstam 1992). Neither the Dutch nor the WWT studies used altimeters on the transmitters fitted to Bewick's Swans, but air

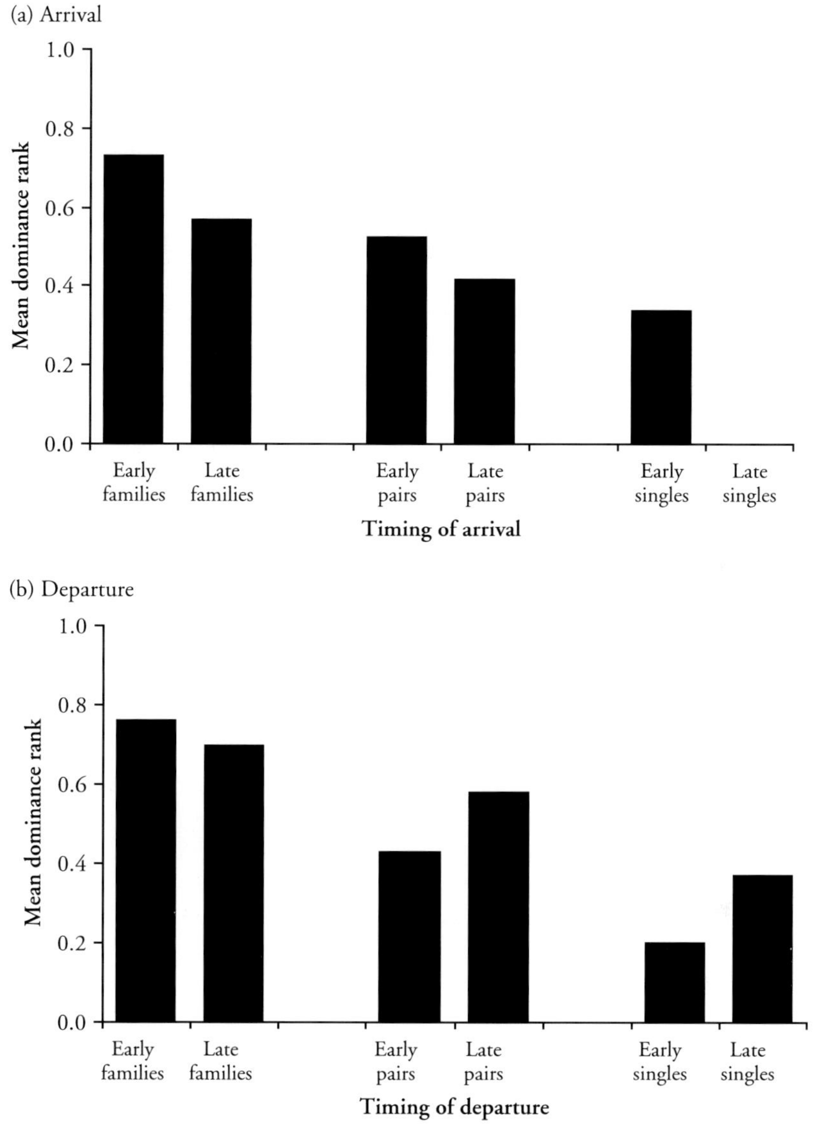

Figure 3.21 Association between dominance rank and the timing of arrival/departure for families, pairs and single swans. From Rees 1988.

temperatures and flight speeds indicated that, like the Whoopers, Bewick's Swans do not generally fly at high altitudes during migration (C. Pennycuick pers. comm.).

Flight speed generally depends upon wind conditions, and again upon the amount of body fat that the birds are carrying. Early studies, which estimated flight speeds by the pacing of swans in cars and aircraft, suggested that swan species can attain an average speed of 55–80km/h in still air (Ogilvie 1972b), and ground speeds of up to 140km/h were reported for Whooper Swans assisted by a jet stream (Owen & Black 1990). More precise measures for Whooper Swans migrating between Britain and Iceland showed that average ground speeds for the crossing varied from 17km/h to 76km/h, although this again depended on wind speed, wind direction and fat loading (Pennycuick *et al.* 1996, 1999). Bewick's Swans similarly reached an average of 64km/h to 70km/h during the Dutch satellite-tracking studies (Nowak *et al.* 1990; Beekman *et al.* 2002).

Bewick's Swans are smaller birds, but are also limited in the amount of fuel (fat) that they can carry to some 24% of body weight, which in turn limits the distance that they can travel. It was originally estimated that they need to feed and replenish their reserves after some 1,450km of non-stop flight, indicating that the birds have to stop two or three times during their 3,000–3,500km journey (Beekman *et al.* 1994a, 1996a). Studies of southbound migration have shown that the birds can fly up to 1,572km non-stop in autumn, and cover up to 2,300km with only short stops, but the importance of major stopovers for refuelling was confirmed (Beekman *et al.* 2002). Drent *et al.* (1981) also suggest, albeit for geese, that movements are timed to ensure that the birds feed on the most digestible and nutritious grasses as they travel to and from their breeding sites, particularly in their selection of sprouting vegetation in the spring. The availability of food resources at regular intervals along the flyway therefore is important for enabling the swans to complete the migrations between the breeding and wintering grounds. In spring they may also be delayed by adverse weather conditions, particularly at more northerly latitudes (Rees & Bowler 1991), which enhances the importance of cold weather refuges along the flyway. Weather is less likely to limit movement to the wintering sites in autumn, due to the generally milder conditions further south. Nevertheless, the swans may linger at northern feeding areas in autumn if conditions permit, such as in the exceptionally mild autumn of 2000, when migration through Estonia was recorded in mid-November (A. Kuresoo pers. comm.).

3.6.2 Orientation and navigation

Young Bewick's Swans remain with their parents during the autumn migration and their first winter, returning with them for at least part of the journey to the breeding grounds in spring (Rees & Bowler 1991). Although yearlings do not appear to consort with their parents on the breeding grounds, not least because the latter become territorial at this time of year, it is not unknown for the young to arrive on the wintering grounds with their parents for at least a second winter. This suggests that they have ample opportunities to learn when, how and where to migrate. On the few occasions when a cygnet has appeared at Slimbridge late in the season without its parents, and after the rest of the wild swans had left, it would make regular flights from the Rushy Pen but keep returning. This behaviour is similar to that of captive-bred geese in wildfowl collections, whose increased frequency of localised flights in autumn indicates migratory restlessness at this time, even though the birds do not move far from the area. Thus, unlike many other avian species, it seems that swans and geese do not have an innate knowledge of the direction in which to migrate, or of the location of areas used by the wild populations in the summer months (Rees *et al.* 2005a).

Bewick's Swan navigation when migrating therefore appears to be learnt from adult swans, although the cues the birds use to navigate have not been tested. It would seem likely that the general direction of flight is determined from celestial cues, particularly since swans wintering at Slimbridge frequently embark on spring migration at dusk, but they may also learn landmarks and chains of wetlands. In this way, traditions could be established and passed on from one generation to the next, although these may have to be modified following changes in the suitability of wetland habitats because of natural or artificial factors. Swans and many goose species therefore differ from ducks; ducklings do not migrate with their parents, and experimental work has demonstrated that they have an innate ability to orientate according to the sun and the stellar landscape (Matthews 1968; Wallraff 1972).

Bewick's Swans follow a much narrower migratory corridor than duck populations, and 'homing' (an ability to find traditional sites) is more prevalent, indicating perhaps that the birds use more integrated navigation mechanisms, such as a combination of the angle of the sun, the stars, the landscape and the earth's magnetic field (Rees *et al.* 2005a). Both topographical and celestial cues are likely to be important, because the birds migrate during both day and night. They are certainly aware of the landscape beneath them; the tendency for new arrivals to land first of all with the main herd (feeding elsewhere in Gloucestershire), rather than alighting immediately at Slimbridge, for instance, shows that they are able to locate temporary wetlands promptly upon arriving in a region. Use of the celestial landscape has not been demonstrated for Bewick's Swans, but the satellite tracking of Whooper Swans migrating between Britain and Iceland has shown that during the overseas crossing the birds continued as long as the sun or moon was above -4° in the sky, and the visibility was greater than 2km (Pennycuick *et al.* 1996, 1999). This indicates that the birds need a visible horizon when flying over water, and that they use primarily visual (celestial and/or landscape) cues to navigate, rather than the magnetic field or inertial systems. Quite remarkably, two birds that were forced by gales from their trajectory recovered their position and eventually made their landfalls in a direct line from their original flightpath, suggesting that they were aware of their position when up to 250km from the main migration route (Pennycuick *et al.* 1996, 1999). The swans' navigation system therefore appears to be both accurate and wide-ranging, whatever the methods the birds use to determine their location.

CHAPTER 4

Food and feeding ecology

'You should know by now, Eileen, that the swans just follow their stomachs.'

David Wilcox (Gloucestershire farmer,
in conversation with the author)

The 20th century saw major changes to the landscape on the Bewick's Swans' wintering grounds. Urban areas and associated infrastructures (e.g. roads and power lines) increased, the drainage of wetlands continued, and farming became more intensive, particularly in the second half of the century. The intensification in agriculture across Europe led to an increase in farmers' use of fertiliser on grasslands and more extensive planting of arable crops such as winter cereals, potatoes and sugar beet. A shift in Bewick's Swan feeding distribution during the second half of the 20th century saw the birds move from wetland habitats to agricultural land. The extent to which this movement onto farmland was driven by a lack of traditional foods, associated with the drainage and pollution of wetland sites, or rather a positive move due to the birds finding the arable crops more nutritious, remains debatable. Whatever the reason for the change in food selection, the increase in population size since the 1960s (Chapter 2) suggests that the swans fare quite well by feeding on arable crops and grassland during the winter months.

This chapter deals mainly with the winter food selection and feeding strategies of the swans feeding on aquatic vegetation in the Netherlands and on farmland in the UK. The swans' behaviour when feeding on aquatic vegetation and the factors affecting their foraging on pondweeds (an important food for the birds in autumn and on spring migration) have been studied in detail by Dutch ecologists. The results have provided a valuable insight into the relationship between the swans and their food supply. Information on the swans' feeding strategies when on farmland, including reasons for variation in feeding patterns between swans in the same flock, are derived mainly from the study of individual swans at Slimbridge, undertaken during the early to mid-1990s (Bowler 1996). Studies of Bewick's Swans along the flyway and in the breeding range have served to emphasise that their choice of food varies in relation to their energy requirements and food availability in different parts of their migratory range, so the more limited published information on seasonal variation in food selection is also described here.

4.1 FEEDING HABITS

Like other swans, Bewick's Swans have a large bill with serrated edges, and this is well suited to the different ways in which they feed. Strong jaw muscles enable the birds to bite firmly and tear vegetation, while the well developed nail on the tip of the mandible and the spiny surface of the tongue are valuable both for grasping and for chopping (Owen & Kear 1972). This is particularly evident when the birds feed on potatoes and sugar beet. The birds' upright stance and strong legs make it easy for them to walk and feed on land, where they also graze on flooded and dry pasture, and probe for roots and stolons. The presence of salt excreting glands, just under the skin above the eyes, enables the birds to feed in coastal and brackish waters as well as in freshwater habitats. When feeding on intertidal vegetation, salt is removed from the bloodstream and excreted in a concentrated solution through the nostrils (Owen & Kear 1972).

The long, muscular and mobile neck enables the swans to take underwater vegetation at a range of depths, with aquatic plants obtained mainly by immersing just the head and neck (head-dipping) or, if the water is too deep for head-dipping, by tipping the body forward (upending). Swans, therefore, are able to reach vegetation down to about 60cm and 85cm below the water surface during head-dipping and upending, respectively (Owen & Cadbury 1975). When head-dipping, Bewick's Swans usually keep their head under water for 15–20 seconds, during which time the bird can swallow continuously, looking up briefly between head-dips to breathe and check for intruders (Owen & Kear 1972). When upending, they may have their heads underwater for up to a minute at a time. They also feed by placing the bill and chin on the surface of the water and sifting particles in the water through the fine serrations along the bill and tongue.

Vigorous paddling or trampling with the feet, which gives the appearance that the bird is rocking from side to side, is frequently seen when the swans are feeding on water and helps to bring plants to the surface. Where underwater pondweed tubers are hidden in the first 30cm of sediment in the lake or riverbed, this trampling enables the birds to dig craters (typically about 1m^2) which expose the tubers. The birds are then able to feed on the tubers, either by head-dipping or upending (Beekman *et al.* 1991; Santamaría & Rodríguez-Gironés 2002).

4.2 FOOD SELECTION

4.2.1 Historical changes in winter diet

Bewick's Swans have traditionally fed on aquatic vegetation in wetland habitats, but there was a major change in their food and habitat selection during the 20th century. In the Netherlands, the swans formerly fed mainly on the below-ground parts of pondweeds (notably *Potamogeton pectinatus* and *P. perfoliatus*), and on other aquatic vegetation including Horned Pondweed *Zannichelia palustris*, hornworts, water milfoil species and stoneworts. They are also known to have grazed on emergent vegetation such as Common Reed *Phragmites communis* and Common Reedmace *Typha latifolia*, and to have taken rhizomes of eelgrass species in tidal basins (from the review in Beekman *et al.* 1991).

The Netherlands has been an important wintering area for Bewick's Swans since the early part of the 20th century, and changes in food choice have been well documented, albeit qualitatively, since then (Brouwer & Tinbergen 1939; Timmerman 1977; Mullie & Poorter 1977; Poorter 1978 and 1981; all reviewed by Dirksen *et al.* 1991). Until 1932 the eastern coast of the Zuiderzee was the only area where the species was seen regularly and in large numbers. Here they fed on eelgrass and pondweeds (Dirksen *et al.* 1991). The building of a 32km long dike (Afsluitdijk) to close the brackish embayment and create Lake IJsselmeer resulted in the disappearance of eelgrass, but the swans continued feeding on the spreading stands of pondweeds from 1932 to around 1940. From 1940 onwards the pondweeds on lake IJsselmeer were lost to land reclamation and eutrophication, but the swans continued feeding on aquatic vegetation on the newly created border lakes during the 1950s and 1960s until pondweeds diminished there as well. In 1968, after feeding on the pondweeds in the Border lakes for a few weeks, the swans suddenly moved to adjacent grasslands, and for several winters thereafter fed mainly on pasture and arable crops. The recovery of aquatic vegetation in the Border lakes, and the colonisation of the recently established Lake Lauwersmeer by pondweeds, resulted in the swans returning to their traditional feeding habits, at least in autumn, from 1975 onwards. But, having depleted the pondweed tubers upon arriving in the Netherlands in autumn, the swans still switch to arable land and grassland as the season progresses (Dirksen *et al.* 1991).

Bewick's Swans did not winter regularly in southern Britain until the 1920s (Chapter 2), and there is no study of their feeding habits in the early years, but it is likely that they fed mainly on soft meadow grasses such as Floating Sweet Grass *Glyceria fluitans*, Marsh Foxtail *Alopecurus geniculatus* and Creeping Bent *Agrostis stolonifera*, which ecological studies found to be their main source of food during the early 1970s (Owen & Cadbury 1975). Reed Sweet Grass *Glyceria maxima* and the starchy roots of Marsh Yellow-cress *Rorippa palustris* also appear to have been taken in late winter in the Ouse Washes, with increasing use of arable land also evident in southeast England by the mid-1970s (Owen & Cadbury 1975). By the early 1990s feeding on farmland was well established; an investigation of the distribution of Bewick's Swans at their feeding sites, undertaken by the Wildfowl & Wetlands Trust during winter 1990–91, found that 60% of the birds were on arable land (mainly on sugar beet and winter cereals, but with some on potatoes), 24% on permanent inland waters, and 13% on improved pasture (usually flooded) over the winter (Rees *et al.* 1997b).

In Gloucestershire, the swans were found to ingest mainly Common Ryegrass *Lolium perenne*, which comprised some 44% of the vegetation found in their droppings. It seemed that the birds did not particularly seek this grass, however, since intake was low in comparison with the quantity of ryegrass available (53% of vegetation in the fields; Rees 1990). Like the swans studied 15 years earlier on the Ouse Washes, the Gloucestershire birds appeared to prefer Marsh Foxtail, which constituted 18% of the vegetation available, but accounted for 32% of grasses ingested. Since Marsh Foxtail is more abundant in the damp depressions favoured by Bewick's Swans, however, it was not clear whether they ingested Marsh Foxtail because they preferred feeding on floodwater (perhaps for safety or because they like to drink whilst feeding), or whether they graze in shallow pools because of the higher density of Marsh Foxtail in these areas. The faecal analyses also found that other soft grasses were favoured, including the meadow grasses (*Poa pretense*, *P. trivialis*) and Timothy Grass *Phleum pratense* (Rees 1990). They avoided Yorkshire Fog *Holcus lonatus*; perhaps the hairs on its leaves make it unpalatable. Interestingly, bearing in mind that clover stolons are favoured by some goose species (Owen & Kerbes 1971; Owen 1980), only very small quantities of White Clover *Triolium repens* were taken, despite constituting 1% of the herbage available in the fields.

There is less information available on any changes in the swans' feeding habits in their breeding range and at northern migratory sites but, since there have been fewer changes in land-use and vegetation cover at the higher latitudes (particularly north of the Baltic), it seems likely that the swans' diet has remained fairly constant in these areas. In more southerly staging sites, however, there has been a major change in the swans' diet, similar to that observed in the wintering range, during the second half of the 20th century. For instance, the Danish staging areas, which are relatively close to the main wintering grounds, have similarly been affected by the intensification of agriculture occurring across western Europe. Denmark receives peak numbers of swans during the autumn and spring migrations, but also holds around 1,000 birds in midwinter. There, the swans showed a similar move from aquatic feeding to field feeding, but not until the late 1970s, five to ten years after they started using agricultural land in the Netherlands (Laubek 1995b).

More intensive farming in Estonia in the late 1970s and 1980s similarly resulted in swans spending more time on pasture and arable crops during migration through the Baltic region, although they continued using aquatic vegetation in coastal waters (Rootsmae 1990). The intensification of agriculture was not sustained, however, and field feeding declined after Estonia gained its independence from the Soviet Union in 1991 (L. Luigujõe and A. Kuresoo pers. comm.). The current situation is rather mixed. Although there has been a general reduction in land management, and this is thought to have reduced the quality and abundance of food at non-wetland sites for swans migrating through Estonia, the change in management has not been even across the country. Thus although the application of fertilisers and other management activity have largely diminished, moderate agricultural activity continues in some areas and maintains grass swards at a level that provides some field feeding for the swans. Moreover, intensive agriculture (including drainage and the use of fertilisers) continues in other areas, and is reducing the availability of aquatic vegetation for the birds (Ots & Kalamees 2000; Kalamees 2000).

4.2.2 Changes in food selected during the course of winter

As well as the major historical changes in the swans' feeding habits, seasonal variation in food selection occurs. Studies from different parts of the wintering range indicate that the timing of any movement from one food source to another varies, probably annually and in relation to food availability. In the Netherlands, the swans feed mainly on aquatic vegetation upon arrival in October and November, before moving to arable land (root crops, then winter cereals) in midwinter, and increasingly onto pasture from midwinter to spring (Dirksen *et al.* 1991). However, the Dutch studies have also shown that the use of aquatic vegetation varies between years (19% of swan-days in 1982–83, dropping to 7% in 1983–84), with the swans spending more time on grassland when feeding on aquatic vegetation declined (Dirksen *et al.* 1991). In both years the swans spent a similar amount of time on arable land, 16% and 17% of swan-days respectively.

The annual variation in the number of swans recorded feeding on aquatic vegetation in the Netherlands is to some degree related to fluctuations of pondweed abundance (Beekman *et al.* 1991). Ongoing studies may indicate whether the amount of pondweed available in one season is influenced by other waterbirds feeding on the leaves and stems of the plant in the summer, or by depletion of the tubers by the swans the previous winter (van Wijk 1988; Beekman *et al.* 1991). Initial findings indicate that the former is the case (Jonzén *et al.* 2002; B.A. Nolet pers. comm.).

The use of aquatic vegetation by Bewick's Swans in autumn was less marked in Denmark and the United Kingdom than by birds arriving in the Netherlands. More than 65% of birds seen in Denmark in October were recorded on stubble fields (Laubek 1995b), and 73% of the swans in the United Kingdom in November were found on arable land, including stubble fields (Rees *et al.* 1997b). The lower level of feeding on aquatic vegetation in Denmark, compared with the Netherlands, may partly be due to the presence of Whooper Swans on the shallow marshes and fjords in autumn. Whooper Swans generally are dominant over the smaller Bewick's Swans (Black & Rees 1984) and also feed on aquatic vegetation, thus potentially depleting the amount of food available for Bewick's Swans in the area. In the United Kingdom, aquatic vegetation favoured by the swans, such as pondweed, stonewort, tasselweed and eelgrass species, may perhaps be less abundant than in the Netherlands. In the United Kingdom and Denmark, the type of arable crop used varies during the winter, with stubble fields selected (where the swans fed on spilt grain following the harvest) in autumn. The birds subsequently transfer to root crops (sugar beet and potatoes) in Britain or brassicas ('OO' strain oilseed rape) in Denmark. From midwinter onwards, those swans on arable land move to shooting cereal fields, but a general movement from arable to pastureland also occurs as the winter progresses (Laubek 1995b; Dirksen *et al.* 1991; Rees *et al.* 1997b; Figure 4.1).

Surveys of Bewick's Swans in Germany in 1995 found that the birds mainly used grassland in both January (87.7% of wintering swans) and March (77.4%), but that the proportion of birds in maize fields and on other cereal crops increased between January and March (from 2.2% to 9.3% for maize; 0.9% to 9.9% for other cereals), whereas there was a decline in the swans' use of Oilseed Rape (from 9.3% to 3.1%) (Degen *et al.* 1996). Subsequent observations found that the swans feed mainly on oilseed rape at wintering sites in Germany, both in Niedersachsen (49%–82% of swans seen close to the River Elbe, Niedersachsen, in winters 1998–99 to 2003–04 were on fields of Oilseed Rape) and in the eastern part of the country, but if the rivers are in flood the swans switch to flooded grassland. Thus in the spring, when flooding becomes more frequent, the proportion of birds in flooded meadows rises (A. Degen pers. comm.).

The reason why swans show a general movement from arable to pastureland during the course of the winter has yet to be confirmed but may be due to depletion of the arable crop,

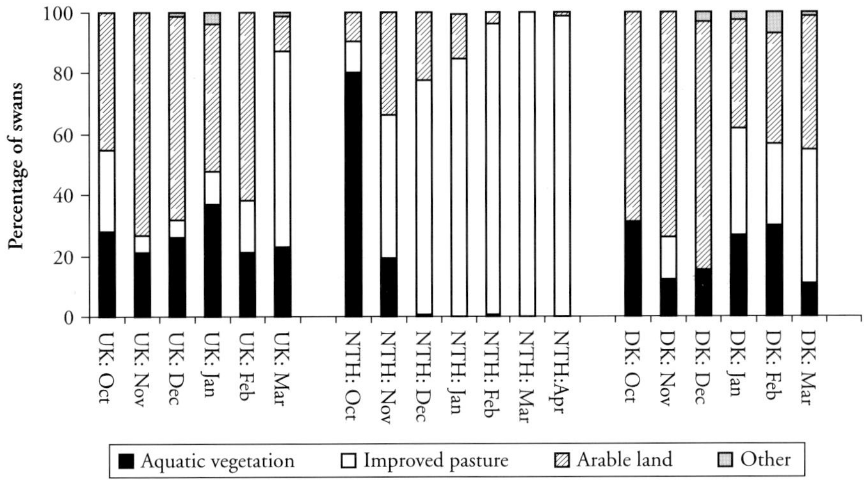

Figure 4.1 Percentage of Bewick's Swans recorded on different habitats in the UK, the Netherlands and Denmark at different stages of the winter. From Laubek 1995b; Rees *et al.* 1997b.

changing food quality in one or the other, a change in the food requirement of the swans, or to a combination of these factors. Certainly the depletion of root crops (notably sugar beet and potatoes) or the loss of these crops due to farming activity (autumn and midwinter ploughing) is one reason, because swans resume feeding on waste root crops where these are provided by a farmer or reserve manager late in the winter season (e.g. in March 1987 at WWT Slimbridge; Rees 1988). However, the switch by both Bewick's and Whooper Swans from winter cereal crops to pasture in late winter indicates that the swans also select grass swards at this time.

Monthly measures of biomass and protein content in the sward confirm that, as might be expected, vegetation quantity and quality (protein levels) both increase substantially between February and March, and again between March and April (Rees 1988). It is not yet clear whether this improvement in grassland habitat coincides with a relative reduction in food quality elsewhere; food availability is less likely to be the reason for the movement to grasslands in spring, since cereal crops are also growing and, except where birds are scared off by farmers, remain available to the birds at this time. Since the swans have to complete a 3,500km migration before embarking on their breeding programme, food with a high energy content is still likely to be important to the birds in late winter, to enable them to put on the body fat needed to complete the flight. High protein intake is also likely to be necessary in preparation for the breeding season, but whether this is most important in winter, during migration or in the breeding range requires further investigation.

Monthly surveys of Bewick's and Whooper Swans wintering in Britain and Ireland made during the 1990–91 winter were used to determine the importance of both habitat and geographic location in influencing feeding site selection by the birds (Rees *et al.* 1997b). Whereas there was a southward shift in the distribution of Whooper Swans (which come predominantly from the Icelandic breeding population) as the winter progressed, over 60% of Bewick's Swans were concentrated in east-central England (notably on the Ouse Washes) from November to February inclusive. Within these areas the percentage of birds recorded on different habitats varied, with an increased used of grasslands later in the season mentioned above. Moreover, there was also a change in the percentage of swans recorded on different crops, with stubbles used in autumn and root crops in midwinter, and with brassicas and winter cereals also used from December onwards (Figure 4.2).

The surveys also found a bias in the midwinter distribution of Bewick's Swan families across United Kingdom swan habitats, in that a higher percentage of juveniles occurred in flocks feeding on permanent inland waters than on pasture or arable land (Rees *et al.* 1997b). Information concerning the range and phenology of food available in the inland waters would help to confirm the reason for families apparently selecting these sites. Meanwhile, it is also possible that cygnet distribution reflects flock sizes on different habitats, rather than families generally preferring to feed on aquatic vegetation. The percentage of cygnets generally is higher in smaller flocks (Section 4.5.1), and average flock sizes are substantially lower on permanent waters (41.6 birds per flock) or improved pasture (38.5) than at the arable sites (90.5) used by most of the swans in the United Kingdom in January (Rees *et al.* 1997b). It should be noted that this pattern is derived from just one winter's study, and that further survey work is needed to determine whether the distribution of cygnets and of Bewick's Swan flocks across habitats are consistent over several years. Given that families are the most dominant group within the winter flocks (Chapter 6) they should be able to hold their own at the prime feeding sites.

4.2.3 Feeding at migratory sites

Food intake in more northerly parts of the flyway has not been studied in detail, but clearly is related to food availability, not least because the vegetation communities are somewhat homogenous at high latitudes. Submerged macrophytes remain an important food at migratory sites. Although movement to cultivated land in Estonia was recorded in the late 1970s and 1980s (Rootsmae 1990), the swans still fed mainly on pondweeds and stoneworts, with the proportion feeding in fields fluctuating substantially between years, and recent major changes in land-use no longer favouring selection of agricultural areas (Luigujõe *et al.* 1996).

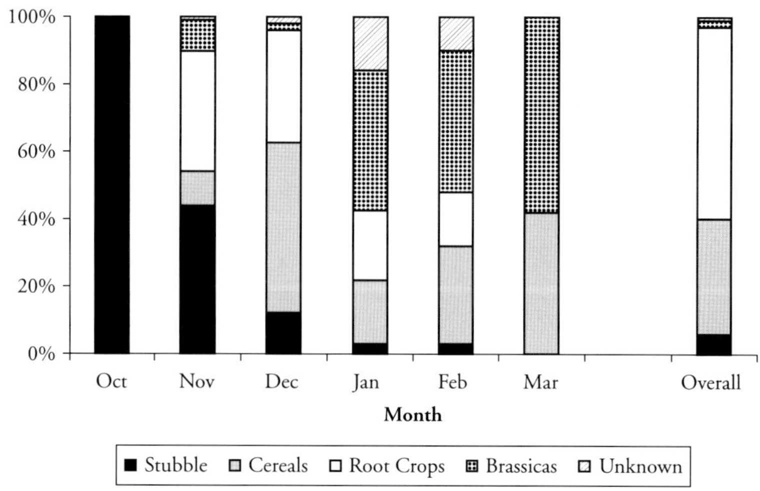

Figure 4.2 Percentage of Bewick's Swans recorded on different arable crops during the 1990-91 winter. From Rees *et al.* 1997b.

Further north, recent studies of Bewick's Swans staging in the White Sea region (Dvina Bay) have found that the birds there feed almost exclusively on submerged waterplants, notably pondweed tubers, eelgrass and stoneworts, usually at low tide when these foods are within reach (Nolet *et al.* 2001b). Some supplemental feeding on emerged food plants such as reed rhizomes and bulrush (*Typha* spp.) leaves occurred around high tide, and it remains possible that the swans also feed on grass rhizomes on flooded pasture, but alternatives to submerged macrophytes are rare at this latitude (65°N) in spring (Nolet *et al.* 2001b). The extent to which the swans feed on different types of aquatic vegetation in Dvina Bay is variable, and apparently dependent on food availability, with swans feeding on stoneworts when ice cover means that no alternatives are available. It seems that the swans may prefer pondweeds, which they move to as soon as the pondweed beds are free of ice, but whether the swans prefer pondweed to eelgrass remains uncertain. In Nolet's study, the swans were able to feed on eelgrass for longer periods (12.8 hours per day) than on pondweed tubers (7.9 hours per day, with an additional 1.2 hours per day on reed rhizomes and bulrush leaves) because eelgrass leaves were within reach during rising and falling tides. There may be nutritional compensation for feeding on eelgrass rather than pondweeds, however, since although eelgrass has lower soluble carbohydrate levels, the young leaves contain about twice as much protein (Nolet *et al.* 2001b).

Recent satellite-tracking studies have shown that Bewick's Swans usually skip the White Sea during autumn migration, or stop there for just a few days (Nolet *et al.* 2001b; Beekman *et al.* 2002; WWT unpubl. data). The difference in staging patterns may be due to the swans' need to replenish their body reserves in spring to enable them to defend their territories and to start laying eggs upon arrival in the breeding range, issues that are not relevant during southward movement in autumn. In autumn, the swans probably arrive earlier on the wintering grounds by spending a little more time in topping up in the Pechora Delta then skipping the White Sea, than by staging and refuelling at the White Sea sites. The advantage to individual swans of arriving early at Baltic staging areas and on the winter grounds may include having primary access to food resources and the opportunity to establish a dominant position in the winter flock (see Chapter 6). It is possible that these factors outweigh the value of refuelling earlier. Moreover, intensive staging in the White Sea in autumn would reduce food availability in the region in spring, because the tuber stock would not grow during the winter period (Beekman *et al.* 1991; Nolet *et al.* 2001b). Thus there appears to be a substantial, albeit indirect, benefit to the birds in missing this potential migratory site in autumn, in that it ensures an undiminished food supply for them in spring.

4.2.4 Food selected in the breeding range

On the breeding grounds, families feed in marshes and the margins of lakes, and graze on tundra, taking tundra sedges *Carex aquatalis* and *C. lachenalii*; grasses *Dupontia fisherii*, *Arctophila fulva* and *Calamagrostis neglecta*; and some herbaceous plants such as *Stellaria humifusum*. Tips of Arctic Marsh-grass *A. fulva* are particularly important in early spring when little else is available. The broader diet in late summer includes *Carex* seedheads, tundra berries (*Empetrum nigrum* and *Rubus chamaemorus*) and the leaves and shoots of pondweeds (*Potamogeton filiformis*), depending on availability (Rees *et al.* 1997a). Analysis of Bewick's Swan droppings, collected as part of the Russian–Dutch Pechora Delta Project (1995–1999), was undertaken by inspecting the epidermal cell structures of vegetation in the droppings through a microscope, and identifying the plant species from the unique patterns of the underlying cells down to the species or genus level. These confirmed that Bewick's Swans mainly ingested sedges and grasses, although dicots (species not known), mosses and berries were also found in the droppings. Pondweeds were eaten mainly by Bewick's Swans moulting in Korovinskaiya Bay (Ubels *et al.* 2000).

The analysis of swan droppings also indicated some changes in the swans' diet during the summer. Sedges proved important throughout the season, but particularly in autumn. Of the grasses, *A. fulva* (roots and leaves) was taken mainly in spring, with other grasses (*Poa* spp., *Dupontia psilosantha*, *Festuca arenaria*, and *Alopecurus pratense*) ingested in relatively low quantities later in the year. Diet also varied with location, however, with the range of food ingested seemingly more diverse in upland areas. There, sedges were marginally less important with cotton grasses, a range of grass species, dicots, berries and horsetails all taken more frequently at upland sites (Ubels *et al.* 2000).

4.2.5 *Occasional feeding on animal matter*

Although mainly vegetarian, the swans have been seen feeding on mud flats in Gloucestershire in spring where only estuarine invertebrates *Corophium* spp. and *Nereis* spp. occur (Bowler 1996; Rees *et al.* 1997a). Bewick's Swans have not been noted feeding on estuarine invertebrates elsewhere, but the conspecific Whistling Swan feeds heavily on estuarine molluscs (*Mya arenaria* and *Macoma baltica*) whilst wintering on Chesapeake Bay in Maryland (Stewart & Manning 1958; Earnst & Bart 1991).

Bewick's Swans' downy cygnets take adult and immature mosquitoes in the breeding range; the parents probably also do so incidentally, but do not appear actively to seek mosquitoes as part of their summer diet. The tadpole shrimp *Notostraca*, however, may form a significant part of the summer diet for breeding birds in far eastern Siberia (Kondratiev 1991b), and it remains to be seen whether invertebrate food, which has a high protein content, has a brief but important role as nutritional requirements change during the life-cycle (Bowler 1996).

4.3 STRATEGIES WHEN FEEDING ON AQUATIC VEGETATION

4.3.1 *Intake of aquatic vegetation*

Dutch scientists have studied Bewick's Swans' feeding strategies at two key stages in their annual migratory cycle: (i) on arrival in the wintering grounds, when up to 7,000 birds congregate at Lake Lauwersmeer in the Netherlands, and (ii) whilst staging at Dvina Bay on the White Sea in Russia in spring. At both Lake Lauwersmeer and at Dvina Bay they feed on below-ground fennel pondweed tubers, in water 30–70cm deep (Nolet *et al.* 2001a; Nolet & Mooij 2002). In both areas, the extent to which the swans depleted the tubers across the study site was determined by taking 10cm diameter cores to a depth of at least 30cm, then measuring the mass of dried tubers per core (Nolet *et al.* 2001a). Thus swan distribution and feeding activity could be related to the amount of food available.

Observations on the Lauwersmeer found that the swans prefer to feed in shallow waters rather than in areas of deeper water in the first instance (Nolet *et al.* 2001a). It was expected that upending, which is necessary in order to reach the tubers in deeper waters, would be more energetically costly than head-dipping, and also that feeding time would decrease as the effort needed to feed increased. Yet the birds spent 1.3–2.1 times longer feeding (with their heads under water) when upending than when head-dipping. They also spent longer periods trampling for food with their feet during the upending bouts, and trampling time increased in deeper waters (Nolet *et al.* 2001a).

The Lauwersmeer study also showed that the swans first take tubers in the sandy sediments in the northern part of the lake, before moving to clay sites at the southern end (Nolet *et al.* 2001a). Tuber density and distribution recorded before and after the swans had been feeding at the Lauwersmeer in autumn confirmed that depletion was greater on sandy than on clay soils, reflecting the swans' use of the sandy areas on arrival and for longer periods. Handling the core samples gave the Dutch researchers further insight into why the swans initially feed in sandy substrates; like the swans, they found that it was easier to extract pondweed tubers from sand than from clay (Nolet *et al.* 2001a).

Recent experimental work has shown that the food intake rates do not differ between birds feeding in shallow and in deep waters if the food is buried to the same depth in the sediment, and if the food is within reach. In the field, however, most pondweed tubers are deeply buried and cannot be reached by the swans in deep water (Nolet *et al.*, in press a). The swans therefore give up feeding in deep water more rapidly than in shallow water, despite high densities of accessible tubers remaining. Assessment of tuber depletion, to the point at which the swans move from one site to another (the so-called 'giving-up density'), found that there was substantial variation in the densities of accessible tubers remaining at the time of the move, and that rich patches were underexploited relative to poor ones. This underuse of rich food patches results in the swans missing some good feeding areas within the pondweed bed, and leaves it looking like a Swiss cheese with lots of holes in it. Nolet *et al.* (in press b) suggest that the patchy food depletion is again due to the depth to which the tubers are buried. Where the swans find food, they feed on the tubers to low levels (the holes) at localised foraging spots but, because the tubers may be out of reach on first checking a new place within the bed, they move elsewhere prematurely and leave the pondweed bed whilst there's still 'cheese' still remaining. Variation in the biomass of tubers remaining when the swans move to a new area (the giving-up biomass) therefore is due more to the relative accessibility of the tubers in the sediment, rather than to feeding being more energetically expensive in deep water (B.A. Nolet pers. comm.).

4.3.2 Searching for food under water

Bewick's Swans use a combination of feeding strategies to ensure that they maximise their food intake rates when feeding on underwater tubers (Nolet & Mooij 2002). The distribution of tubers within a feeding site tends to be patchy, and the swans have no visual cues to identify areas of high tuber density. In these conditions they alternate between localised intensive searching and longer distance searching. When searching locally for tubers the swans tend to move quite slowly and to follow a sinuous path, turning regularly to feed in the tuber patch, but on moving between patches they swim or fly more rapidly and directly until another clump of food is found. Computer simulations confirmed that, provided that the tubers occur in clumps, the swans are able to achieve a greater food intake and higher energy gains by using a combination of these two feeding strategies than if they were to use a single method of searching, or to move at random (Nolet & Mooij 2002). Thus the swans appear to use the optimal tactics for feeding in this type of environment.

4.3.3 Timing of switching between habitats

After feeding on pondweed tubers for 2–4 weeks in autumn, Bewick's Swans on the Lauwersmeer switch to other habitats, predominantly sugar beet left after the harvest

(Beekman *et al.* 1991; Nolet *et al.* 2002). Although the factor most likely to affect the timing of this change is the depletion of the tubers, other variables such as risk of predation, food quality or changes in nutritional requirements may also be responsible.

The timing of the swans' switch from feeding on pondweed tubers in the Lauwersmeer to feeding on sugar beet in fields around the lake therefore was investigated. The investigation was to determine whether the switch is regulated primarily by the birds achieving a higher net daily energy intake at the new habitat compared with the old one (Nolet *et al.* 2002). Since the swans feed on sugar beet only during the day (they fly to the Lauwersmeer to roost), and an earlier study had shown that the swans also feed on pondweed tubers at night (van Eerden *et al.* 1997), the possibility that the swans adopt a mixed feeding strategy (i.e. feeding on both tubers and beet in the same day) was taken into account. Energy expended during the day by swans on the different habitats was calculated from the amount of time that the swans spent in each activity (recorded during field observations) and from the energy cost of each activity (derived from heart-rate measurements in captive swans; also see Section 4.6). Energy intake each day was estimated by multiplying the time spent foraging with calculations of digestible (metabolisable) energy, in the form of carbohydrates, in the food ingested (Nolet *et al.* 2002). Pondweed tubers proved to be highly digestible, with 90% of the available energy extracted by the swans (Nolet *et al.* 2002).

In the first year of observations (1995) the movement from tubers to beet was indeed induced by the depletion of pondweed tubers, but in the second year (1996) it was due to high water levels on the lake. Differences in daily energy expenditure were relatively small (2.3 x basal metabolic rate, or BMR, when feeding on pondweed; 2.0–2.5 x BMR when on beet), but the daily energy intake declined substantially as the tubers were depleted or became inaccessible due to water level rises (Nolet *et al.* 2002). Most swans switched habitats when the net energy gain in the day achieved by staying on the pondweeds fell to just below that gained from feeding only on beet. However, the analyses indicated that the swans would attain a substantially higher energy gain by feeding on both beet and tubers, and there was evidence both from van Eerden *et al.* (1997) and from the time-budget data that mixed feeding does occur in the Lauwersmeer area. Thus it seemed that the swans switch to the beet fields long after they would first benefit from doing so due to energy gain alone (Nolet *et al.* 2002).

There are a number of possible reasons for the delay in changing habitats, which are considered in detail by Nolet *et al.* (2002). These include protein requirements, since tubers contain more protein than sugar beet (14.8% versus 6.6% of organic matter; van Eerden *et al.* 1997), and protein levels in grass swards are also thought to affect site selection by Bewick's Swans feeding on pasture (Rees 1990; Section 4.4.1). Perceived predation risk (from foxes and wildfowling) and disturbance levels may be higher on agricultural land, making this less attractive to the swans whilst they are still obtaining sufficient food and energy from the tubers alone. The birds may also be reluctant to change sites, and particularly to switch to a site that involves twice-daily flights between the feeding and roosting areas, but it should be noted that the energy cost of these flights was accounted for in the comparison of aquatic and field feeding (Nolet *et al.* 2002).

4.3.4 Effect of swan grazing on their food supply

Although pondweed tubers play an important role in providing a food source for Bewick's Swans in autumn and at spring staging sites, how grazing by swans and other waterbirds affect the distribution and abundance of the vegetation is still poorly understood. In temperate regions, the shoots (ramets) of Fennel Pondweed emerge from mid-April, with green growth occurring from mid-April to mid-September, and the tubers are produced from July to September. During the winter, when no growth occurs, the plant survives as underground tubers.

The relationship between the swans and their food supply is actually quite complex; by feeding on the tubers in autumn, the swans clearly reduce tuber and ramet density for the following spring, but simulation models suggest that this thinning of the tuber-bed serves to enhance tuber growth rate later in the year (Jonzén *et al.* 2002). Thus it seems that, theoretically, by feeding on pondweed in autumn the swans may have a positive affect on their food supply 12 months later, since they do not rely on the Lauwersmeer for food throughout the year. The impact of feeding by other waterbirds would also affect this cycle, however, and there is some evidence that increased grazing of ramets by Eurasian Coots *Fulica atra* during the summer months has reduced tuber densities the following autumn (L. Santamaría unpubl. data in Jonzén *et al.* 2002). Idestam-Almquist (1996) also found that waterfowl numbers had a major influence on the abundance of Fennel Pondweed in the Baltic.

The annual cycle of pondweed growth and senescence means that the effects of swan grazing on pondweed abundance would vary with the time of year (Jonzén *et al.* 2002). For this reason, intensive feeding in the White Sea in spring does not damage food availability for the following spring, because the tubers regenerate during the summer months. Intensive feeding by the swans at the White Sea in autumn would have a detrimental effect on food supply the following spring, however, and make the site less suitable for refuelling during spring migration, because the tubers do not grow over winter (Beekman *et al.* 2002). Consequences of feeding at the White Sea in autumn (which might occur should the swans encounter persistent head winds during migration), could result in a longer staging period at the site the following spring (and thus late onset of the breeding season). Alternatively, it could result in the birds putting on less body fat during spring migration (Beekman *et al.* 2002). Either scenario would likely impinge on the success of the breeding season in that year, should the birds arrive late or be in poor condition on reaching the breeding grounds (Chapter 5).

4.4 STRATEGIES WHEN FEEDING ON FARMLAND

4.4.1 Movement between feeding areas during the winter

The previous sections describe the swans' strategies for optimising their intake of aquatic vegetation during migration and on arrival in the wintering range, but the birds also feed on farmland for much of the winter. Different variables therefore may determine feeding site selection on terrestrial habitats, particularly since (unlike when feeding on pondweed tubers) the birds are able to see the range of food available in the vicinity. Researchers at the Wildfowl & Wetlands Trust have investigated factors affecting the swans' distribution at feeding sites across Gloucestershire during the 1986–87 winter (Rees 1990), then again during the mid-1990s (Bowler 1996), to determine factors affecting their selection of fields for feeding. Livestock farming was the main type of agriculture in the area, so the land was mostly put to pasture, and the swans chose between a range of grass fields during the winter months.

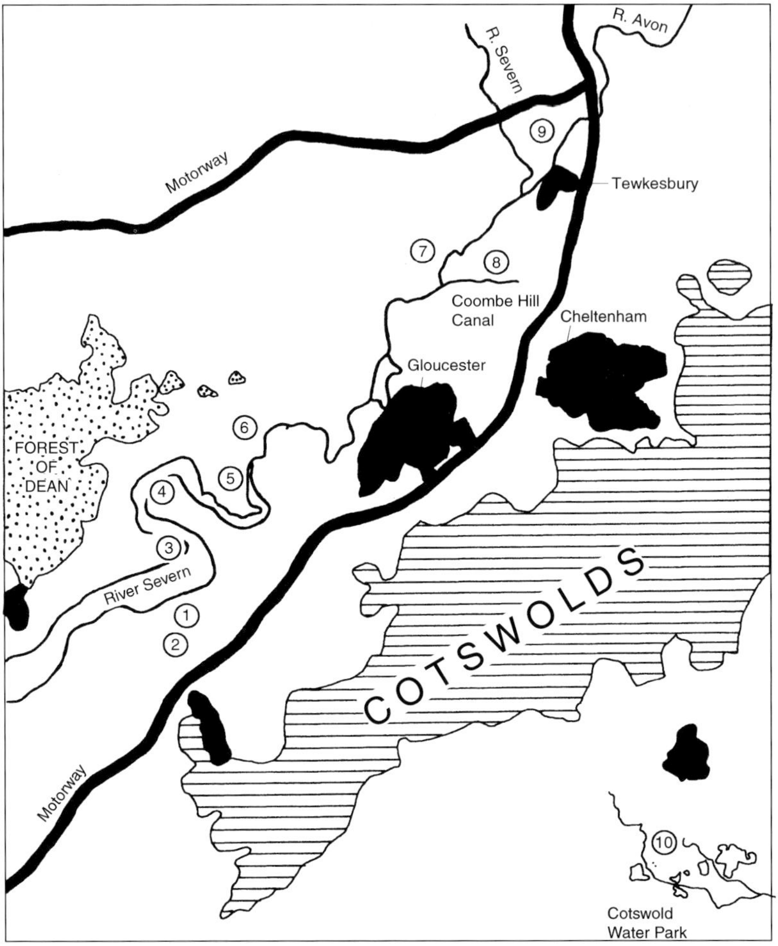

Figure 4.3 Location of feeding sites regularly used by Bewick's Swans wintering in Gloucestershire in the mid-1980s: 1. New Grounds; 2. The Moors; 3. Awre; 4. Arlingham; 5. Rodley; 6. Walmore Common; 7. Ashelworth/Hasfield Hams; 8. Coombe Hill; 9. Avon Meadows; 10. Cotswold Water Park. From Rees 1990.

Both studies of Bewick's Swan distribution in Gloucestershire found that the birds fed together in a small number of large flocks in autumn, but by January they were more dispersed, with the birds feeding in smaller groups at different sites across the county. Generally they returned to Slimbridge to roost, but occasionally remained overnight at Walmore Common (some 10km from Slimbridge on the other side of the River Severn; Figure 4.3) if the fields there were flooded. In 1986–87, the few birds that had arrived by the end of October all frequented the Wildfowl & Wetlands Trust's reserve at the New Grounds at Slimbridge. In November and December they again used the reserve but also fed regularly at two other sites (Rodley and Walmore Common), and by January and February they had dispersed more widely to a greater number of the Gloucestershire sites whilst continuing to use several fields on the reserve at Slimbridge (Figure 4.4a,b). In March 1987 most of the swans returned to the New Grounds and concentrated on one field

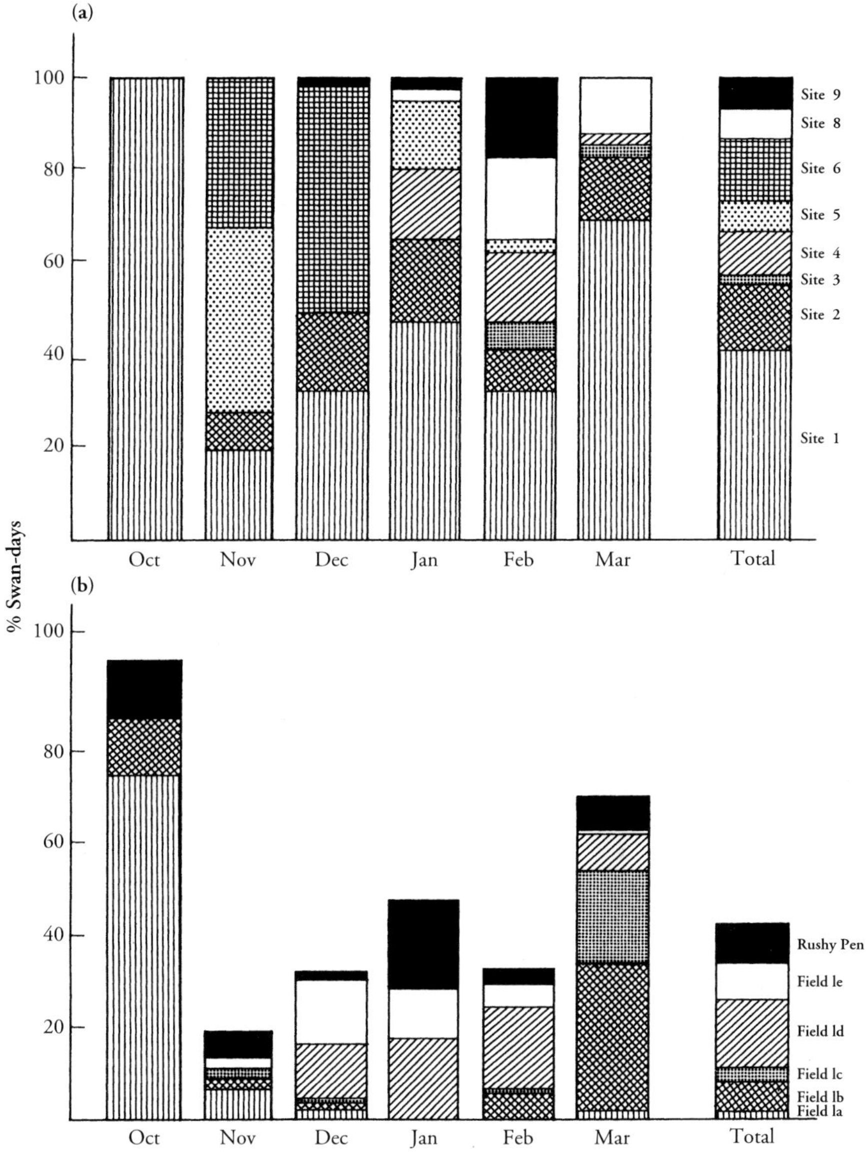

Figure 4.4 Distribution of Bewick's Swans each month: (a) throughout Gloucestershire; (b) at the New Grounds, Slimbridge. Names and locations throughout Gloucestershire are given in Figure 4.3, with Site 1 = the New Grounds. From Rees 1990.

where potatoes had been spread along the edge of a permanent pool, which clearly appealed to the birds but effectively terminated the study of feeding distribution in that year! The study in the mid-1990s, whilst finding that dispersal to a wider range of sites occurred slightly earlier (in December and January), also found that they used fewer fields at the end of the season (February and March) despite potatoes not being provided in these years. Thus although the pattern of increasing dispersal across several fields in midwinter may be at least partly due to depletion of the vegetation, coupled with the swans' need to maintain their food intake rates, it could also reflect seasonal variation in the number of birds in the area (Bowler 1996).

The extent to which Bewick's Swans used different fields in Gloucestershire (measured as the number of swan-days recorded for each field each month) was considered in relation to habitat variables (also measured monthly) in fields traditionally used by the birds during the winter season. Feeding site selection by the birds was found to depend on both the quantity and quality of food in each field each month. The number of swan-days recorded in those fields traditionally used by the birds was associated primarily with the biomass of green grass available (Rees 1990), but with protein content proving a significant secondary factor after biomass had been taken into account (Rees 1990). Moreover, when other grazing species of waterfowl (European White-fronted Geese and Eurasian Wigeon) were also present, site usage was more closely correlated with maximum sward length than with biomass, perhaps because the grazing activity of the geese and Eurasian Wigeon caused some thinning of the sward. There was no significant association between the depth of flooding and site selection, although swans were more likely to be seen on flooded rather than non-flooded fields. Subsequently, the number of swan-days recorded for each field each month was found to be associated not with the depth of water on the land but with the area of flooding (Bowler 1996). After initially increasing with the area of flooding, swan use actually declined on heavily flooded fields, presumably because the swans were unable to reach vegetation in deep water.

Further assessment of Bewick's Swan distribution in Gloucestershire found initially a negative relationship between the percentage of swan-days per field and grass biomass at the site (Bowler 1996), which contrasted with the positive association found earlier (Rees 1990) and contradicted conventional rules concerning optimal foraging strategy. However, by the time that this later study was undertaken, in the early 1990s, Bewick's Swans wintering at Slimbridge had segregated into two roost groups with different distribution patterns (see Chapter 6), and on investigating habitat selection for each roost group it transpired that the birds were selecting fields for different reasons. Swans roosting within the Wildfowl & Wetlands Trust's enclosures (the 'Grounds' roost) spent less time feeding in the fields and appeared to select daytime sites with both low biomass and low protein levels close to the roost, apparently having gained sufficient food from the grain distributed in the pens. The low biomass and protein levels at fields close to the roost were thought to stem from the less intensive agriculture practised on the refuge compared with that on neighbouring farmland. Swans from the Rushy Pen roost, on the other hand, spent more time feeding during the day, travelled further and selected fields on the basis of high protein content as well as percentage water cover. This suggested that they selected their feeding sites more critically than swans from the Grounds roost, perhaps because they gained less food from the distribution of grain at the roost site, due to the number of other waterbirds frequenting the Rushy Pen (Bowler 1996).

4.4.2 Variation in time spent feeding

Several studies of Bewick's Swans' feeding activities have given differing information concerning diurnal and seasonal variation in feeding rates, which may be linked to the different conditions (i.e. sites, day length, weather conditions, energy requirements, food being ingested and food abundance) in which the birds were observed. The earliest study of the swans' feeding patterns in winter found that the proportion of the birds feeding

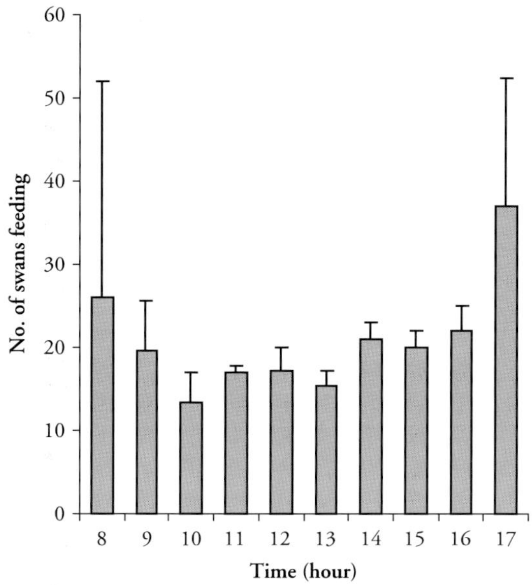

Figure 4.5 Proportion of adult swans recorded feeding for each hour of the day, during flock scans of Bewick's Swans in fields in Gloucestershire for the 1990–91 to 1992–93 winters inclusive. Derived from Bowler 1996.

increased during the day when the birds were on water (Owen & Cadbury 1975). When feeding on pasture, however, feeding activity peaks in the morning and late afternoon if the birds are undisturbed (Bowler 1996; Figure 4.5). Although they usually fly to feeding areas in the morning and return to the roost at dusk, swans may continue feeding at night under a bright moon, particularly if on bays, lakes or floodwater, where they are safer from predators than they are on dry land (Rees & Bowler 1991; Bowler 1996). During summer, while at the breeding grounds, feeding continues in bouts throughout the 24 hours, but with longer rest periods during the Arctic 'night' (*c.* 22:00–04:00 hrs; Rees *et al.* 1997a).

During the 1990s, adult Bewick's Swans in the fields around Slimbridge spent 33% of their time feeding (Bowler 1996), compared with 54% recorded for swans at the same site during the mid-1980s (68.5% before and 48.5% after 1 January; Rees & Bowler 1991), and 60% for Bewick's Swans on the fens at WWT Welney (Cadbury 1975; Scott 1980a). Cygnets spend significantly more time feeding than adults (39% and 33% respectively, in Bowler 1996), with comparatively less time devoted to other activities. The discrepancy in the amount of time that the swans spent feeding in the two Slimbridge studies may at least be partially due to neither study controlling for any variation in the weather conditions. There may also have been some inconsistency in the activities recorded as feeding. For instance, the second study recorded feeding during flock scans only if the swans had their heads down to graze or were evidently ingesting food items just collected; steps between pecks were not included. Certainly, more detailed continuous monitoring of the foraging behaviour of individual swans has shown that, during feeding bouts, the birds spend 71.9% of their time with their heads down feeding; 17% with the head up being vigilant; 7.8% moving between feeding patches; and 3.3% in other activities (Bowler 1996).

These estimates of the amount of time that Bewick's Swans spend feeding are similar to figures reported for Whooper Swans at Rostellan Lake, Co. Cork, which spent 40% of their time feeding in winter (O'Donoghue & O'Halloran 1994); for Whooper Swans in fields in central Scotland (around 60% of time spent feeding in the day, Rees *et al.* 2005b); and also to the activities of the more sedentary (non-migratory) Mute Swans when on water. The latter spent 42% of the day feeding whilst moulting in Gloucestershire (Ryley & Bowler 1994) and 36% of the day feeding when wintering at Cork Lough, Ireland (Keane & O'Halloran 1992). Bewick's Swans at Slimbridge may generally spend less time feeding when in the fields during the day because of the provision of grain at the roost site. Their feeding activity was certainly lower than that of other grazing wildfowl in winter: 90% feeding for European White-fronted Geese (Owen 1972), 70% for Pink-footed Geese *Anser brachyrhynchus* (Giroux & Patterson 1995), 75–85% for Barnacle Geese *Branta leucopsis* (Ebbinge *et al.* 1975), 74% for Eurasian Wigeon (Mayhew 1988) and 60% for Gadwall *Anas strepera* (Paulus 1984). Variation in the amount of time spent feeding is also likely to be due to the effects of other factors, however, such as feeding habitat, prevailing weather conditions and the time of year. For instance, although the swans at Matsalu Bay, Estonia, spent 59% of their time feeding (and 24.5% sleeping) over a ten-day period in spring 1988, only 25.5% of their time was spent feeding (compared with 38.2% sleeping and 12.3% preening) on the day when large numbers of birds arrived in the area. This difference was thought to be due to the birds needing to rest following a long flight to the migratory site (Rees & Bowler 1991).

It might be expected that, as suggested for geese (Drent & Prins 1987), the rate at which swans ingest their food would vary with food availability, and that they would increase their feeding rate or change their diet in response to a reduction in food supply. In his study of White-fronted Geese wintering at Slimbridge, Myrfyn Owen found that the geese increased their pecking rate as the food supply diminished towards the end of the winter (Owen 1972). Bewick's Swans wintering at the same site spent a greater proportion of their time feeding in early February, when food supply was lowest (Bowler 1996; Figure 4.6). Further assessment of factors affecting feeding activity found that, in addition to the

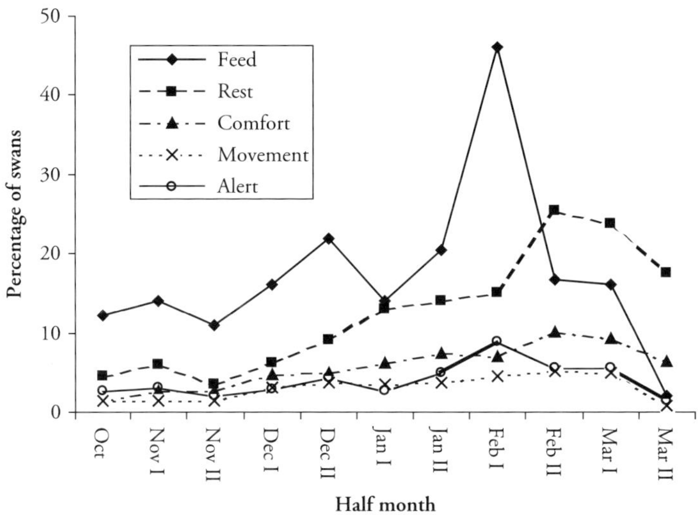

Figure 4.6 Proportion of adult swans recorded feeding each month, during flock scans of Bewick's Swans in fields in Gloucestershire for the 1990–91 to 1992–93 winters inclusive. Derived from Bowler 1996.

significant effects of time of year and time of day, feeding activity increased significantly with water cover and with protein content of the sward in both adults and cygnets. Biomass of vegetation and flock size did not seem to affect the percentage of birds recorded feeding, but flock size was positively correlated with the proportion of adults involved in comfort activities (preening or bathing) and social activities (mainly aggressive interactions, Bowler 1996).

The day length during the study of the swans' feeding activities in Estonia in 1988 (21 April–3 May) was about 15.4 hours, giving an average of 9.1 hours per day spent feeding when on Matsalu Bay (Rees & Bowler 1991), although it is likely that individuals differ in their feeding activities. The mean day length in Gloucestershire in autumn is just 8.6 hours so, even using the higher estimates of feeding activity at the site (from Rees & Bowler 1991), the average length of time that the birds spent feeding was about 5.9 hours per day from late October to the end of December, and 4.9 hours per day from 1 January to early March (the latter based on an average day length of 10.1 hours). Both are appreciably lower than the 9.1 hours per day feeding at Matsalu Bay (Rees & Bowler 1991). The comparatively high proportion of daylight hours spent feeding in Gloucestershire in autumn may reflect the swans' need to replenish their nutrient reserves, following the energy demands of autumn migration, by maximising their food intake during the short periods of daylight on arrival in the wintering range. They certainly put on weight at this time; weights recorded for Bewick's Swans wintering at Slimbridge increase during November and early December, but reach a plateau towards the end of the year (Evans & Kear 1978; see Chapter 3).

The author monitored the feeding activity of Bewick's Swan pairs seen staging at Matsalu Bay in spring to determine whether there was any difference in feeding intensity between members of a pair. It was considered that females might need to feed more intensively in order to prepare for egg-laying and incubation, but that breeding males would also need to gain condition prior to defending their breeding territories. Whilst head-dipping for food, paired males and females had their heads down feeding for the same amount of time (69.8% and 69.2% of the total observation period, respectively), and neither sex looked up significantly more frequently than the other during feeding bouts (Rees & Bowler 1991). This result differs quite markedly from studies of the feeding behaviour of Bewick's Swans in winter, which show that males with cygnets are more alert throughout the winter, apparently to protect their young (Bowler 1996; Section 4.5.2). During the spring migration, however, it seems that males may be making up for lost time since paired females do not feed for longer periods than their partners at Matsalu Bay, despite the advantage that might accrue to the females in developing fat deposits prior to the breeding season. Moreover, males did not spend more time alert than their mates, although they did spend significantly more time than females in aggressive encounters with other swans (Rees & Bowler 1991).

Paired swans looked up for significantly shorter periods than single swans whilst feeding, which probably enhanced their food intake during feeding periods and indicates that benefits to the pair in remaining close to each other continue throughout migration. Juveniles showed the most intensive feeding activity, and had their heads underwater for 73.6% of the focal-bird observations on this age group at Matsalu Bay. They also looked up for longer periods than paired birds (but not singles) whilst feeding, however, perhaps to ensure that they were not separated from their parents. There was no evidence to suggest that the parents rejected their cygnets during the spring migration, although this may happen north of Estonia, closer to the breeding grounds. It also remains possible that sex differences in the feeding activity of paired birds may occur closer to the breeding range, or would be evident in a longer study in Estonia.

4.5 INDIVIDUAL DIFFERENCES IN FORAGING STRATEGIES

John Bowler also studied the effects of pairing and breeding success on the swans' foraging strategies by monitoring their distribution in Gloucestershire in the 1989–90 to 1992–93 winters. In particular, he considered whether the social classes differed in their feeding site selection and also whether single swans, paired birds and family parties had different feeding patterns.

4.5.1 Biases in feeding distribution for the different social classes

Different social classes were each more likely to be seen feeding in some fields than in others, with single swans, for instance, occurring in greater numbers than expected on the saltmarshes (the Dumbles at Slimbridge), but the reasons for this bias in distribution were not always obvious. The proportion of families identified in flocks generally decreased with distance from the Slimbridge roost sites, whereas the proportion of single swans increased, indicating that those without young tended to roam further afield. There was no obvious trend for pairs without cygnets.

Flying is energetically costly, which puts a premium on those good-quality feeding sites closest to the roost. Thus families, which tend to be dominant in the social hierarchy (see Chapter 6), appear to establish themselves at feeding sites closer to the roost. Although families were not observed displacing other swans from the fields, the proportion of cygnets in the feeding flocks increased with decreasing flock size, suggesting that family parties may deter other birds from using the same feeding sites (Bowler 1996). Other studies have also shown that flock size is closely correlated with the percentage of young recorded in the flock. Flocks of less than 20 birds have a higher percentage of juveniles than larger groups, and this is true for both Bewick's Swans and Mute Swans in Britain (Rees *et al.* 1997b), reinforcing the view that smaller groups consist of just one or two family parties.

4.5.2 Feeding rates of parents and non-breeders

The continuous monitoring of the foraging activity of individually identified Bewick's Swans, alluded to in Section 4.4.2, described several factors such as the swans' age, sex and breeding status that influence their food intake rate (Bowler 1996). It was noted earlier that, on average, swans spent 71.9% of their time with their heads down during feeding bouts, 17% of the time being vigilant, 7.8% moving between food patches and 3.3% in other activities. There was a marked sex difference in feeding rate, however, with females spending significantly more time actually feeding during the feeding bouts, and they also had a higher peck rate but slower step rates than males. Consequently, females spent less time in vigilance, movement and social interaction.

Breeding males were found to have slower peck rates, to be more vigilant and to be more frequently involved in social interactions than other males, probably because they were protecting their mate and offspring from interference from the other swans at the site. Females with young were more vigilant than paired females without young, indicating that they shared responsibility with their mates for the protection of their cygnets. Males spent less time being vigilant, and fed more intensively, when grazing on well-fertilised swards,

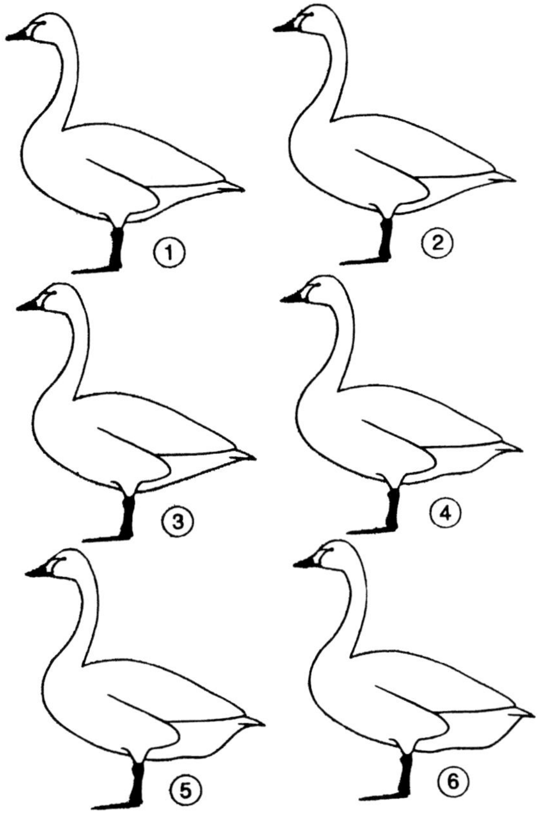

Figure 4.7 Abdominal profile scores for Bewick's Swans, used for assessing body condition in the field. 1 = very concave; 2 = concave; 3 = straight; 4 = convex; 5 = very convex; 6 = sagging. From Bowler 1994.

but conversely invested more time in protecting their mates and any offspring from intraspecific feeding competition when food quality (in terms of protein content) was low. Peck rates by both sexes decreased as sward length increased. If one assumes that bite size increased with sward length, this may reflect an increase in 'handling' (i.e. chewing and swallowing) time as the volume of food ingested increased (Bowler 1996).

John Bowler developed an abdominal profile index for Bewick's Swans (Bowler 1994), similar to the one described for geese (Owen 1981), which enabled him to monitor changes in body mass for individuals seen in the field, and thus to investigate the consequences of variation in food intake rates on birds of different sex, dominance rank (for further details on dominance see Chapter 6) and breeding status. The scale reflected the shape of the abdomen under the tail and was scored from 1 to 6, from 'very concave' to 'sagging' (Figure 4.7).

The swans' abdominal profiles varied between years, and also with time of year (measured in half-months during the winter) and social status. Nevertheless, females had significantly higher abdominal profile scores than males throughout the winter. Moreover, the effects of dominance and breeding status worked in opposite directions for the two sexes: low-ranking single males had higher abdominal profile scores than higher-ranking paired males, whilst low-ranking single females had lower abdominal profile scores than

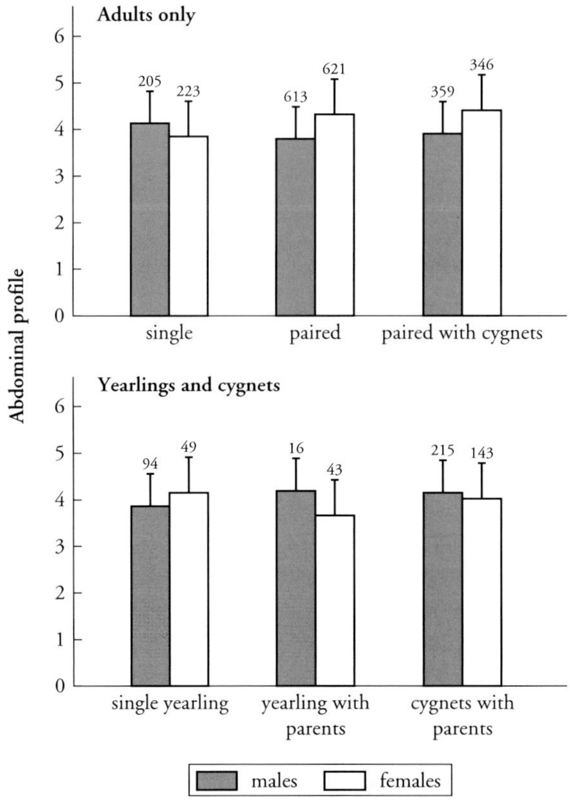

Figure 4.8 Mean abdominal profile scores plotted against social class (histogram SD), with the number of individuals involved in each category indicated. From Bowler 1994.

higher-ranking paired females (Bowler 1994; Figure 4.8). Thus, breeding females and their cygnets and, to a lesser degree, paired females without cygnets, not only benefited from the protection afforded by their mates by being able to feed more intensively (as indicated by the faster peck rates and slower step rates), but this was also reflected in their improved body condition during the winter. The abdominal profile of paired females increased, and the condition of their mates decreased, with increasing dominance rank of the pair, again indicating the benefits to females and the costs to males of the time that males spend in vigilance and social encounters. Cygnets of dominant pairs were also in a better condition than those of lower-ranking birds, but there was no evidence that yearlings associating with their parents were in a better condition than single yearlings (Bowler 1994).

4.6 ENERGY REQUIREMENTS

Estimates of energy expenditure for captive Bewick's Swans have indicated that feeding on pondweed is more costly than feeding on the leftovers of the sugar beet harvest. Birds spent 38.9W of energy in a day while active on the pondweed bed, compared to 28.3W per day when in sugar beet fields (Nolet *et al.* 2002; 1999 data in Table 4.1). The most notable difference between the two feeding sites was the relative high energy expenditure during

Table 4.1 Energy costs (mean ± SE(n)) per day of different behaviours based on heart rate measurements, with the number of individuals given in parentheses. Swimming while intermittently dipping the head into water was recorded as 'sampling'. Estimates of the costs of flying are given in the text. From Nolet *et al.* 2002.

Behaviour	Cost (W)	
	Pondweed	**Sugar Beet**
1999		
Resting	19.4 (1)	
Active	38.9 ± 0.7 (3)	28.3 ± 2.0 (2)
2000		
Resting	22.2 ± 5.9 (3)	21.7 ± 8.2 (3)
Preening	33.8 ± 9.4 (5)	26.2 ± 8.6 (4)
Drinking	26.7 ± 6.4 (5)	24.1 ± 7.1 (4)
Foraging		
Feeding (head-dipping)	27.4 ± 7.3 (5)	
Feeding (upending)	38.6 ± 17.6 (2)	
Bite-feeding	44.2 ± 21.6 (3)	26.1 ± 6.5 (4)
Surfacing	30.8 ± 8.0 (4)	
Trampling	50.9 ± 10.1 (5)	
Sampling	23.9 ± 6.1 (5)	
Interactions		
Threatening	42.5 ± 10.0 (2)	24.8 ± 1.1 (2)
Avoiding	28.3 ± 10.5 (2)	68.7 (1)
Calling	31.5 ± 11.0 (5)	61.9 ± 17.9 (4)
Wing-flapping	47.1 ± 11.5 (5)	76.5 ± 21.9 (4)
Looking	35.4 ± 8.8 (5)	27.7 ± 7.0 (4)
Walking		52.1 ± 15.1 (4)
Swimming	28.2 ± 9.0 (5)	

trampling (i.e. using the feet to dig for pondweed tubers), but this was balanced by swimming being less energy demanding than walking (Nolet *et al.* 2002). Preliminary estimates of tuber consumption by Bewick's Swans feeding on *Potamogeton* upon arrival in the Netherlands in the autumn put the daily intake at around 283 (254–318) grams dry weight per bird per day which, given that the calorific content of the tubers is 17kJ per gram (dry weight), indicated an average energy intake of 4,810kJ/swan/day (Beekman *et al.* 1991). This high-calorie diet, together with the digestibility of the tubers (>90% of the energy extracted, see Section 4.3.3), helps the birds replenish the fat reserves depleted during the autumn migration and prepares them for colder weather during the winter. Nolet *et al.* (2002) also initially estimated the energy expenditure during flight at 402.7W from formulae developed by Pennycuick (1989), but calculations derived from Norberg (1996) give a lower value of 204.2W, which may be closer to the real cost (B.A. Nolet pers. comm.).

Although studies of energy intake and expenditure have not been undertaken elsewhere in the wintering range, monitoring of Bewick's Swan body condition at Slimbridge has confirmed that the swans gain weight rapidly from their arrival to reach a maximum in late December to early January. This is then followed by a midwinter decline when natural food supplies are low, but there is a further increase in body fat prior to migration at the end of March (Bowler 1994; Chapter 3). In the winter when Bewick's Swans left Slimbridge for the continental mainland in January, and reached western Germany before returning to the United Kingdom in February, it was estimated that the return journey of at least 1,600km in 14 days represented an energy expenditure of about 7.3% of an adult bird's body weight. At Slimbridge this weight takes approximately 22 days to accumulate (Evans & Kear 1978; Evans & Rees 1978; Evans 1982). The swans duly finally departed on their spring migration after spending a further 15–26 days (average 23 days for eight swan units) at the site.

The importance to the swans of having extra body reserves on arrival at the breeding grounds has still to be determined, but if the body reserves for egg-laying are primarily laid down at late migratory sites and then transported to the breeding grounds (so-called 'capital-breeding'; Drent & Daan 1980; Klaassen *et al.* 2001), the short period of refuelling on the White Sea could be crucial for the breeding success and thus the viability of the European Bewick's Swan population. Arriving on the breeding grounds with good body reserves may also be important for survival, particularly for breeding pairs, since they remain on their territories and therefore have fewer feeding opportunities during incubation.

Observations of the Bewick's Swans' behaviour at Dvina Bay on the White Sea in mid- to late May 1995 and 1996 illustrated not only the importance of the site to the swans but also the significance to individuals of arriving early at the site. A total of 17,972 swans (>60% of the flyway population) were seen flying into the Dry Sea part of Dvina Bay in May 1996. Individual birds were estimated as staying between one and six days at the site, although two swans tracked by satellite staged for 15 days and 16 days respectively, suggesting that the swans use a wider part of Dvina Bay or the White Sea than the Dry Sea alone (Nolet & Drent 1998). Dvina Bay is tidal, and the swans forage on the pondweed tubers only during low tide. As the staging period progresses and more swans arrive, the food in the shallows is depleted and the swans move progressively into deeper water to feed. There they have less time for feeding before the tide turns and puts the tubers out of reach, to the extent that the swans have 6.0 hours feeding time per tide on 20 May compared with just 3.3 hours per tide by 28 May (Nolet & Drent 1998). Measures of tuber densities before and after grazing by the swans suggested that, in deep water, the density and thus the food intake rate might initially have been higher than in shallow water. Nevertheless, this did not compensate for the lack of feeding time. Nolet and Drent calculated that swans arriving early at Dvina Bay are able to completely replenish the fat deposits used during migration in two weeks, but that swans arriving one week later (i.e. as soon as the first patches of tubers become available in the Dry Sea in mid-May), suffer a declining intake rate due to the depletion of the tuber bed, and therefore are able to replace only 44% of their body reserves. An even later bird, arriving one week after this, would be able to replace only 10% of its reserves, indicating that late migration could have serious consequences for the individuals involved.

Studies of the energy requirements of Bewick's Swans at other stages in their migratory and reproductive cycles are still to be completed, but Dutch scientists are making a detailed assessment of how the energy demands (and thus food requirements) change with the time of year. They are paying particular attention to key stages in the swans' life-cycle, including the period before egg-laying and incubation, the time prior to the annual moult when the birds are unable to fly in search of food, the swans' preparation for migration, and how the swans respond to adverse weather conditions. An important question being addressed is whether there is a bottleneck in the optimum nutritional requirements of these long-distance migrants with, for example, the quantity and quality of food available at the spring staging sites regulating reproductive success during the summer months. Although energy calculations and satellite-tracking studies indicate that the swans have a potential flight range of some 1,800-2,300km before needing to feed and replenish their reserves during migration (Beekman *et al.* 1996a, 2002; see Chapter 3), whether the swans need additional energy intake at spring staging sites in preparation for the breeding season or whether they develop the reserves for egg-laying and incubation in the vicinity of the breeding territories has yet to be determined (M. Klaassen pers. comm.).

4.7 FEEDING COMPETITION OR COEXISTENCE

Since the relatively large Whooper Swans dominate the smaller and lighter Bewick's Swans at sites where both species occur, it is thought that Whooper Swans may displace Bewick's Swans in these areas (Black & Rees 1984). However, there is no conclusive evidence that competition between the two affects their distribution except at a local level.

The extent to which the three swan species that winter in Britain use the same sites was assessed; first, the number of sites with single-species and mixed-species (Bewick's, Whooper and Mute Swan) flocks was recorded in northwest England and east–central England (where several hundred of each species are found in midwinter) during the 1990–91 winter. This was compared with the expected frequency of occurrence of these flocks, calculated from the known distribution of swans across sites in these two regions (Rees *et al.* 1997b). Results showed that Mute Swans were more likely to be found in single-species flocks, probably because they tend to be more widely dispersed. They were also less likely to use terrestrial habitats, and were more likely to occur in urban habitats, whereas the number of single-species Whooper Swan and Bewick's Swan flocks did not differ significantly from the expected frequencies. Moreover, records of Whooper Swans and Bewick's Swans present at the same site were higher than expected, although this was not statistically significant (Table 4.2, from Rees *et al.* 1997b). It seems, therefore, that Whooper Swans do not displace Bewick's Swans from fields in these regions, but more detailed behavioural studies of interspecific interactions are needed to determine whether Bewick's Swans and Whooper Swans form truly mixed flocks, or whether conspecifics group together and use different food resources from other species using the same site. The number of Bewick's Swans wintering on the Ouse Washes remains high, despite the increase in Whooper Swans at the site (Figure 4.9a). However, the drop in numbers at Martin Mere as Whooper Swan counts continue to rise (Figure 4.9b) could be due to Whoopers monopolising the best feeding and roosting areas, as well as to Martin Mere being at the edge of the Bewick's Swans' wintering range.

Whereas the effects of interspecific competition for food on the swans' distribution is still unclear, their movements to feeding areas within a wintering site seem to be influenced by food depletion both by conspecifics and by other birds using the same resources. Section 4.4.1 above, for instance, shows that Bewick's Swans wintering in Gloucestershire disperse over a wider range of sites as the winter progresses — as the food supply diminishes — which implies that a reduction in the food available due to grazing by other animals will also affect feeding site selection. Similarly, Bewick's Swans feeding in the Onnerpolder, in the Netherlands, avoid pastures grazed by sheep, partly because the sheep graze the pasture to the extent that there is very little food left for the birds (J. Nienhuis, unpubl. data), and a similar pattern is now being found at a national scale (J. van Gils, unpubl. data). A study of the grazing succession of Eurasian Wigeon, European White-fronted Geese and Bewick's Swans at WWT Slimbridge reinforced the view that the choice of feeding sites by one group

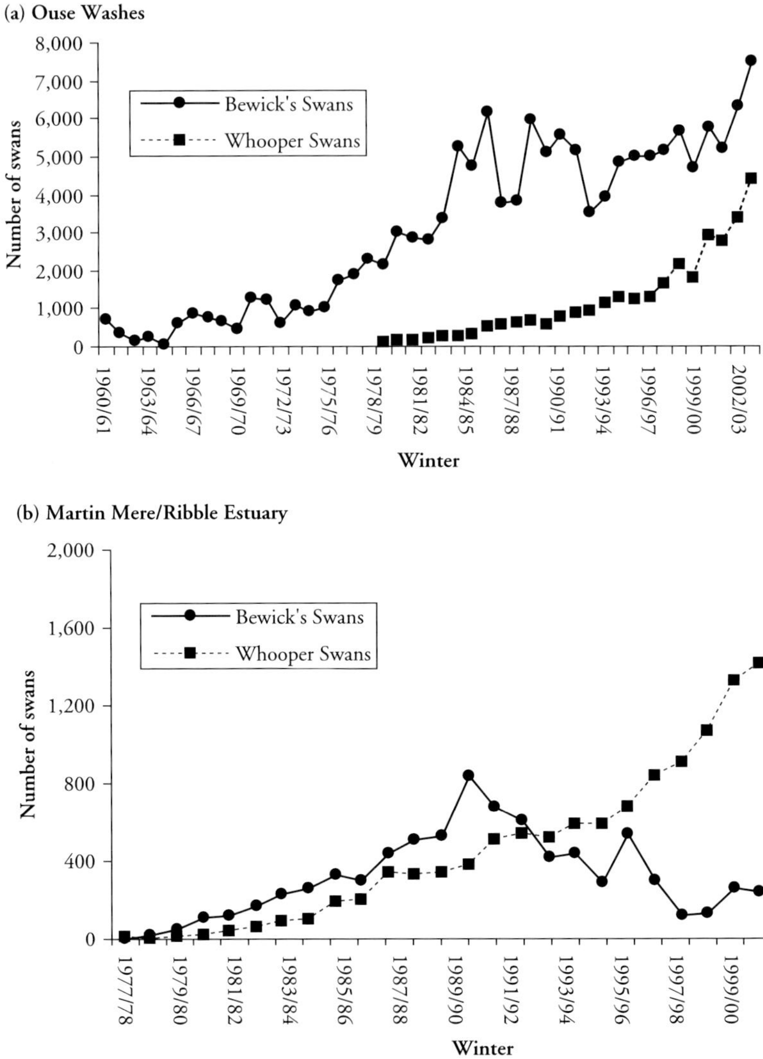

Figure 4.9 Maximum Bewick's and Whooper Swan counts each winter at (a) the Ouse Washes, (b) Martin Mere.

may be influenced by the activities of others in the area, and indicated that in some cases this need not be a negative effect. A comparison of the percentage of swan-days recorded in one month with the percentage of goose-days and of wigeon-days recorded in the following month showed that, although there was no association between the distribution of swans and wigeon, geese did not select fields that had been grazed intensively by swans earlier in the winter. Yet, on the other hand, wigeon appeared to prefer fields that had been used intensively by geese the previous month (Rees 1990). Thus it seems that grazing pressure by the swans depletes the vegetation to such an extent that the geese do better by moving to other fields, but that the grazing activity of the geese helps to reduce the sward length to a bite size more suitable for the shorter-billed wigeon.

Table 4.2 Comparison of the number of sites observed with single or mixed species flocks, with the expected frequency of such flocks occurring. From Rees *et al.* 1997b: Flock types W+B+M = mixed Whooper, Bewick's and Mute Swan flocks; W+M = mixed Whooper and Mute Swan flocks; B+M = mixed Bewick's and Mute Swan flocks; W+B = mixed Whooper and Bewick's Swan flocks; W = single species Whooper Swan flock; B = single species Bewick's Swan flock; M = single species Mute Swan flock. %W, %B and %M represent the percentage of sites at which each species was recorded (and thus the probability of each species occurring at any one site). p values are for χ^2_1 tests: * = <0.01, ** = <0.001, n.s. = not significant.

Flock type	No. sites in NW England			No. sites in EC England		
	Observed	Expected	p	Observed	Expected	p
W+B+M	9	9	n.s.	12	18	n.s.
W+M	8	29	**	3	18	**
B+M	1	11	*	7	29	**
W+B	9	5	n.s.	25	18	n.s.
W	25	17	n.s.	21	19	n.s.
B	7	6	n.s.	36	30	n.s.
M	53	35	n.s.	57	29	**
%W		45.4%			37.9%	
%B		23.2%			49.7%	
%M		63.3%			49.2%	

Other studies have similarly shown that the feeding activity of one species may produce suitable habitat for another; for instance, although Brent Geese *Branta bernicla* and Brown Hares *Lepus europaeus* compete for a common food supply (van der Wal *et al.* 1998), the geese do benefit from the hares' grazing activity, in that the latter additionally feed on swards of the shrub *Atriplex portulacoides* and prevent it from spreading over the goose feeding areas (van der Wal *et al.* 2000). Eurasian Wigeon seen feeding on water quite often are amongst or behind swans (mainly Mutes, but also Whooper Swans), where presumably they benefit from vegetation disturbed or brought up by the swans from deeper water (C. Spray pers. comm.). Grazing successions have also been described for large herbivores in the Serengeti National Park, Tanzania, with feeding site selection by one group being influenced by the movements of other animals in the area (Bell 1971; Sinclair 1979; Maddock 1979). The Serengeti clearly is very different in scale and habitat to the pastures of Gloucestershire, but it seems that the same principles apply. Again, there was little evidence for overt aggression between swans, geese and wigeon, but more detailed studies of the rate of food intake for birds feeding adjacent to members of their own or other species are needed to determine the extent to which they compete or coexist.

CHAPTER 5

Breeding biology

So doeth the swan her downy cygnets save,
Keeping them prisoners underneath her wings.

William Shakespeare (*Henry VI, Part I*)

For most of the 20th century very little was known about the Bewick's Swan's summer haunts in the High Arctic. As late as 1990, the Russian tundras were largely closed to Western ornithologists, and the vast, sparsely populated terrain made it difficult even for Russian scientists to undertake detailed studies in the field. The few Russian publications in English on the swans' breeding biology (including the seminal work of Dementiev and Gladkov, *The Birds of the Soviet Union*) focused mainly on numbers and distribution, rather than on behavioural ecology. The Anglo-Soviet expedition to the Yuribey region of the Yamal Peninsula in 1978 (Chapter 2) found just 83 Bewick's Swans and no pairs on nests, which was not encouraging for future research programmes, at least in that particular area (Scott 1985 and pers. comm.). It was not until the advent of *glasnost* and *perestroika* that British, Dutch and Danish scientists were able to join Russian colleagues on joint expeditions to the Arctic, and that fieldwork undertaken by Russian scientists during the 1970s and 1980s became better known in the West.

The first of several international expeditions to the Russian Arctic in the late 20th century was made in 1991 by the author and Dafila Scott (daughter of Sir Peter Scott, and illustrator of this book) from the WWT. We were accompanied by Russian colleagues Yuri Shchadilov, from the All-Russian Research Institute for Nature Protection (Yuri having done the hard work in organising the expedition within Russia), and Yuri Mineyev, from the Komi Science Centre. Biochemist Yuri Morozov took a few weeks' holiday from the Institute of Endocrinology to act as interpreter, and has been a regular member of the team ever since. According to Russian friends, we were the first British people to reach Nar'Yan

Mar (at the mouth of the River Pechora) since the explorer and ornithologist Henry Seebohm visited the area at the end of the 19th century and, since the city of Nar'Yan Mar (base camp) was closed to Westerners for most of the 20th century, we were minded to believe this.

Like other natural historians of the Victorian era, Seebohm's interest was in taxonomy, and his records are of birds shot for close inspection in the hand, not of their distribution across the region. Yet his expedition to the Pechora River basin in 1875 has provided valuable information on species present in this area more than 100 years ago (Seebohm 1901). Whilst preparing for the journey in Arkhangelsk, Seebohm sought further information concerning the mighty Pechora River, and was told by the Samoyede people that there were two species of swan on the Pechora: "the larger one common and breeding there, the smaller one rare and appearing only in autumn". This he found to be substantially correct, and he succeeded in collecting the skin and eggs of a Bewick's Swan that had been breeding on an island in the mouth of the Pechora River Delta. Following a boat trip to the north, he then found swans nesting on open tundra presumably (from the description, although the site is not named) in the current Nenetskiy National Nature Reserve. One hundred and thirty years later, the situation is much the same: Whooper Swans occupy the delta region and Bewick's Swans travel a little further north to breed on the open tundra and islands (notably Lovetskiy Island) in the Pechora Sea. Our own expeditions now aim to determine the main reasons for variation in Bewick's Swan survival and breeding success, in what is now known to be an important breeding site not only for the swans but for a wealth of other wildfowl and waders.

From 1992 onwards several international expeditions visited the region, working sometimes collaboratively and sometimes in parallel, depending on the aims of the different studies. In Nar'Yan Mar, our paths also crossed those of American and Scandinavian geological survey teams, who were assessing oil and mineral extraction opportunities. The first of several Dutch ornithological studies in the area was initiated in 1992, when Jan Beekman (from the Netherlands), with Pelle Andersen–Harild (Denmark), commenced his study of Bewick's Swan energy requirements during breeding and migration. Broader ecological studies of the state nature reserve and Pechora Delta, undertaken during the late 1990s by a Dutch–Russian team, led by Mennobart van Eerden in collaboration with specialists from the Komi Science Centre, provided a valuable description of the structure and dynamics of the Pechora ecosystem (van Eerden 2000). Their work has been resumed, and aims to develop a management plan for the reserve within the PRISM project. Dutch researchers are particularly interested in the Pechora region, not only for its integral ecological value and relatively pristine condition, but because its similarities to the Rhine Delta area make it a valuable case-study for informing wetland restoration work in the Netherlands (van Eerden & Lavrinenko 2000; Chapter 1). Scientists from the Norwegian Polar Institute have also been active in the area, particularly in monitoring seabird concentrations along the Barents Sea coast (A. Glotov, pers. comm.). Meanwhile, the Anglo-Russian team was joined by Anna Belousova in 1992, who went on to gain her Candidate of Science degree with a thesis on Bewick's Swans' breeding biology (Belousova 1999).

The main study site for Bewick's Swan research is centred on the abandoned fishing settlement of Khabuicka (named after the nearby Lake Khabuicka) within the then Nenetskiy *zakaznik* (hunting sanctuary). The region, including the Korovinskaiya Bay, received further protection upon being designated a state nature reserve (the Nenetskiy

Figure 5.1 Field observations are made from a vantage point on the derelict fish barn at Khabuicka.

zapovednik) by the Russian government in December 1997 (Kotkin 2000). Khabuicka lies on the coastal plain just to the north of the Pechora River Delta, on the shore of the Pechora Sea, some 100km north of the city of Nar'Yan Mar. There, vantage points from the roofs of the two huts and a fish barn make it possible to locate swans on nests at up to 1km over the flat terrain, facilitating visits to the nests to record clutch and brood sizes (Figure 5.1). In due course, 'Khabuicka' became the name for the core study area, covering about 30 hectares of low-lying terrain with slightly elevated relief, dotted with pools and small lakes and intersected with numerous channels. For a number of years authors (including this one) referred to Khabuicka and the surrounding area as being on the Russki Zavorot Peninsula, but this seems to be a misconception. The Russki Zavorot Peninsula is actually the spit at the northernmost end of the Malozemelskaya tundra, and Khabuicka is a little further south, in part of the Malozemelskaya tundra known by the old name of Zakharin Bereg.

The rapid improvement in communication with Russian scientists from the late 20th century onwards led not only to joint expeditions to the Nenetskiy *zapovednik* and Pechora Delta region, but to the results of Russian studies of swans in other parts of the Arctic becoming known in the West. Kondratiev's observations on Bewick's Swans' nesting biology in far eastern Russia were presented at the Third International Swan Symposium held in Oxford in 1989, but only in recent years have other studies been described. These include those undertaken by Evgeny V. Syroechkovsky, Konstantin Litvin and Elena Gurtovaya on Vaygach Island; the results of 40 years of monitoring swan numbers and nesting distribution on the Lena Delta by the Lena-Nordenskiöld Biological Station in Yakutia; and the surveys of swan distribution in Chukotka by Evgeny E. Syroechkovski Jnr. Vaygach Island is a High Arctic breeding area north of the Russian mainland, so the six-year study of Bewick's Swans' nesting biology at this site has, by comparison with observations made further south, provided insight into the effects of latitude on the swans' breeding programme.

This chapter synthesises the results of recent studies to give an overview of factors affecting the swans' breeding biology. It considers conditions encountered by the swans at different stages of the breeding season and discusses how a combination of variables, both environmental and relating to the individual birds, may influence the breeding success of the population as a whole. Thus, following decades when the reasons for variation in breeding success were based almost entirely on observations of individuals in winter (the notable exception being Ernst Poorter's 1991 monograph on the correlation between Arctic weather conditions and the percentage of young recorded in the wintering range), the hazards facing the birds during the summer months, and how these may be ameliorated by the traits and habits of the individual swans, can now be described for the first time.

5.1 PAIR FORMATION

Pair formation does not usually occur in the winter, at least not in the UK. In 40 years of studying Bewick's Swans at Slimbridge, during which time nearly 4,000 pairs were recorded, there have been no instances where both members of the pair were seen separately at the site in the winter prior to being first recorded together. Swans must therefore pair off elsewhere, either at spring migratory sites, on the breeding grounds, or during summer in the non-breeding flocks. Moreover, since one would expect that a small number of birds from Slimbridge would pair 'by chance' in the spring or summer, the fact that this has not been recorded suggests that they may actively choose unfamiliar mates, perhaps to reduce the possibility of inbreeding (Bateson 1983). On arrival in the breeding grounds, established pairs embark promptly on territorial defence and nesting, whereas non-breeders congregate into non-breeding flocks where courtship is frequently observed (albeit not yet studied in detail), characterised by head-turning, stomping (with straight neck and erect neck feathers) and wing-flicking displays. Thus it seems that some if not most pair formation occurs in the breeding range.

First pairing occasionally occurs when the birds are two years old (i.e. during the swan's third summer), but more usually pair bonds are established at age three or four years, with females generally pairing earlier than males. Bewick's Swans tend to pair one year later, and also to breed later, than both Whooper Swans and Mute Swans (Rees *et al.* 1996; Chapter 7). These differences have been attributed to the comparatively long migrations that Bewick's Swans undertake, and to the shortness of the Arctic breeding season, which makes it rare (though not unknown) for Bewick's Swans to both find a new mate and produce a brood in a single season. Although siblings may associate with each other during their second or third winter, there are no cases of sibling pairing recorded to date. Kinship recognition mechanisms therefore may be in place to reduce the possibility of inbreeding (e.g. Choudhury & Black 1994), which would be important in site-loyal species such as the Bewick's Swan. Certainly, swans seem to recognise each other easily, since paired birds identify and greet one another effusively in a flock of several hundred birds, even after a temporary separation of several months, and offspring regularly rejoin their parents at traditional wintering sites in their second and subsequent years (further information provided in Chapter 6). Factors affecting mate selection are considered in greater detail in Chapter 7.

5.2 MATING

As with pair formation, copulation has not been recorded at wintering sites in the UK, whereas Whooper Swans have been observed mating at WWT centres in late winter and early spring. Copulation does occur at spring migratory sites as far south as Estonia (Rees & Bowler 1991), and may be quite frequent both there and in the White Sea region since it requires standing water, which is often absent upon arrival at the breeding territories (Bowler 2005b). Copulation observed in early to mid-June on the breeding areas involved only non-breeding pairs which had established a territory but did not then build a nest. Although it is unlikely that these pairs would embark on nest-building and breeding so late in the season, and there is no record of them doing so, presumably mating helps to reinforce new or existing pair bonds.

Although the swans use a range of overt neck-pumping and wing-flapping displays in defence of their territory in summer, and also to establish dominance in the wintering flock (see Chapter 6), pre-copulatory display is more subtle, limited to ritualised head-dipping movements by both sexes. This display is often initiated by the female, who repeatedly dips her head and neck into the water, wings lowered and with the head and neck feathers erect. The male joins in, wings slightly raised, and the pair head-dip and run water over their backs side by side until the movements become synchronised, whereupon the male mounts for *c.*10–15 seconds, partially submerging his mate and grasping the back of her head or neck with his bill (Evans 1975; Cramp & Simmons 1977; Bowler 2005b). After mating, the pair rise up on the water with necks stretched upwards, facing each other with wings fully extended and calling softly, in a gentler version of the triumph display used following aggressive encounters with other swans (Chapter 6). They then turn a partial circle before settling to bathe and preen or feed. The whole process takes 50–90 seconds and may be repeated several times daily on arrival on the breeding territories (Bowler 2005b).

5.3 TIMING AND DURATION OF THE BREEDING SEASON

Bewick's Swans arrive on the breeding grounds in mid- to late May, or in early June in late years or further north, depending on local conditions. Non-breeders arrive later than breeders in nesting areas, and these may include some two- or three-year-old birds, which are former offspring summering close to their natal site. Swans marked with yellow leg rings (denoting birds of known age) were seen in non-breeding groups at Khabuicka (J. Bowler pers. comm.) and, although it was not possible to read the rings, this suggests that they had been caught on nearly territories as cygnets in previous years. Pairs with previous experience of the breeding sites appear to embark on their breeding programme whilst the territories are still largely covered with ice and snow, whereas new pairs are more likely to wait for the thaw (see Section 5.15). Eggs are laid in late May to early June and hatch from late June to early July. Generally only one clutch is laid in the wild, although second clutches have been recorded for Bewick's Swans in captivity (WWT unpubl. data).

Movements from the tundra breeding grounds begin in early September, non-breeders tending to leave before birds with young, and the swans congregate in large numbers at pre-migratory sites. Yuri Mineyev's observations during the 1970s and 1980s found that most families remain on the breeding grounds until mid- to late September (Mineyev 1991), presumably waiting for the young to fledge and to be strong enough in flight to embark on migration. Mass departure from the breeding grounds in European Russia usually occurs towards the end of September (Rees *et al.* 1997a; Chapter 3), although this depends on the timing of the onset of winter. For instance, in his observations of the onset of autumn migration for Bewick's Swans across European Russia, Yuri Mineyev noted that birds congregate in pre-migratory flocks (e.g., in Bolvanskaya and the Korovinskaiya Gulf) in mid-September (13–20 September), and that birds with broods join the flocks in the last 10 days of September, coinciding with the appearance of ice on small lakes and streams. Almost no swans remain in the eastern parts of the European breeding range (Yugorsky and east Bolshezemelskaya tundra) by the end of September. The last birds usually depart from breeding areas further west in the first half of October as ice forms on the coastal bays (Mineyev 1990).

Until recently it was thought that only very few Bewick's Swans remained in the breeding area until mid-October, but in an exceptionally mild autumn for this part of the Arctic in 2003, mass migration occurred about 10 days later than normal (Andrei Glotov pers. comm.). Satellite tracking of swan migration in this year confirmed that all the swans with transmitters left the tundra for the northern reaches of the Pechora Delta in early September and remained on the delta until 15–19 October before heading south to the wintering range. In general, the timing of mass migration is linked directly to the timing of the freezing of the water over the entire region (with ice formation spreading from east to west), thus reducing feeding opportunities for the birds (A. Glotov pers. comm.). In most years the first birds are passing along the Baltic coast and reaching wintering sites in the Netherlands by early to mid-October. Observations on the Lena Delta, in the eastern part of the breeding range, found similar arrival and departure patterns, with swans arriving in the second half of May (the earliest date was 15 May) and departing in mid-September (the latest date that swans were still present being 4 October; Pozdnyakov 2002), despite the area being at a higher latitude than the main breeding areas in European Russia.

Interspecific variations in the distribution of long-distance migrants, and the length of the migrations undertaken, have been associated with differences in body size, particularly where species are closely related. Thus Bewick's and Whistling Swans breed at higher latitudes than the larger Whooper and Trumpeter Swans. Whooper Swans, which breed on taiga lakes across Eurasia, may gain more from wintering close to their breeding grounds than the smaller Arctic-breeding Bewick's Swans. This gain is made by reducing the energy costs of migratory flight and by the longer ice-free periods at lower latitudes, enabling their comparatively slow-growing cygnets to fledge before the onset of winter. There is some limited overlap in the breeding distribution of Bewick's and Whooper Swans, with the two species occasionally using adjacent breeding territories at the northern edge of the shrub tundra zone (Y. Mineyev pers. comm.), although this depends on spring weather conditions enabling Whooper Swans to breed further north than normal.

5.4 TERRITORIAL DEFENCE

Although the evidence for prolonged parental care is well established, with parents and offspring remaining together throughout the cygnets' first winter, and departing together on the spring migration, adults generally return to the breeding territories without their previous year's brood. The mechanism for the break-up of the family unit in spring is still not certain, but it appears that the offspring become increasingly independent, since

Figure 5.2 Aerial chase over a Bewick's Swan territory.

although parents were not seen being aggressive towards their cygnets whilst staging in Estonia, the cygnets were spending long periods apart from the adult birds and ventured further from them at this time (Rees & Bowler 1991). More detailed behavioural studies in spring and early summer are needed to determine whether some adults become more aggressive towards their previous year's offspring whilst staging on the White Sea or as they arrive on their breeding territories. At Khabuicka, although only adult pairs occupy the nesting territory, yearlings are often nearby in non-breeding flocks. There, a yearling may occasionally visit a breeding territory and sit with a pair near the nest (A. Belousova pers. comm.), but on Vaygach Island the few yearlings seen attempting to join territorial pairs in late springs were driven off by the adult swans (Litvin *et al.* 1999). Thus, although most breeding pairs arrive on their territories without their previous year's offspring, in some cases the break-up of the family unit may not occur until the pairs separate from the flock and reoccupy their nest sites. Offspring also may rejoin their parents in their second and third winters, with one bird ('Hannah') seen consorting with her parents at Slimbridge when six years old (Chapter 6).

Pairs are highly territorial when nesting and usually remain on their territories until the downy young have fledged. They are very active in defence of their breeding territory, particularly upon arrival at the nest site, where the male or both members of the pair frequently display (wings-extended-and-flapping) at swans flying over them, and also take off to chase away potential intruders. Bewick's Swans are extremely agile in the air, particularly in comparison with the larger and heavier Whooper Swans, and will dive and swerve to peck at the tail feathers of any other swans over their territory, like dragons in the sky (Figure 5.2). Once the other swans have been displaced to 100–200m from the territory boundary, the swan that did the chasing (usually the male) returns to perform an aerial triumph display (wings fully extended and flapping, with loud calls) over the nest and whilst landing, and may then take off again to make a victory flight around the boundaries of the territory, calling, with slow wing-beats and neck outstretched (Bowler 2005b).

Observations from mid-June onwards suggest that paired birds already on a territory generally succeed in defending it, although they may occasionally lose, particularly if a second pair makes a determined bid for it. It is possible that occupancy of a particular territory may be less secure for those arriving very early at the breeding grounds (in late May to early June), but there are very few behavioural records at this time of the season. Successful pairs can be merciless towards their weaker opponents if the latter do not move away. Following one fight where an occupying pair lost its territory to intruders, both members of the new pair stood on the prone body of the defeated male and repeatedly pecked at his neck, back and wings for more than three hours. Fortunately he was able to hide his head and upper part of his neck under a log, which may have saved his life. Immediately after the attack he was unable to fly, but swam to Lake Khabuicka to recover, and was also re-sighted in Western Europe the following winter (Bowler 2005b; A. Belousova pers. comm.).

Except when defending territories, and despite some bickering within flocks, Bewick's Swans tend to be gregarious in summer as well as in winter, with non-breeding birds gathering in flocks of up to several hundred on lakes, marshes and coastal bays during the breeding season and to moult (Shchadilov & Orlov 1987; Krechmar *et al.* 1991). After moulting they congregate in larger premigration groups on shallow coastal waters (Mineyev 1991). The biggest concentration, of over 15,000 birds, was observed on the Korovinskaiya Bay in September 1992 (Mineyev 1994; Beekman *et al.* 1994a, 1996b). Parents and their newly fledged offspring leave the breeding territory together and remain together throughout the autumn migration to the wintering grounds, joining other families, pairs and single swans in large flocks at the migratory and wintering sites.

5.5 PROPORTION OF BIRDS THAT ATTEMPT TO BREED

Aerial surveys made over the coastal plain (part of the Malozemelskaya tundra) north of Korovinskaiya Bay in the early 1980s and again during the 1990s found that a high proportion of Bewick's Swans (54%–62%) do not occupy breeding territories in spring, but remain in non-breeding flocks (Shchadilov *et al.* 2002). This is similar to observations made of Whooper Swans in Iceland, where more than 60% congregate in non-breeding flocks in spring and summer (Garðarsson & Skarphéðinsson 1984; Rees *et al.* 1991; Einarsson 1996), and a large non-breeding component has also been reported for Whistling Swans in North America during the breeding season (Bart *et al.* 1991; Ritchie *et al.* 2002). Since some 61%-89% of the adult Bewick's Swans seen at Slimbridge in winter are paired (average 73% over a 15-year period; Evans 1979a), it seems that some paired birds do not make a prolonged attempt, if any, to establish a breeding territory, but quickly join the non-breeding flocks in spring and early summer.

Even for pairs occupying territories, there is a substantial annual variation in the proportion that build nests and thus apparently attempt to breed. In the years 1991, 1998 and 1999, 65%, 71% and 20% of territorial pairs, respectively, had nests in June (Shchadilov *et al.* 2002). The proportion of territorial pairs still attempting to breed in July was similar (20%–64% still incubating or with downy young in July for the years 1996–1999;

Shchadilov *et al.* 2002). Similarly, on Vaygach Island the percentage of breeding pairs among territorial pairs varied from 22% in 1997 (a late spring) to 70% in 1988 (Litvin *et al.* 1999). This mirrors observations made of the swans in the wintering range, with the proportion of paired birds with cygnets at Slimbridge varying from 12%–66% (Evans 1979a). The most likely reason for this variation is the impact of the Arctic climate at different stages of the breeding cycle. In cold springs (the Arctic spring being the second half of May and June), fewer pairs breed, and those that do lay smaller clutches. If the summer is cold, or there is an early onset of winter weather, cygnet survival is likely to be lower, although this has not been studied in detail.

Although there is still little information on cygnet survival shortly after hatching (but see Sections 5.12 and 5.13), and especially during the autumn migration, it seems likely that there is substantial annual variation in survival determined by food availability and the timing of the onset of winter. Further studies are needed to see which factors impinging on the breeding cycle have the greatest effect on Bewick's Swans' breeding success. These may vary across the breeding range, depending on latitude, migration distances and local predator levels.

5.6 VARIATION IN BREEDING DENSITY

Breeding density varies between years and in different parts of the Russian Arctic, with average densities of 1.1 nesting pairs per km^2 around Khabuicka on Zakharin Bereg in the Nenetskiy State Nature Reserve, 0.22 nests per km^2 on Vaygach Island and 0.20 nests per km^2 in the Chaun lowlands of Chukotka (Table 5.1). Although the years and the size of the study areas are not directly comparable, nevertheless this reinforces earlier reports

Table 5.1 Variation in Bewick's Swan breeding density across the Russian Arctic.

Location	Years	Size of study/survey area (km^2)	Mean density (nests/km^2)	Range of annual densities for study period (nests/km^2)	Reference
Khabuicka, Nenetskiy Nature Reserve	1991–2000	25	1.10	0.48–1.60	WWT and All-Russia Research Institute unpubl. data
Malozemelskaya tundra	1977–79, 1982, 1986 & 1988	Not known		0.27–1.40	Mineyev 1991
Bolshezemelskaya tundra	1973–79	Not known		0.02–0.18	Mineyev 1991
Yugorskiy peninsula	1981–84 & 1987	Not known	0.18	0.08–0.66	Mineyev 1987, 1991
Vaygach Island	1986–88 & 1995–97	80	0.22	0.12–0.36	Syroechkovsky *et al.* 2002
Chaun lowlands, Chukotka	1975–77 & 1980–84	200	0.20	0.13–0.25	Krechmar & Kondratiev 1986, 1991b

(e.g. Mineyev 1991) that Zakharin Bereg has one of the highest breeding densities of Bewick's Swans in northeast European Russia, and probably one of the highest throughout the range. The proximity of the Zakharin Bereg to the Pechora Delta is likely to be a main reason why this is a preferred site for Bewick's Swans and other Arctic-nesting species. American researchers have already shown that Whistling Swans breeding in Alaska are found at higher densities close to major river deltas than on mainland tundra; it is thought that these areas may provide better food resources and longer periods of open water than non-delta habitats (review in Ritchie *et al.* 2002). Although Bewick's Swans are territorial – not colonial – breeders, nests can be rather close together in optimum habitats. Thus, at Khabuicka the average distance between nests was between 500m and 600m each year, but some nests were as little as 200m apart. The pairs defended territories of, very roughly, 0.2km^2 during incubation (Bowler 2005b).

Surveys of Bewick's Swan density on the northeast part of the Malozemelskaya tundra (i.e. including the Nenetskiy State Nature Reserve) and also further east on the Bolshezemelskaya tundra during the early 1980s found that density varied both between areas and across different habitats within the two regions (Table 5.2; Shchadilov *et al.* 2002, derived from Shchadilov & Orlov 1987). Densities both of territorial birds and of all Bewick's Swans recorded were significantly lower on the Bolshezemelskaya tundra than on the Malozemelskaya tundra. Although densities appear higher on the maritime tundra than on the inland tundra at both sites, differences between habitats were not significant. Combining observations made on the Bolshezemelskaya and Malozemelskaya tundras may in this case have resulted in regional variation overriding any effect due to habitat. More extensive monitoring of Bewick's Swan distribution in relation to habitat is necessary to determine not only the main habitat variables influencing breeding density, but also the reasons for the variation in Bewick's Swan distribution across the breeding range.

A comparison of the 1980–81 surveys of the northeast part of the Malozemelskaya tundra with those made 10 years later (from 1991 to 1999) found no significant increase in the density of territorial Bewick's Swans over this period. In contrast, there was an increase in the Bewick's Swan population wintering in northwest Europe during the 1980s and early 1990s (Beekman 1997, Delany *et al.* 1999), and also an increase in the density of territorial Whooper Swans just a little further south on the Pechora Delta over the same period (Table 5.3; Figure 5.3a,b; Shchadilov & Orlov 1987; Shchadilov *et al.* 2002). There is increasing evidence, however, that the Bewick's Swan population declined in the mid- to late 1990s (Delany *et al.* 1999; SOVON Ganzen- en Zwanenwerkgroep 2001; Chapter 2), which may account for the lower Bewick's Swan breeding densities recorded after 1996 (Figure 5.3a).

When considering annual variation in breeding density, rather than long-term trends, the importance of spring weather conditions is evident. Section 5.2 describes the marked variation between years in the proportion of territorial pairs that build nests and thus attempt to breed. Some of the variation may be due to the age structure of the population, with young birds being territorial but perhaps not breeding when occupying territories for the first time. However, observations made both at Khabuicka and on Vaygach Island have shown that spring weather is a key factor in influencing breeding density, and that this has knock-on effects for the success of the whole of the breeding season. Over a six-year period on Vaygach Island, a High Arctic breeding area north of the Russian mainland, highest nest densities and subsequent breeding success were recorded in years when the swans arrived on an average date and the weather remained relatively mild thereafter (Syroechkovsky *et al.*

Table 5.2 Variation in Bewick's Swan density by region and habitat in 1980 and 1981. [a]Single swans, pairs without nests and pairs with nests were all classified as territorial swans. Number of birds counted given in parentheses. From Shchadilov *et al.* 2002, derived from Shchadilov and Orlov 1987.

Region	Habitat	Area surveyed (km^2)		Density (individuals/km^2): Territorial Swans[a]		Grouped swans		Overall density (total swans)	
		1980	1981	1980	1981	1980	1981	1980	1981
Bolshezemelskaya tundra									
	Maritime tundra	148	161	0.29 (43)	0.36 (58)	0 (0)	0.55 (89)	0.29 (43)	0.91 (147)
	Inland tundra	571	459	0.14 (80)	0.17 (78)	0 (0)	0.27 (124)	0.14 (80)	0.44 (202)
	Pechora Delta	105	44	0.22 (23)	0.40 (18)	0.19 (20)	0 (0)	0.41 (43)	0.40 (18)
Malozemelskaya tundra									
	Maritime tundra	86	75	0.50 (43)	1.04 (78)	2.09 (180)	3.21 (241)	2.59 (223)	4.25 (319)
	Inland tundra	256	245	0.51 (131)	0.80 (196)	0.39 (100)	0.62 (152)	0.90 (231)	1.42 (348)

Table 5.3 Bewick's Swan density (swans/km²) and breeding status, recorded during aerial surveys of the Nenetskiy region (including Zakharin Bereg). The total number of birds counted *(n)* is given in parentheses. [a]In 1980 and 1981, density is given for singles and pairs combined. From Shchadilov *et al.* 2002.

Date	Area surveyed (km²)	Density of individuals observed as				Overall (*n*)
		Singles	Non-breeding pairs	Breeding pairs	Flocks	
1980	342	0.51[a]			0.82	1.33 (454)
1981	320	0.86[a]			1.23	2.08 (667)
June 1991	362	0.38	0.72	1.36	2.86	5.3 (1,923)
July 1996	7.5	0.13	3.20	0.80	0	4.13 (31)
July 1997	22	0.18	0.41	0.82	0.18	1.60 (36)
June 1998	22	0.14	0.45	1.09	0	1.68 (37)
July 1998	22	0.14	0.45	0.82	0.14	1.54 (34)
June 1999	11	0.36	0.54	0.18	0.27	1.36 (15)
July 1999	22	0.09	0.45	0.18	0	0.72 (16)

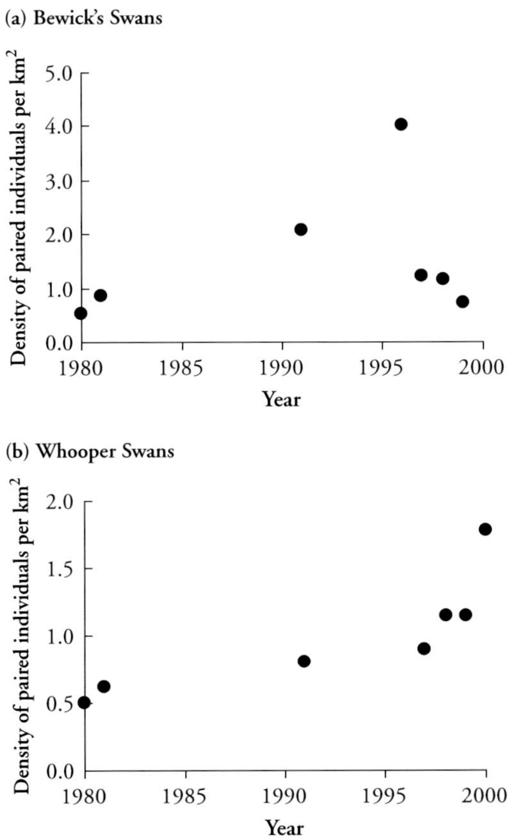

Figure 5.3 Density of (a) territorial Bewick's Swan pairs (with and without a nest) on open tundra in the Nenetskiy State Nature Reserve, and (b) territorial Whooper Swan pairs on the Pechora Delta, from 1980–2000. From Shchadilov *et al.* 2002.

2002). A return to cold weather during laying and incubation may result in nests on Vaygach Island being abandoned, lowering nesting density in otherwise early springs (Syroechkovsky *et al.* 2002; see Section 5.12).

5.7 NESTING

Bewick's Swans breeding on the tundras of Arctic Russia nest in habitats similar to those used by Whistling Swans in North America. The Bewick's Swans favour the open maritime tundra of the north Russian coast, nesting in moss–lichen, moss–sedge and sedge tundras, and low-lying marshes, where numerous pools and lakes are intersected by river channels. Non-breeding birds congregate on shallow lakes and marshlands, in the same areas as the breeding pairs. The swans are rarely found in shrub tundra during the breeding season, and are recorded in the forest–tundra and taiga zones only during migration; they also tend to avoid mountainous regions (Mineyev 1991). On Vaygach Island, a relatively high proportion of nests (26%) were situated in dry areas with willow bushes (*Salix* spp.) (Syroechkovsky *et al.* 2002). This differed from the situation on the Bolshezemelskaya and Malozemelskaya tundras, where the swans generally prefer to nest in sedge and moss–sedge habitats (Mineyev 1991; WWT unpubl. data). Even on the hillier Vaygach Island (which reaches 150m above sea level), however, most bred in sedge meadows or moss–sedge terrain (Figure 5.4).

Bewick's Swans are extremely shy during the breeding season, and freedom from disturbance is an important factor in influencing breeding distribution and breeding success. Long-term monitoring of Bewick's Swan numbers and distribution in the Lena Delta, an important breeding area for the species east of the Ural Mountains, showed that breeding density was lowest in the southeastern part of the delta, adjacent to the main commercial shipping route linking the Lena River with the Laptev Sea (Pozdnyakov 2002). At Khabuicka, on the Malozemelskaya tundra, the swans appear to prefer nesting on slightly raised mounds, hummocks or low ridges (still <5m above sea level in this flat, open landscape), often south facing, which are thought to be the first areas to become snow-free in spring and not liable to flooding (see Section 5.17). These mounds also have the advantage of providing a view over several kilometres of open tundra, enabling the swans

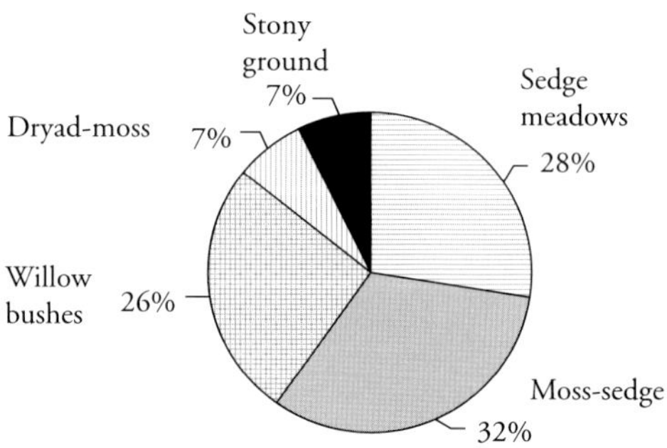

Figure 5.4 Proportion of Bewick's Swans recorded on different habitats on Vaygach Island ($n = 77$ nests). From Syroechkovsky *et al.* 2002.

to monitor the movements of Arctic Foxes and other predators. The swans leave the nest to hide if humans approach to within a few hundred metres, in some cases initially lying flat behind the nest with their neck stretched out horizontally (Figure 5.5), then retreating to a pool or lake if the intruders get closer.

The swans build the nest mound from vegetation found in the immediate vicinity, so it may consist of sedges, grasses, moss, *Empetrum* and even dwarf willow or birch. The male passes material to the female sitting on the nest, and both members of the pair may continue adding to the mound during incubation, particularly following nest relief when the swans change over. The mound itself is large, some 1.0–1.5m in diameter at the base and 15–50cm high, and the depression in the cup is normally 10–30 cm deep (WWT and All-Russian Research Institute unpubl. data). The cup is lined with softer grasses, and also with small amounts of down. The mean distance of nests from the nearest water body at Khabuicka was 27.7m (range < 1 m to 150 m, $n = 87$ over a five-year period), and the water bodies closest to the nests had a mean area of 90,600m^2. The distance from the nest to the nearest pool or lake therefore appears to be greater at Khabuicka than in some other parts of the breeding range (e.g. usually <15 m from water in the Chaun lowlands of Chukotka; Krechmar & Kondratiev 1986), and also in comparison with Whooper Swans and Mute Swans which generally nest adjacent to water. The distances from the nests to water at Khabuicka did not vary significantly between years, so did not appear to be a seasonal anomaly (WWT unpubl. data). Since the nests on the ridges tended to be further from water those in low-lying areas at Khabuicka, the greater distances may reflect the relative importance of a slightly elevated site in this species in an otherwise flat and open landscape, such as on the maritime tundra of Zakharin Bereg.

The importance of nest-site elevation was considered in further detail by coding nest site topography at Khabuicka as being either (i) 'ridge' (i.e. the nest was located on raised dry *Empetrum*-dominated tundra), (ii) 'hummock' (a raised hummock in low-lying marsh), or (iii) 'low-lying' (the nest located in low-lying *Carex*-dominated vegetation but not on a raised hummock), and monitoring the use of these sites during the 1990s. The proportion of pairs nesting on ridges, hummocks and at low-lying sites varied considerably between years. Although each type of location was used every year, generally the swans were most likely to nest on ridges. This was the most frequently used location type in all years except

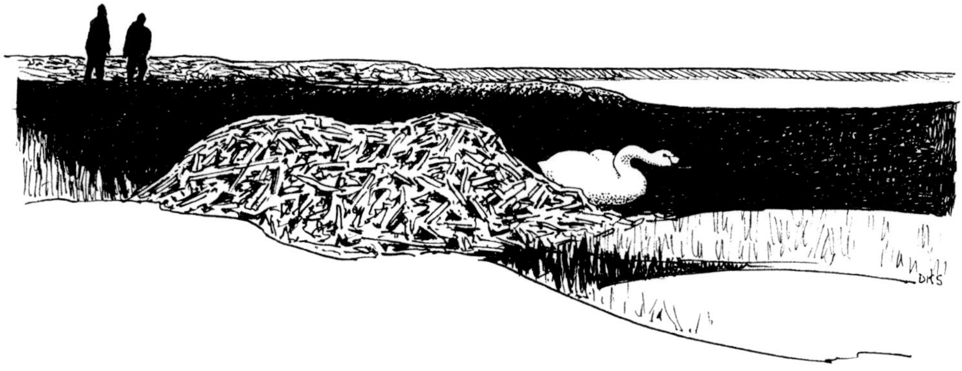

Figure 5.5 Bewick's Swan hiding behind its nest.

Table 5.4 Number of nesting pairs in the Khabuicka study area using ridge, hummock and low-lying nest sites in the years 1991–1995 inclusive, and the timing of the spring thaw. "Not known" indicates nests that were seen but not visited.

Year	Ridge	Hummock	Low-lying	Not known	Spring thaw
1991	12	8	6	7	Average
1992	6	17	5	5	Early
1993	11	10	1	0	Late
1994	6	6	1	20	Late
1995	10	10	9	9	Average

1992, which had a very early spring thaw (Table 5.4). The use of the lower sites, particularly hummocks, was more in evidence in the early and average springs.

Since swans nesting on ridges commence incubation slightly earlier than those in lower-lying areas (Section 5.10), it is suggested that a more rapid thaw of the snow at ridge sites may be one reason why they are selected by the birds. Of four pairs known to have switched territory during the study, and where the switch was not due to being displaced by an incoming pair (see Section 5.8 below), two moved their nest site from a hummock to a ridge, one involved moving from one ridge to another, and one involved nesting for the first time at a hummock nest site following a year of holding a territory and not nesting. The sample size is small but it does suggest that pairs attempt to improve their nest-site quality when they change territory, and that at least on the flat maritime tundra at Khabuicka, ridges with brood-rearing pools not overly distant (< 150m) from the nest were the preferred nest sites. Elsewhere, Bewick's Swans on the Lena Delta also usually nest on the elevated (but not precipitous) banks of channels and small lakes, and on the banks of polygonal pools (Pozdnyakov 2002).

Nests may be reused over several years, with new material added each spring, or a new nest may be built within the same territory. Of 92 nests examined on Vaygach Island where age was established, 57% were new, and the ratio of old to new nests varied between years (Syroechkovsky *et al.* 2002). There were significantly more new nests in late springs than in other years, and the ratio of old to new nests also correlated with the mean onset of laying date in that year, perhaps reflecting differences in the availability of old nests for swans in early and late springs. It was suggested that, since old nest mounds become frozen during winter, they did not thaw in time for breeding in cold springs, but that in milder years the pairs were able to reuse them (Krechmar & Kondratiev 1986; Syroechkovsky *et al.* 2002). Monda *et al.* (1994) conversely suggest that existing Whistling Swan nest mounds thaw more quickly, and that they have a larger buffer of thawed soil than the surrounding tundra. Theoretically, this should make the old nests available to the birds at an earlier stage in the breeding season, but it is not clear whether Whistling Swans use the old nest sites in preference to building new ones, or if other factors affect the swans' tendency to reuse old nest sites. Whether the territory is being occupied by the same pair or a new pair may also be relevant; this has yet to be investigated in detail. Russian colleagues observed that two apparently inexperienced pairs, new to the Khabuicka study area, made very primitive nests consisting simply of a circle of sedges (without a base) in which they laid their eggs. Both were abandoned, although the swans remained in the vicinity (A. Belousova pers. comm.).

Although nest refurbishment usually occurs at the start of the breeding season and during incubation, swans at Khabuicka were seen adding material to their nests after the brood had hatched. Such behaviour may aid brooding on cold 'nights' in the first few days after the eggs have hatched, when the parents continue to bring the cygnets to the nest. Later addition of nest material seems unlikely to improve nesting success for that year, but may serve instead to enhance nesting success in subsequent years by increasing the size of the nest mound. Nest mounds may be stabilised, and thus more permanent, through colonisation by vegetation which is fertilised by the faeces of the swans themselves and of other species using the mound as a vantage point outside of the breeding season. The formation of permafrost cores during the winter months may also help this process (Monda *et al.* 1994; J. Bowler pers. comm.).

5.8 RETURN RATES AND NEST-SITE FIDELITY

Throughout the 1990s, WWT and the All-Russian Research Institute monitored the occupancy of territories by Bewick's Swans in a 25 km^2 study area at Khabuicka, to ascertain the annual variation in the occupancy of the territories and to determine whether consistency of occupancy is related to the breeding success of the pairs on those territories. Return rates for pairs identified by natural and artificial markings was monitored, with a view to establishing the frequency of turnover of pairs on the territories. Only a small number of swans recorded in the study area had been marked in the wintering range ('Dina' and possibly 'Croquet' from Martin Mere, and 'Panier' and 'Aintree' from Welney), but parents and their cygnets were caught and ringed on territories at Khabuicka during the early to mid-1990s to support the recognition of individuals by their bill pattern markings. By mapping the location of nests and territorial pairs (without nests) each spring, and checking each bird for rings and bill patterns, it was thus possible to determine whether known individuals returned to the same or to a different territory each year. Observations soon showed that Bewick's Swan pairs occupy the same breeding territories over several summers (Shchadilov *et al.* 1998). Of 184 pairs where both male and female were identified between 1991 and 1995, two (1.1%) were pairs where both members returned to the same site in all five years. A further eight pairs (4.3%) returned for a minimum of four years, 22 (12.0%) for at least three years and 39 (21.2%) for a minimum of two years. The remaining 113 pairs were seen on territories that they had not occupied in a previous season, and were thought to be newly formed pairs.

Individual pairs generally returned to the same territory used in the previous year, unless ousted by an incoming pair. Of 111 pairs returning to Khabuicka where at least one member of the pair had been recorded there previously, 92% returned to the same territory that they had occupied in a previous year. Of the nine that changed territory, two involved birds which returned alone having lost their mates, three were pairs which were forced off their territory by another pair, two were pairs which had been absent in the previous year, and the remaining two moved into adjacent unoccupied territories without the loss of the original territory to another pair. In all three instances where an incoming pair displaced the established pair from their territory, the new pair did not attempt to build a nest on the territory until the following year. Just holding a territory therefore may be important for the swans' longer-term breeding success, though new territorial pairs (i.e. that did not nest in the first year) were not subsequently more likely to nest than established pairs who had been on the territory for at least two years (WWT unpubl. data).

The Khabuicka study found that the loss of a clutch or suboptimal breeding conditions did not necessarily deter the swans from returning to the same site. The territory closest to the huts was occupied by a female swan ('Dina') for six years, during which time she had two mates. Despite efforts to keep disturbance levels to a minimum, 'Dina' was frequently displaced from the nest by human movement to and from the camp, and she only succeeded in raising one cygnet over the six-year period. Nevertheless she kept returning, maintaining her fidelity to her territory at Khabuicka until finally moving to another breeding site or joining the non-breeding flock in summer 2001. She is still seen regularly in the wintering range, most recently on the Ouse Washes in January 2004.

5.9 THE LAYING PERIOD

The onset of egg laying varies between years and appears to be influenced mainly by the timing of the ice and snow melt on the breeding grounds. On Vaygach Island the timing of the first clutch over a six-year period ranged from 28 May (recorded in 1995) to 17 June (in 1997). Early laying dates were associated with above average spring temperatures for the region, whereas delayed onset of laying was recorded when the break-up of river ice did not occur until mid- to late June (Syroechkovsky *et al.* 2002; Figure 5.6). Further south, at Khabuicka on the Russian mainland, the onset of laying appears to be a little earlier, ranging from 23 May in 1995 to 6 June in 1994 and 1997 (WWT and All-Russian Research Institute, unpubl. data). The timing of breeding appears to be similar for Bewick's Swans taking the eastern flyway. The birds usually arrive in the Lena Delta during the second half of May (the earliest arrival was 15 May in 1995 and 2000; the latest arrival was 30 May for the south of the delta and 9 June for the north in 1996; Pozdnyakov 2002) and, in the easternmost part of the range, nesting at Chaun Bay commenced between 29 May and 6 June over a nine-year period (Krechmar & Kondratiev 1986).

The eggs are initially white in colour but stain to brownish yellow after a few days in the nest. They are almost equally rounded at the ends, with mean dimensions of 102.7 x 66.4 mm (89.0–117.4 x 60.5–72.0) (n = 342, for eggs measured near Khabuicka; WWT and All-Russia Research Institute unpubl. data; Bowler 2005b). The average mass of freshly laid eggs recorded on Vaygach Island ranged from 237.2 g (n = 32 clutches) in 1995 to 245.2 g (n = 76) in 1988, with egg mass tending to be higher in early and medium springs than in late springs (Syroechkovsky *et al.* 2002). Eggs of swans on Vaygach Island may also weigh less than those nesting further south; mean egg mass for swans in the Nenetskiy State Nature Reserve in 1996 was 247.6 g (n = 115; Buzun & Grigoryan 2004), compared with 240.6 g (n = 69) on Vaygach Island in the same year (Syroechkovsky *et al.* 2002), although further comparisons between the sites are needed to confirm this observation. The mass of eggs recorded for captive Bewick's Swans (mean = 290 g, range = 252-326, n = 44; Cramp & Simmons 1977) appears to be heavier than for swans in the wild.

Clutch size varies across Arctic Russia, and also varies between years within relatively small study areas, due to annual variation in weather conditions and food availability. Nevertheless, clutches of 2–5 eggs are the norm, although occasionally 6 eggs have been recorded. At Khabuicka, average final clutch size was 3.4 eggs (range = 1–6 eggs in 214 clutches) in 1991–2000, which is higher than the estimate of 3.1 eggs over a broader range of northern European Russia in earlier years (range = 1–6 eggs in 73 clutches, Mineyev 1991), but lower than those recorded on Chukotka (5-year mean of 3.8 eggs; Kondratiev 1991).

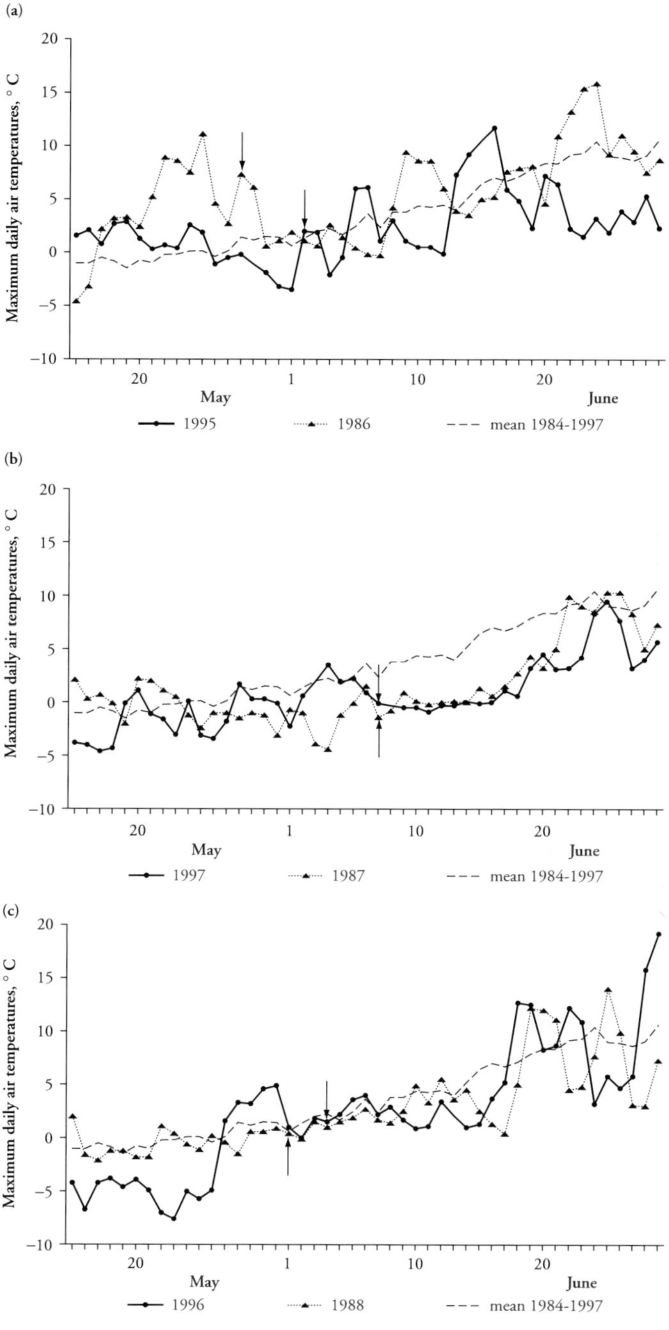

Figure 5.6 Mean daily temperatures in: (a) early springs (dotted line = 1986, solid line = 1995); (b) late springs (dotted line = 1987, solid line = 1997); and (c) normal springs (dotted line = 1988, solid line = 1996). Dashed lines show the mean daily maximum air temperatures for 1984–1997. Arrows show the date of onset of egg laying in each year. Air temperature data were recorded at the Bely Nos weather station. From Syroechkovsky *et al.* 2002.

Table 5.5 Breeding parameters for Bewick's Swans nesting on Vaygach Island. Mean values are given throughout. Sample sizes are given in parentheses. +++ = early spring, ++ = medium spring, + = late spring. From Syroechkovsky *et al.* 2002.

Year	1986+++	1987+	1988++	1995+++	1996++	1997+
Mean clutch initiation date (days from 1 June)	8.7 (13)	11.6 (5)	9.1 (25)	1.6 (10)	9.4 (24)	12.3 (6)
Percentage of old nests	47% (15)	17% (12)	46% (24)	63% (8)	50% (26)	33% (6)
Nest density (nests/km^2)	0.19	0.12	0.31	0.21	0.36	0.13
Distance (km) to nearest nest	0.97 (18)	1.73 (13)	1.08 (32)	1.29 (10)	0.86 (27)	1.98 (6)
Nesting success (%)	66% (9)	66% (12)	94% (18)	86% (7)	71% (17)	20% (5)
Mean clutch size	3.0 (18)	2.4 (14)	2.8 (29)	3.3 (11)	2.9 (27)	2.4 (5)
Modal clutch size	3	3	3	4	3	3
Calculated fresh egg mass	245.0 (52)	235.9 (32)	245.2 (76)	237.2 (32)	240.6 (69)	231.0 (12)

Clutch sizes on Vaygach Island are markedly lower, however, with a mean of 2.8 eggs (range = 1–5 eggs in 104 clutches) in the years 1986–1988 and 1995–1997 (Syroechkovsky *et al.* 2002), perhaps reflecting the higher breeding latitudes and later onset of egg laying. Annual variation in clutch sizes is evident, with annual means of 2.6 to 3.9 eggs at Khabuicka in 1991–2000, and of 2.4 to 3.3 eggs on Vaygach Island in 1986–1988 and 1995–1997. In 1996, average clutch size on Vaygach Island was significantly smaller than clutch sizes recorded that year in mainland Russia (Syroechkovsky *et al.* 2002). It seems likely that swans nesting at higher latitudes would have smaller clutch sizes, because of the shortening of an already brief breeding season, but further data are required to determine whether this tendency is consistent over several years.

Clutch size varies with the onset of laying date; generally more eggs are laid in early clutches than in those laid later in the year, and clutches tend to be both early and to include more eggs in years with an early spring thaw (Table 5.5). Other variables such as pair identity, whether the pair had been seen breeding at the site in previous years ('breeding experience'), proximity of the next nest and nest topography, does not seem to influence the number of eggs laid (WWT unpubl. data). The association between onset of laying date and clutch size is consistent for several studies in the breeding range (Syroechkovsky *et al.* 1991, 2002; Pozdnyakov 2002) and is also well documented for other species (e.g. Monda *et al.* 1994 for Whistling Swans; Birkhead *et al.* 1983 and Walter *et al.* 1991 for Mute Swans). Moreover, the largest clutches generally contain the largest eggs. The decrease in clutch size for birds that start laying later in the season may help those laying late to hatch their eggs at a similar time to the rest of the population. The time between laying each egg has been estimated at 48 hours in captivity (Evans 1975), and also for the first three eggs for clutches in the wild (albeit with a longer gap of up to 72 hours reported for the last eggs of a clutch; Krechmar & Kondratiev 1986). This means that incubation may start some eight days after onset of laying for a clutch of five eggs, compared with after four days for a clutch of three eggs. Alternatively, this trend could have evolved in response to the energy constraints of Arctic-breeding, such as swans breeding late in the season being in poorer condition and therefore having limited nutrient reserves for egg production, rather than to increase hatching synchrony *per se*, although hatching date is important for ensuring that the cygnets have enough time to fledge and migrate before ice and snow return to the tundra.

5.10 INCUBATION BEHAVIOUR

The female swans incubate the eggs, which take 29–30 days to hatch. Initial observations found that both parents sat on the nest, indicating that both sexes incubated the eggs (Kondratiev 1985). Measuring the temperature of artificial eggs placed in the nest reinforced the view that the males take an active role, and that they are responsible for 20%–44% of incubation during the day (Krechmar & Kondratiev 1986). Egg temperatures varied from 33.0–25.0°C, however (mean = 30.1°C; Krechmar & Kondratyev 1986), and detailed behavioural observations suggest that only the female settles on the nest sufficiently to heat the eggs via her brood patches (Bowler 2005b). Further work therefore is needed to confirm the precise role of the male in incubation, although he certainly sits on the nest when the female leaves to feed, thus reducing chilling and protecting the eggs from predation. Nest occupancy by the male is more frequent on warm days, whereas the female is more likely to incubate on cooler days and at night (J. Bowler pers comm.). Since the male generally sits whilst the female is away, the covering of eggs with down and nest material, noted in other species (including Whooper Swans), is not usual in Bewick's Swans.

Observations made at Khabuicka during the 1990s found that, despite annual variation in weather conditions, the mean date for starting incubation was quite consistent, ranging from 30 May in 1992 to 9 June in 1997 (WWT unpubl. data). The prompt start to the breeding season is likely to be crucial for Arctic breeders, for ensuring that the parents have sufficient time to raise their brood, and even this ten-day span may prove important if a late spring coincides with the early onset of winter. Incubation commences once the last egg is laid. Swans nesting on ridges initiated incubation, on average, one day before those nesting on hummocks, which in turn initiated incubation one day earlier than pairs with low-lying nests (Table 5.6). Nest topography was significant not only in determining the timing of incubation, but its influence proved more important in some years than in others (Bowler *et al.* in prep.). Snow generally disappears earlier from ridges than from lower-lying areas in the spring, whilst low-lying areas are more prone to flooding during and immediately after the spring thaw. This places a premium upon the selection and defence of ridge nest sites in years with an average or late spring thaw. Conversely, in early years (e.g. 1992) pairs were able to start laying early at low-lying and hummock nest-sites as well, so the advantage of holding a ridge nest-site was reduced for that year.

5.11 PREDATION

Swan eggs and young are particularly at risk from predators such as Arctic Foxes, Wolverines *Gulo gulo*, and probably also White-tailed Eagles *Haliaeetus albicilla*. These strike mainly when the adults are away from the nest or if the brood is dispersed, since the adults are capable of defending their young at least against the foxes. Skuas *Stercorarius* sp. and *Catharacta* sp. are a constant threat to swan eggs. Other potential predators include Herring Gulls *Larus argentatus* and Glaucous Gulls *Larus hyperboreus*, Snowy Owls *Nyctea scandiaca* and Rough-legged Buzzards *Buteo lagopus*. One nest on Vaygach Island was predated by a Polar Bear *Ursus maritimus* (Syroechkovsky *et al.* 2002), but this can be considered unusual. Brown Bears *Ursus arctos* usually occur to the south of the swans' breeding grounds and, when they do venture into the tundra (sometimes occurring at Khabuicka) it is in the summer to feed on berries. The jury is still out on whether

Table 5.6 Date of onset of incubation in the Khabuicka study area for nests in different topographical positions. See text for definitions. SE is the standard error for the incubation dates. Year and nest topography were both significant in determining the timing of incubation ($p<0.001$ and $p<0.01$, respectively), and a significant Year*Topography interaction showed that topography was more important in some years than in others.

Year	Nest topography	Mean date incubation commenced	Number of nests	SE (days)
1992	Ridge	1 June	2	0.98
	Hummock	30 May	7	0.86
	Low-lying	1 June	1	–
1993	Ridge	6 June	4	0.91
	Hummock	7 June	5	0.51
	Low-lying	–	0	–
1994	Ridge	7 June	3	0.58
	Hummock	8 June	5	0.60
	Low-lying	11 June	1	–
1995	Ridge	30 May	9	1.14
	Hummock	2 June	8	0.91
	Low-lying	3 June	9	0.67
Mean date 1992–1994	Ridge	3 June	4	1.93
	Hummock	4 June	4	2.02
	Low-lying	5 June	3	3.06

domestic Reindeer *Rangifer tarandus* have a harmful effect on Bewick's Swan breeding success, and the risk may vary across the breeding range. Reindeer have been reported eating goose eggs (Syroechkovsky & Krechmar 1981) and trampling on waterfowl nests (Egorov 1965; Pozdnyakov 2002), and there is a suggestion that swan breeding density declines with increasing Reindeer numbers in the Lena Delta region, although this trend did not prove statistically significant (Pozdnyakov 2002). However, the animals do not seem to have a major effect on swan nests in the Nenetskiy State Nature Reserve or on Vaygach Island (Syroechkovsky *et al.* 2002; WWT and All-Russian Research Institute unpubl. data), so whether there is a conflict between Reindeer and swans, and whether this depends on the number of Reindeer herds and the timing of their movements across the tundra remains uncertain. Further consideration of this issue should take into account not only the Reindeer themselves, but also the effect on the swans of people with children and dogs associated with the Reindeer herds.

At Khabuicka, Arctic Foxes were identified as being the chief predators of eggs and young cygnets. In the years of most intensive monitoring (1992–1994 inclusive), they accounted for three out of five total clutch/brood losses by 31 July, two out of four partial clutch losses, and may have been implicated in instances where eggs were taken by gulls or skuas, for example by distracting the parents away from the nest. Eggs were also seen taken by Herring Gulls (three nests) and Arctic Skuas *Stercorarius parasiticus* (two nests). During the same three years, a further 14 of the study pairs lost all of their eggs and/or cygnets, and 25 pairs lost one or more eggs and/or cygnets to unknown agents. Three eggs were possibly taken from one nest by a Wolverine in 1995, and eggs or newly hatched cygnets were also taken by a Wolverine from a separate territory in 1999 (A. Belousova and C. Liggett pers. comm.).

Cycles of lemming *Lemmus* sp. abundance have been linked to the breeding success of Arctic-nesting geese, and they may also affect swans in some areas although this has to be confirmed. For instance, predation pressure on eggs and downy young by Arctic Foxes is thought to be higher when populations of their preferred prey, lemmings, are low, particularly if a peak in lemming numbers the previous year resulted in an abundance of foxes in the area. The breeding success of Bewick's Swans at Khabuicka was certainly reduced in years when there were three or four active Arctic Fox dens (with cubs) in the immediate vicinity (WWT unpubl. data). However, breeding success on Vaygach Island remained high in a year when lemmings were scarce and Arctic Foxes were abundant, despite Barnacle Geese and Tundra Bean Geese *Anser fabalis rossicus* having greatly reduced breeding success in low-lemming years (Syroechkovsky *et al.* 1991). The difference between the two studies may result from differences in the relative nesting densities of the Bewick's Swans at the two sites, and in the relative nesting densities of other prey species including geese. At Khabuicka, Bewick's Swans always nest at considerably higher densities than White-fronted Geese and Tundra Bean Geese combined, whereas on Vaygach Island swan nest densities are always considerably lower than for all goose species combined. Arctic Foxes on Vaygach Island therefore may find it more profitable to rob the nests of the smaller geese, which are less able than the swans to defend their eggs and young.

5.12 HATCHING

Hatching is somewhat asynchronous, and takes place over a 24–72 hour period, from the first egg piping to the last cygnet emerging. Eggshell from the first egg to hatch is sometimes removed by the female and carried up to 100m from the nest. The other eggshells are left in the nest, but the adults cover them with nest material upon leaving (Bowler 2005b). At Khabuicka, average brood size in the first five days after hatching was three cygnets during the 1990s (range = 2.8-3.4), compared with an average clutch size of 3.4 eggs over a similar period (1991–2000; see Section 5.9). Yuri Mineyev's extensive monitoring of Bewick's Swan nesting success across a wide expanse of European northeast Russia found an average 2.7 cygnets per family in the region in the first days after hatching (from 485 clutches), compared with an overall average clutch size of 3.1 eggs per clutch (Mineyev 1991).

Intensive fieldwork at Khabuicka in 1995 found that 91.8% of eggs hatched (n = 98 eggs in 25 clutches). The Vaygach Island study gave similar results, with 89.8% of eggs hatched (n = 137 eggs in 45 clutches where the exact number of cygnets that hatched was known, for all years of the study). Further investigation of the 14 eggs that failed to hatch on Vagach Island found that in two (1.5%) cases the cygnets were dead in shell (perhaps due to slow embryo growth), five (3.6%) were rotten (perhaps not fertilised), two (3.6%) were predated, presumably by gulls after the nest was checked, and five (3.6%) disappeared from the nests (one from each nest) for reasons not known (Syroechkovsky *et al.* 2002).

Longer-term but less detailed monitoring at Khabuicka between 1992 and 2000, to determine hatching success for a larger sample of clutches, found that 151 (54.9%) of

275 clutches surveyed produced at least one cygnet, but the outcome of a further 103 was not known. Twenty-one (7.6%) of the clutches were known to be totally lost during the study although, given the large number of clutches where hatching success was unclear, the rate of total clutch loss to predators is likely to be higher (WWT and All-Russian Research Institute, unpubl. data).

The proportion of cygnets that hatched at Vaygach Island varied between years (Table 5.5) and this, together with the variation in nesting density and clutch size, indicates that Bewick's Swan breeding success is dependent on several parameters related to weather conditions at different stages of the breeding programme, as has been demonstrated for Arctic-nesting geese (Owen 1980; Syroechkovsky *et al.* 1991). Although the precise cause of total nest failure is not well established, it is suggested that a main reason may be that the swans abandon their nests prematurely if cold weather strikes during incubation (Syroechkovsky *et al.* 2002). Nesting success (measured as the percentage of eggs that hatched) for 69 clutches monitored through to hatching on Vaygach Island was lower in years with cold weather during laying (1986 and 1995) or late springs (1987), or both (1997) than in milder seasons (1988 and 1996; Table 5.5) and, although it was not possible to determine the number of abandoned nests amongst the unsuccessful ones, at least in some cases the swans had left before the predators attacked because cold eggs remained in the nest for at least 2–3 days before being destroyed by gulls or Arctic Foxes (Syroechkovsky *et al.* 2002).

A lack of breeding experience may also influence hatching success. There is little data for this to date in Bewick's Swans, but an apparently inexperienced pair at Khabuicka left the nest with two cygnets when a third was within half a day of hatching (A. Belousova pers. comm.). Pairs generally see all their viable eggs through to hatching before leading the brood to water.

5.13 BROODS AND PARENTAL CARE IN SUMMER

The downy young are active within a few hours of hatching, although they may remain in the nest for up to three days whilst the rest of the clutch hatches. The cygnets feed themselves from the outset, although they may also take vegetation brought to the surface of the water or pulled from the banks by the adult birds, particularly when small. Both parents are actively involved in rearing the chicks; the female broods the young and both parents, but particularly the male, defend the territory from incoming swans and other grazing wildfowl. The male also drives off predators such as Arctic Foxes, whilst the female leads the young to hiding places in vegetation at the edge of the pool.

Previous breeding experience and co-operation between members of a pair (Section 5.17) may be particularly important for cygnet survival during and immediately after hatching. On one occasion, an Arctic Fox distracted the male parent by jumping around erratically, and managed to steal a newly hatched cygnet. The male started calling until the female left the nest, and the pair then appeared to search for the missing cygnet. This was not a good tactic, since it left the fox to take a second cygnet and an egg directly from the nest (A. Belousova pers. comm.). More usually, both members of the pair stay close to the nest throughout the hatching period, so that the cygnets can hide in the nest if a predator approaches. Similarly, families remain close together on leaving the nest for water for the first time.

Adoption of young birds from a neighbouring family may occasionally occur in areas of high nesting density. At Khabuicka, eight cygnets were seen climbing the bank of a channel, followed by their two parents, when maximum clutch size in the vicinity was of only six eggs. Although the downy cygnets are brooded to protect them both from predators and from chilling in the first few weeks after hatching, they are not carried on their parents' backs in the manner noted for Mute Swans and Black-necked Swans. Families are more mobile once hatching is completed than during the incubation period, and may move into territories of adjacent pairs at this time. Parental care continues to the first winter and spring, as discussed in Chapter 6, but offspring from previous years generally do not accompany their parents when the latter arrive on the breeding territories (Litvin *et al.* 1999; Section 5.4).

Brood sizes decline during July as a result of predation and adverse weather conditions. As might be expected, clutch size is correlated with brood size at both the beginning and at the end of July ($r_s = 0.577$, $p < 0.01$, $n = 45$ and $r_s = 0.345$, $p < 0.05$, $n = 45$, respectively), but there was no correlation between clutch size and the number of eggs and/or cygnets lost by the end of July ($r_s = -0.015$, $p > 0.1$, $n = 47$). Preliminary analysis of breeding success data recorded at Khabuicka has shown that variation in brood size at the end of July, by which time the cygnets are about 3–4 weeks old, is attributable to the previous breeding experience of the parents. Pairs which included one or more members which had nested at least once before at Khabuicka, had larger brood sizes by the end of July than inexperienced breeders; 2.02 ($n = 47$) and 1.74 ($n = 23$) for experienced and inexperienced breeding pairs respectively. Although there was some annual variation in brood sizes by the end of July, these differences were not significant: 2.12 ($n = 26$) in 1992, 1.85 ($n = 33$) in 1993, 2.19 ($n = 32$) in 1994 and 2.0 ($n = 16$) in 1998. The smallest broods occurred in 1993, the year with a late spring thaw, but the range of brood sizes recorded meant that they did not differ significantly from those seen in other years. Further analysis found a significant interaction between breeding experience and year, indicating that the effect of previous breeding experience upon brood size was more important in some years than in others (WWT unpubl. data). This supported observations made in the wintering range, which also found that the number of years that a pair had been together had a greater influence on breeding success (i.e. the number of cygnets associating in winter) in some years than in others (Rees *et al.* 1996; Chapter 7).

5.14 CYGNET SURVIVAL

The unpredictability of the weather in early springs, coupled with a possible reduction in the availability of food and its quality, appears to diminish any benefits to the swans of starting their breeding programme early. Certainly post-hatching mortality in cygnets at Khabuicka seems to be high in the first two weeks after hatching. Of the 91.8% of eggs that hatched at Khabuicka in 1995 (98 eggs in 25 clutches), average brood size was 3.91 on hatching, 3.91 at 5 days and 3.36 at 10 days (11 families monitored for 10 days after hatching). Thus, of 43 cygnets that hatched for the 11 families, 86.1% survived to 10 days old. For a larger number of families monitored less frequently (i.e. at >5-day intervals), but over a longer period, average brood size was 2.6 by the time the cygnets were about 5–6 weeks old ($n = 73$ broods) from an average final clutch size of 3.6 eggs ($n = 97$ clutches) between 1992–1994.

These figures give a general indication of the perinatal losses suffered, but hatching success and cygnet survival in the first few days after hatching varies between years, depending on the weather conditions. Brood size at hatching also varies between years, in line with the annual variation in clutch size described in Section 5.9. The proportion of successful nests (i.e. at least one cygnet hatched) was lowest and the number of eggs or cygnets lost per clutch highest at Khabuicka in 1992 (mean egg or cygnet loss of 1.96, $n = 24$ clutches), which saw an early spring thaw and three active Arctic Fox dens in the study area. By comparison, the number of eggs or cygnets lost per nest was lower in 1993 and 1994 (1.22, $n = 9$ for both years respectively), which had late spring thaws and only one active fox den in the vicinity.

The data are currently too patchy to determine whether there is a meaningful association between the location of the nest site and the predation of eggs or cygnets, but the initial indication for swans at Khabuicka is that rates of egg or cygnet loss are, on average, highest for low-lying nests, intermediate for nests on hummocks and lowest for nests on ridges (Table 5.7). Pairs nesting on ridges lost significantly fewer eggs or cygnets than those nesting on hummocks in 1993 and, when all the data were combined, the loss of eggs or cygnets from ridge nests was significantly lower than that from low-lying nests (Table 5.7; Bowler *et al.* in prep.). The rearing of young from nests on ridges therefore was significantly higher than at lower-lying nests in these years although, given that ground-based and aerial predators are able to reach all the sites, the biological reason for this is unclear. Of 61 experienced breeding pairs, 41% nested on ridges, 42.6% nested on hummocks and 16.4% in low-lying areas, suggesting that experienced pairs may select the more elevated locations. Whether their choice of nest location improves their chances of breeding successfully, or whether it is

Table 5.7 Comparison of the numbers of eggs and/or cygnets lost by 31st July from nests in different topographical locations in the Khabuicka study area. [a] Includes a low-lying nest site from 1995, to improve the sample size. *In 1993, significantly fewer eggs/cygnets were lost from ridge nests, compared with hummock nests. ** Over the whole study period, significantly fewer eggs/cygnets were lost from ridge nests, compared with low-lying nests. All other comparisons of egg/cygnet loss in relation to nest topography were not significant.

		No. of eggs and/or cygnets lost per nest	
Year	Nest topography	Mean (± SE)	Number of nests
1992	Ridge	2.00 (± 1.00)	4
	Hummock	1.67 (± 0.36)	15
	Low-lying	2.80 (± 0.80)	5
1993	Ridge*	0.80 (± 0.29)	10
	Hummock	2.00 (± 0.31)	9
	Low-lying	3.00 (–)	1
1994	Ridge	1.00 (± 1.00)	2
	Hummock	1.50 (± 0.56)	6
	Low-lying[a]	1.50 (± 1.50)	2
All	Ridge**	1.13 (± 0.34)	16
	Hummock	1.70 (± 0.23)	30
	Low-lying	2.50 (± 0.60)	8

their breeding experience that at least partially explains the improved productivity at these sites, remains to be determined.

Cygnets on the Nenetskiy State Nature Reserve fledge at 60–70 days, with most fledging occurring at around 10–15 September, and the last young thought to become fully winged by 20–25 September (J. Beekman pers. obs. in Rees *et al.* 1997a). Reports of cygnets fledging sooner, at 45–55 days (Kondratiev 1991b), may perhaps refer to the time taken for the birds to gain their body contour feathers, rather than being capable of flight, although nesting latitude and habitat are known to affect cygnet growth rates. For instance, the estimated age at fledging for Whooper Swan cygnets from peatlands and oligotrophic lakes were, on average, 18 days and 10 days later, respectively, than cygnets from eutrophic lakes (Knudsen *et al.* 2002). Bewick's Swan cygnet mortality in the weeks prior to fledging and during the autumn migration is not yet known but is also likely to be high. Mean brood sizes recorded during winter range from 1.6 to 1.9 cygnets per family (Rees *et al.* 1997a), or maybe as high as the 2.1 cygnets per family seen at Slimbridge (Evans 1979a), and this does not take into account pairs that lost all their young earlier in the season. Even for cygnets that reach the wintering grounds, the probability of returning for a second year is quite low, with just 66.2% seen for a second winter (Scott 1988). The survival rate for adult birds wintering at Slimbridge is, for comparison, estimated at 87.1%, with no differences between the sexes (Evans 1979a; Scott 1988).

5.15 MOULTING PERIOD

Adult Bewick's Swans undergo a complete moult of their flight feathers every year, and individual swans are flightless for about three weeks, with the main moulting period ranging from the last ten days of July to mid-August. Pairs with cygnets moult slightly asynchronously (up to ten days apart, Rees *et al.* 1997a), with female parents generally moulting their flight feathers before the males. This asynchrony within breeding pairs enables one parent to be fully winged for a larger proportion of the moult period. Not only are they therefore capable of defending the mate and brood from predators, they may also be able to reassemble the family if adults and young should become separated. Non-breeders moult at the same time as the breeding pairs; both breeders and non-breeders have been caught for ringing whilst flightless in early to mid-August. The tail and body moult starts later, as in other waterfowl species, with the replacement of tail feathers complete by October and of body feathers by February (J Beekman, unpubl. data in Rees *et al.* 1997a).

Swans are at higher risk both of predation and of more limited access to food whilst flightless, so many non-breeders move to the safety of large water bodies where they congregate to feed on aquatic vegetation, whilst territorial and breeding pairs remain on the breeding territories. Others may roam across the tundra, feeding on sedges, grasses, and particularly the berries of Cloudberry *Rubus chamaemorus*, Cowberry *Vaccinium vitis-idaea* and Crowberry *Empetrum nigrum* that are ripening at this time. They can see potential predators at great distances across the flat landscape, are adept at lying low in even the shortest vegetation and, if driven to it, are very rapid runners. So they can quickly hide or escape to the numerous lakes and river channels to avoid land-based predators.

5.16 WEATHER CONDITIONS AND BREEDING SUCCESS: A SYNTHESIS

Earlier parts of this chapter have described the number of ways in which the weather conditions can influence Bewick's Swan breeding success. Spring weather influences breeding density and the timing of laying, and the latter in turn determines clutch sizes. Nesting begins later in years when the thaw is later, nesting density is lower in these years, and clutch size and egg mass are also lower. It is thought that spring weather, including the occurrence of cold snaps, is the single most important factor affecting Bewick's Swan breeding success on Vaygach Island (Syroechkovsky *et al.* 2002), perhaps because the swans have a tighter timescale for breeding at these high latitudes than those on the Russian mainland. Further south, there may still be time for swans that start breeding late, in mid- to late June, to hatch and raise young if conditions are good later in the year, and temperatures do not drop consistently below freezing until late September or early October.

The preliminary observations at Khabuicka suggest that cygnet survival after hatching was in fact higher in years with late spring thaws, which coincided with fewer active fox dens in the vicinity (Section 5.14), but further information is needed to determine whether the breeding density of Arctic Foxes is consistently related to spring weather conditions. Even if it is, the reduced predation rate may not compensate for the smaller number of breeding swans and reduced clutch sizes in years with colder springs.

Non-breeding Bewick's Swans in the Nenetskiy State Nature Reserve have been seen feeding intensively on the berries of Cloudberry, Cowberry and Crowberry in August. Given the apparent importance of this food source to non-breeders, it may also be important for cygnet survival in some years, but the level to which fruiting varies with weather conditions, the extent to which the birds rely on these tundra berries in addition to their staple diet of Arctic sedges and grasses, and the extent to which cygnets feed on the berries have yet to be determined. Preliminary analyses of five swan droppings collected in autumn (i.e. August to early September) on the Nenetskiy State Nature Reserve suggested that berries comprise a major part of the diet for only some swans, and that less than 10% of the Bewick's Swans' autumn diet was of tundra berries (Ubels *et al.* 2000). However, only a few autumn droppings were analysed, and there was some evidence that the swans' diet in spring (late May to June) and summer (July) varied with location (Ubels *et al.* 2000; see Chapter 4). There is therefore still much to learn concerning the environmental factors affecting Bewick's Swan survival and breeding success in the second part of the breeding season.

5.17 BREEDING EXPERIENCE AND BREEDING SUCCESS: A SYNTHESIS

Studies of Bewick's Swans in the wintering range have shown that reproductive success improves with the number of years that a pair has been together (Rees *et al.* 1996; Chapter 7), but just how pair duration and previous experience of the breeding grounds helps to improve the breeding success of a pair can only be determined by studying known swans over several years at their nest sites. It is possible that established pairs arrive on the breeding grounds in better condition (and thus fitter to breed) than less experienced pairs, in which case breeding success may be influenced by conditions encountered before the start of the breeding season. However, since there is currently no information on factors affecting fitness at this time (e.g. mate guarding whilst feeding at the staging sites), or even whether there is significant variation in the swans' body condition on arrival at the nest sites, this section focuses on information gained from the breeding grounds.

Observations of known individuals at Khabuicka found that experienced pairs (i.e. those that included at least one bird which had previously nested at the site) started incubation significantly earlier than inexperienced pairs. The only exception was in 1992, a year with an exceptionally early spring thaw, when both experienced and inexperienced breeders started incubation at around the same time (30 May). Annual variation in spring weather conditions also influenced the timing of incubation, so that the importance of previous breeding experience was more marked in some years than in others. Further investigation found that experienced breeding pairs not only started laying their eggs earlier, but that they laid larger clutch sizes than inexperienced breeders, and were also significantly more likely to nest on ridges (Section 5.14). Conversely, of 20 inexperienced breeders where nest topography was recorded 10% nested on ridges, 55% on hummocks and 35% at low-lying sites. The higher reproductive success of experienced breeding pairs was most pronounced in years with an average or late spring thaw, when ridges became free of snow earlier than lower-lying areas, and which, in addition, were less prone to flooding during the thaw. There were no significant differences between inexperienced and experienced breeders for more detailed measures of habitat quality in their territories, including distance of the nest from the nearest water body (means = 14.5 m, n = 8 and 28.9, n = 27, respectively), size of nearest water body (72.9 m^2, n = 8 and 103m^2, n = 27, respectively), or in the biomass (1.36 g/m^2, n = 8, 1.21 g/m^2, n = 34) and protein content of vegetation (14.98 %, n = 9 and 15.34%, n = 32) at the feeding areas (Bowler *et al.* in prep.).

One of the hypotheses for the increase in breeding success with pair duration, initially observed for pairs seen in the wintering range, is that knowledge accrued from previous mutual breeding attempts improves co-operation between pair members (Black *et al.* 1996; Rees *et al.* 1996). Firstly, if clutch size is associated with laying date as in other species, experienced breeders may select nest-sites that allowed them to initiate egg-laying earlier than inexperienced breeders, within territories that contain more abundant, higher quality food. Additionally, improved cooperation between pair members may reduce the risk of the loss of eggs and hatchlings, for example by improved co-ordination in the timing of changeover for guarding or incubation at the nest, in a species in which both members of the pair undertake nest-sitting activities. The exchange of nest duties has not been specifically investigated, but anecdotal information suggests that it may be significant in influencing breeding success at more than an individual level. In just the few pairs monitored at Khabuicka, one pair where the male tried to push his mate from the nest abandoned the nest after a few days. In a second case, where the female did not return to resume her nest-sitting duties, the eggs were predated (A. Belousova pers. comm.).

If coordination of responsibility for parental care within the pair is a factor, experienced breeding pairs might be expected to lose fewer eggs and cygnets than inexperienced breeders, and achieve higher rates of hatching and fledging success. There was insufficient information for both experienced and inexperienced pairs using each of the habitat categories at Khabuicka to confirm whether breeding pairs were more successful in each, but the higher proportion of experienced pairs on the more productive ridge sites suggests that their previous knowledge of the area enables them to return early and to occupy these sites.

Nevertheless, experienced breeders generally lost fewer eggs or cygnets during the summer than inexperienced breeders. It remains to be seen whether this is due to territory quality (e.g. food supply) or to parental cooperation in maintaining vigilance once the young have hatched, thus reducing the risk of losing the hatchlings.

5.18 HABITAT AND OTHER FACTORS INFLUENCING BREEDING

The effects of broad-scale habitat differences on Bewick's Swan breeding density and breeding success have yet to be determined. Although density appeared to be higher on open maritime tundra than on inland tundra across the Malozemelskaya and Bolshezemelskaya tundras (Shchadilov *et al.* 2002), regional differences may have been more important than habitat effects. Bewick's Swans nesting in northwest Vaygach Island, about 100km north of mainland Russia, experience rather different habitats to those in the open tundra of the Nenetskiy State Nature Reserve. The island is hilly (up to 150m above sea level) and the upper slopes are rocky; in its lowlands, shallow lakes are surrounded by wide marshy plains. Thus, although there is preliminary evidence that Bewick's Swan breeding success is lower on Vaygach Island than further south in mainland Russia (Syroechkovsky *et al.* 2002), the extent to which this is due to habitat or to latitudinal differences has yet to be determined. Persistent winds on Vaygach Island make for dense snow cover in winter, which melts very slowly in spring, particularly in lowland areas, so the rich lowland vegetation is not available to the swans for feeding until relatively late in the breeding season (the end of June) (Syroechkovsky *et al.* 2002). Food supply therefore may be poorer and more limited for the swans on arrival at Vaygach Island than in other breeding areas, which in turn may have consequences for nesting success, but this can only be a subject of conjecture until there has been a detailed habitat study and knowledge of the condition of returning birds.

Although, on a more local scale, nest location and breeding experience have been found to influence the onset of incubation and breeding success, and these variables seem to be more important in some years than in others, potential components of success such as vegetation abundance and protein content of the sward on the breeding territories did not prove significant on initial investigation. More detailed habitat monitoring is necessary, both on a broad scale to describe habitat variables influencing breeding density, and at a fine-grained level to determine the characteristics of good quality breeding territories. A confounding factor in such studies is the apparently overriding effect of weather conditions, which determines timing of arrival on the breeding grounds, the proportion of pairs that attempt to breed each year and the success of these pairs in that breeding season.

Studies of Bewick's Swans in the wintering range have found that male body size also influences breeding success, in that larger males tend to bring more young to the wintering grounds. It is thought that this is because larger males are more successful at defending breeding territories and protecting their young (Scott 1988; Chapter 7). At Khabuicka, female body size seemed more important, in that the larger females (determined by their skull and tarsus lengths) generally laid larger clutches, with the combination of female and male body size also being significant (WWT unpubl. data). However, since the birds caught and measured in summer were all territorial or breeding pairs, the importance of male size in holding territories on arrival in the breeding range was not tested, and this could account for the discrepancy with observations in the wintering range. Larger females may be able to accumulate higher fat reserves for the egg-laying and incubation period in absolute terms than smaller females, perhaps whilst staging in the White Sea region, and therefore be able to produce more eggs. This would be particularly beneficial in years with a late spring thaw, when additional feeding immediately prior to, and during, the egg-laying period may not be possible. In accordance with this hypothesis, the relationship between female size and clutch size was not seen in 1992 when the spring thaw was unusually early (J. Bowler pers. comm.).

5.19 OVERVIEW OF FACTORS AFFECTING BREEDING SUCCESS

Early analyses of how conditions on the breeding grounds affect Bewick's Swan breeding success have shown that annual variation in the percentage of juveniles recorded in the wintering population was attributable to weather conditions during the short Arctic breeding season (Poorter 1991). However, the detail of how climate and other variables interact has only recently been described. This chapter has illustrated annual variation in the percentage of birds that attempt to breed, in clutch size and in hatching success, all of which are influenced by nesting habitat and the breeding experience of the pair, but with weather conditions on the breeding grounds in spring and early summer having an overriding influence on the breeding success of the population as a whole.

The substantial annual variation in the percentage of juveniles recorded in the wintering grounds (e.g. ranging from 5.1%–25.6% at Slimbridge between 1970–71 and 1995–96 inclusive; Evans 1979a; WWT unpubl. data) thus reflects the whole range of factors and conditions encountered by the birds during the breeding season, including spring weather conditions across the Arctic breeding grounds, the proportion of birds that attempt to breed, their clutch sizes, predation levels, cygnet mortality on autumn migration, and the characteristics of the pair that enable them to maximise their breeding success at each stage of the breeding season.

Although weather conditions may affect breeding success of the population as a whole, at Khabuicka it seems that by the acquisition and successful defence of ridge nest sites within a territory containing appropriate brood-rearing pools that are not overly distant (< 150m) from the nest, experienced pairs are able to breed more successfully than inexperienced breeders, particularly in years with a late spring thaw. This hypothesis is consistent with the observation that the annual variation of an individual pair's breeding success, as measured in the wintering grounds, is correlated with the duration of that pair's previous mutual breeding experience.

The breeding range is also limited not only by latitude (notably, extreme weather conditions at high latitudes) but by the absence of land northwards, and perhaps by competition with Whooper Swans at the edge of the shrub tundra zone to the south. Longitudinally, the swans' breeding distribution may be influenced by the distances the birds have to migrate (either west or east) to reach suitable wintering grounds. Thus breeding areas in the more central parts of the Russian Arctic may be less attractive to the birds, since they would have further to migrate than swans at more westerly nest sites (if part of the northwest European population), or than those breeding further east (for the eastern population). Conservation of breeding areas with currently high nesting densities therefore is of high priority for the future welfare of the species. Given that the swans nest in a very narrow band across the Russian Arctic, and that the effects of global warming are predicted to have the greatest impact on habitats at high latitudes, the implications of climate change for Bewick's Swan habitats should also be addressed. Finding a solution to any such changes may, of course, be much more difficult to achieve.

CHAPTER 6

Social behaviour in winter

'... wild swans (and wild geese) are very special birds because their society is based on a permanent pair bond and a family life which keeps the young with their parents until breeding time comes round again. It took me a little time after I had learnt these things before I decided to give up shooting.'

Peter Scott (*Observations of Wildlife*)

When Bewick's Swans leave their breeding grounds, families and territorial pairs become more gregarious and join non-breeding birds in flocks at the migratory and wintering sites. Overt physical and vocal displays are used to defend their territories in summer, but displays become even more evident in winter as the birds attempt to maintain family cohesion and to establish their dominance rank in the wintering flock. Our understanding of the social life of the swans in winter and the significance during these months of their ritualised displays owes much to the work initiated by Dafila Scott at Welney (Scott 1978) and to studies undertaken by John Bowler at Slimbridge (Bowler 1996). Scott's work provides an important insight into the costs and benefits to the swans of prolonged parental care and long-term pair bonds. Bowler built on this to consider whether position in the peck order affected the distribution of individual birds within a wintering site, the likelihood of individuals returning to the site, and also their subsequent pairing and breeding success. This chapter aims to provide an overview of life within the Bewick's Swans winter flocks, particularly how the birds behave towards each other, and the consequences of this for the individuals involved.

6.1 FLOCK STRUCTURE

When Bewick's Swans congregate in wintering flocks, the largest and most stable single unit is the family. This usually consists of the parental pair with one to five cygnets, and sometimes offspring from previous years. Unlike Whooper Swans, which become independent more rapidly, Bewick's Swan yearlings (i.e. birds in their second winter) quite regularly rejoin their parents on the wintering grounds, even if they arrive on different dates. Grown offspring only occasionally associate with their parents on reaching adulthood (i.e. from their third winter onwards), however, and if they do it is usually prior to pairing. The oldest offspring seen consorting with its parents was 'Hannah', who was brought to Slimbridge as a cygnet by her parents 'Forsythe' and 'Findlay' in winter 1973–74, and was still associating with them six years later in 1979–80. In 1980-81 she returned with a mate, 'Frederick', and the pair had two cygnets of their own in winter 1980–81. On rare occasions, offspring may join the parental group when accompanied by a mate of their own, particularly when the extended family embarks on aggressive disputes with other swans in the flock. An example was the swans 'Pampas' and 'Pamper', who joined 'Pampas's parents ('Pampisford' and 'Brinkley') and their cygnets, to form a very dominant group at Slimbridge in the winters 1979–80, 1980–81 and 1981–82. It was only in 1982–83 that 'Pampas' and 'Pamper' brought their own offspring to the site.

Wintering flocks consist of all social classes, and include not only families (parents with cygnets), but pairs without offspring, single adults and yearlings. The ratio of birds or (for family groups and pairs) units from the different classes within the flock is not static and may vary between years. This is to be expected as it reflects annual variation in the breeding success of the population, with the ratio of parents to pairs without young changing accordingly (see Chapter 7), and with the number of yearlings obviously reflecting breeding success in the previous year.

In addition to the main social classes, some more ephemeral groupings have been noted. Siblings occasionally associate with each other, and at Slimbridge where most of the detailed observations of individuals have been made, this usually occurs in their second or third winters at the site. This association is limited to cooperating in aggressive encounters and in maintaining proximity to each other on the lake or in the fields. There are no known cases of siblings courting, or any evidence of them breeding successfully. Groups of three birds ('trios') – and sometimes even a larger group of four birds – aged 16 months or more, may support each other in aggressive interactions. These groups usually consist of a pair joined by one of their offspring from a previous year, but sometimes there is no obvious relationship between the three individuals. Temporary associations between two birds that are not a definite pair also occur during the winter months. They display together, perform triumph ceremonies together following an aggressive encounter and, like families, pairs and trios, they maintain spatial proximity to each other in the flock. Whereas members of other units associate with each other throughout the winter, however, temporary associates tend to be seen together only for a few days or weeks, and are apart at other times. They also generally arrive and depart on different dates at the start and end of the winter, so appear to be migrating separately.

6.2 DOMINANCE RANKS

6.2.1 Aggressive encounters and other displays

Aggressive encounters between singles, pairs or family units are common in the wintering flock, and result in the establishment of a dominance hierarchy. These encounters may consist of a head lowered threat or peck, or the swans may use a range of more prolonged displays, which are usually performed in the same order and are of increasing intensity, with opponents facing each other and calling loudly. The series, described in detail by Scott (1978; Figure 6.1) generally commences with a neck-stretching movement, and continues as follows:

Neck-stretching and head-pumping;
Neck-stretching and pumping with wings half-open but not flapping;
Wings half open and flapping;
Wings fully extended and flapping;
Wings fully extended and still.

Finally, one opponent may lunge at the other and the two birds, usually male, may grapple together, pecking at the base of the neck and beating each other with their wings. The encounter may be curtailed at any stage by one of the parties retreating, so the swans only occasionally become involved in physical combat. Detailed monitoring of the interactions between Bewick's Swans at Slimbridge in winter 1982–83 found that, of all the encounters recorded, those reaching the first main stage (i.e. neck-stretching and head-pumping) were most common, accounting for 30.7% of encounters, followed by simple head-low threats or pecks (18.4%) and wings half open and flapping displays (18.1%). Only 8.2% of encounters resulted in the opponents grappling with each other, but once the opponents had reached the stage of fully opening their wings, there was a 50% chance that the encounter would proceed to physical combat (Rees 1988).

Male swans tend to be more actively involved in aggressive encounters than their mates or dependent offspring, but family members appear to lend support during an encounter by approaching and joining in the display (Scott 1980b). In general, although more than 75% of families, pairs and trios identified at Slimbridge during a season were seen to be involved in aggressive encounters, fewer than 25% of single birds present were recorded in overt displays (Rees 1988). Subordinate single swans tended to simply give way to the more dominant groups in the herd. This submissive behaviour, in which a subordinate bird turns or moves directly away from a more dominant individual or social unit, was often accompanied by a particular head carriage with the bill tilted slightly up and the head leaning back. This possibly enables them to see behind, and so monitor the aggressor's behaviour and determine whether they are still under threat (Scott 1978).

Aggressive encounters are frequently followed by triumph displays, in which the winning swan returns to his or her mate or family and displays towards them, using similar actions to those used to win the encounter. Triumph displays therefore also consist of neck-stretching, wing-flapping and loud calling, depending on the intensity of the encounter, and are generally performed by both members of the pair and also other family members. Whilst aggressive encounters have been studied in some detail over the years, much less attention has been paid to the function of the triumph displays, despite these also being a highly overt form of behaviour. Dafila Scott, however, observed that, although triumph displays are usually performed immediately after winning an aggressive encounter (with 80% of neck-stretching and head-pumping displays being followed by a triumph display), triumph displays also occur when members of a pair are reunited after a temporary separation ('greetings triumph'), and also occasionally when pairs came in to land ('landing triumph') (Scott 1978). Greetings triumphs are seldom seen and usually occur when the temporary separation has been prolonged. The more usual greeting on meeting following a temporary separation the courtship 'bow'. Landing triumphs only occur occasionally in the wintering

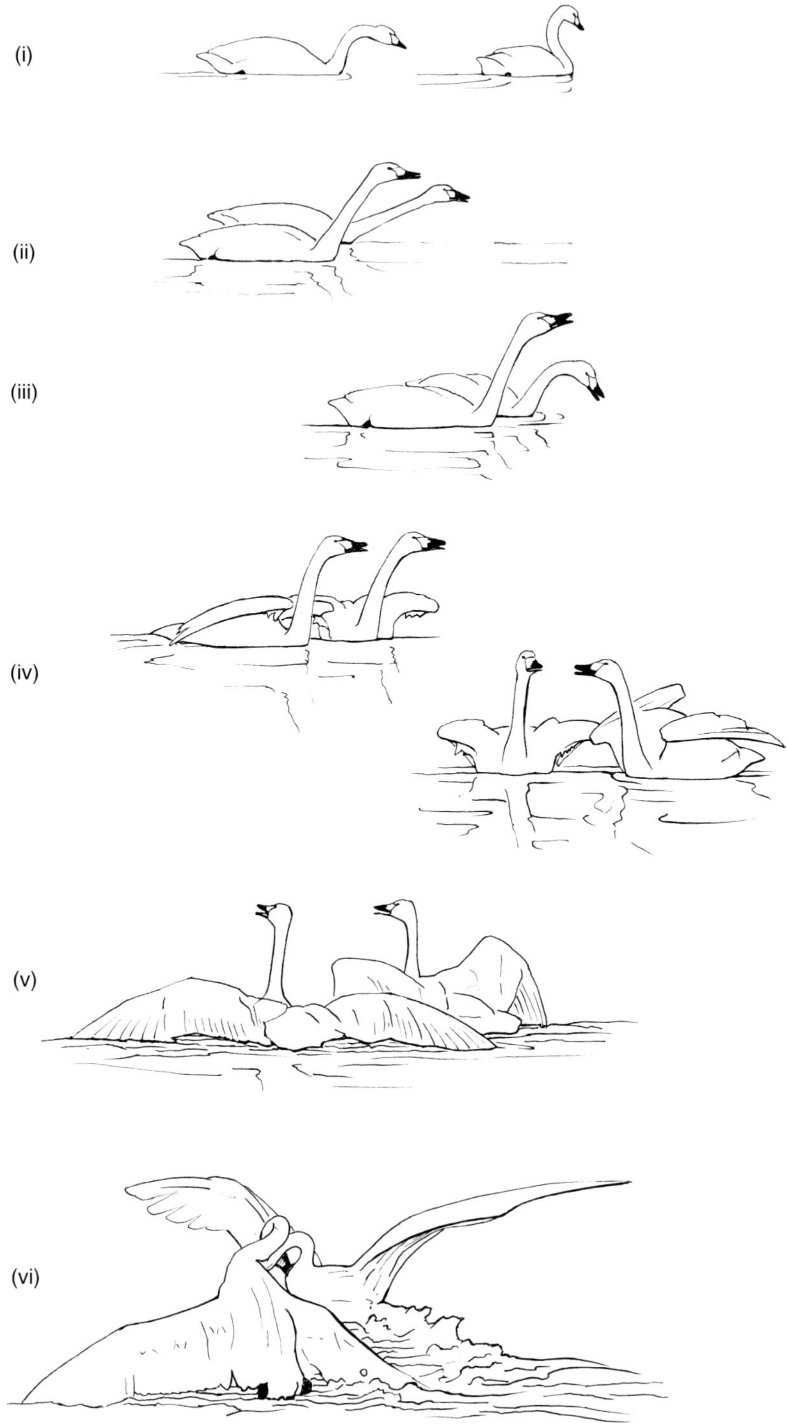

Figure 6.1 Range of displays used by Bewick's Swans in aggressive encounters: (i) head horizontal threat; (ii) neck-stretching; (iii) head-pumping; (iv) half-open wing display; (v) spread-wing display; and (vi) physical combat.

flock, but may be more common whilst swans are defending territories during the breeding season. In landing triumphs, the vocalisations sometimes begin before touchdown and continue for several seconds (up to 20 seconds) after landing. There are two possible outcomes to a landing triumph: either the display will fade if there is no response from adjacent swans, or it provokes another pair to display. When the latter occurs, either the triumphing pair are chased off (and the triumph curtailed) or the challengers are themselves defeated.

Dafila Scott (1978) suggests that the function of triumph displays, and particularly landing triumphs, is to demonstrate fighting ability, either directly to a defeated opponent or indirectly to swans of lower rank in the vicinity. She found that triumphing birds are generally of higher rank than average, especially those performing landing displays. Triumphs are more frequent when the nearest birds are of lower dominance rank. If the nearest birds are of higher rank than those landing, there is a higher probability of the triumph being curtailed. Individuals tend to triumph when the nearest swans are at a medium distance, so not when they are very close (<3 swan lengths away) or far off (>12 swan lengths away). It should be noted, however, that this hypothesis does not rule out a potential alternative or additional function of triumph displays, which is that paired birds reinforce their pair bond by displaying together.

Bewick's Swans use a wide range of other social displays. These include the pre-flight signal of head-bobbing—when the neck is alternately bent and stretched with increasing frequency while calling, immediately prior to flight. Pre-flight signals serve to ensure that actions within a social group are coordinated and that members of a pair or family take off together. As described in detail earlier (Chapter 3, Section 3.4.2), males lead the movement of the pair in the autumn, and are the first to open their wings for flight at this time, but flight initiation by the female becomes increasingly common prior to departure on the spring migration. This change suggests a shift in responsibility between members of a pair for leading movement as the winter progresses (Rees 1987).

Members of a pair, rejoining after a temporary separation, appear to locate each other in the herd by voice, then greet each other with a 'bow' of ritualised head-turning with

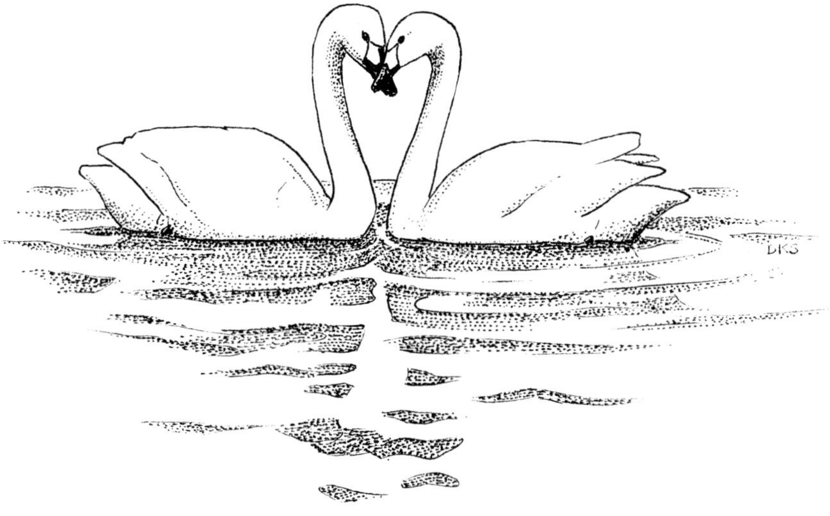

Figure 6.2 A courtship bow.

straightened necks. This is usually accompanied by a slight ruffling of the neck feathers and a slight raising of the wings, with the swans remaining close together, their bodies facing obliquely (Scott 1978; Bowler 2005b; Figure 6.2). In addition to the formal displays, paired birds and family groups also use less intense head-bobs and greetings calls as a means of staying in contact within the flock.

The frequency of aggressive displays, and particularly the frequency of encounters that proceed to physical combat, is usually high in October and November as the swans arrive at the wintering sites and establish a dominance hierarchy. The frequency then diminishes during the winter (Scott 1978; Figure 6.3). The frequency of fights did not decline from January onwards in Scott's study, but at this time there was a small increase in the number of new arrivals (perhaps due to a mid-winter influx of swans from the continent), and thus a need to re-establish the social hierarchy. The frequency of aggression, and of all types of interaction, also varies with the density of the flock. This is particularly apparent when comparing birds at a provisioned site with those occasions when the same birds are on more natural habitats. For instance, interaction rates are higher when the swans are feeding in front of the observatory at Welney than when they are on farmland in the surrounding area. A comparison of flocks of differing densities on the same habitat (fenland) also found that there were higher interaction rates when the birds were feeding closer together (Scott 1978, 1980a).

6.2.2 *Establishing dominance relationships*

Several studies of dominance relationships in Bewick's Swans have been made over the years, all of which required some measure of the dominance rank of the social units in the wintering flock. This measure has been achieved by recording the outcome of aggressive encounters between known individuals — the identification of which is made much quicker and easier through bill-pattern recognition. Encounters occur mainly when the swans are on water, so it would be difficult to determine which swans are involved by reading their leg rings. Dominance rank can be calculated at the end of the winter as the number of social units beaten as a proportion of the total number of units fought over the winter season (Scott 1978; Rees 1988; Bowler 1996).

Comparison of the ranks recorded for the different units shows that families are the most dominant, winning more than 70% of their encounters with other social groups. They are followed by trios, then paired birds, and with single swans generally being subordinate in the flock. The relative success of the small number of trios reinforces the results of similar studies on geese, which suggest that position in the social hierarchy is related to the size of

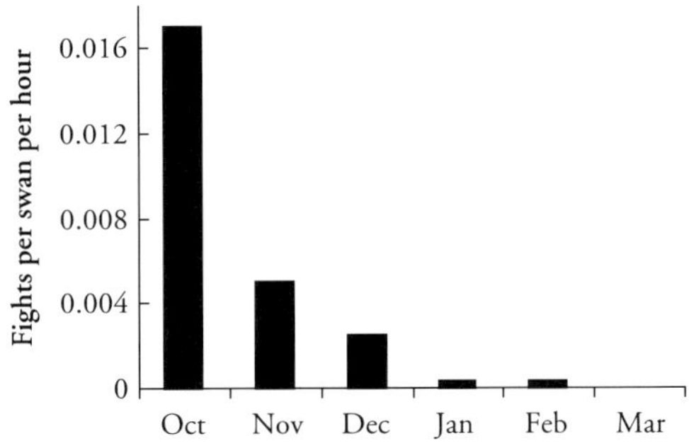

Figure 6.3 Decline in the frequency of aggressive displays recorded on the fens and in front of the observatory at Welney during the 1975–76 winter. From Scott 1978.

the unit (Black & Owen 1989). Although initial studies at Welney did not find a relationship between family size in swans (i.e. parents and the number of their cygnets) and the dominance achieved (Scott 1978, 1980b), intensive monitoring of the smaller flock at Slimbridge confirmed that the dominance rank achieved by two-parent families increases linearly with the number of cygnets that they brought to Slimbridge that winter. Moreover, the small number of single-parent families, where one of the adults had died or become lost between breeding and the family's arrival at Slimbridge, ranked between the two-parent families and pairs without cygnets (Bowler 1996). In this particular study, the two groups of trios recorded both consisted of yearlings (i.e. sub-adult birds), and these unexpectedly ranked between pairs and single swans.

The success of the different social units, and also of individual birds, was extremely constant over the winter season. This held true irrespective of whether the birds were feeding in the fields or at provisioned sites in front of the observatories at WWT Welney and WWT Slimbridge. The only apparent difference between the field and observatory results was the overall success of different sized units, with pairs beating families less frequently and singles beating pairs less frequently in the field (Scott 1980a). Since most swans are not found at provisioned sites, this suggests that class differences in dominance rank are well defined in the population as a whole. Nevertheless, the situation is not static, and swans do occasionally beat a unit that has been successful against swans with a higher dominance rank.

Given the close association between social class and dominance ranking, it is perhaps not surprising that families, trios and pairs are all more likely to be observed in encounters with their own social class than with one of the other classes (Rees 1988). This suggests a tendency for swans to compete more frequently with birds close to them in the social hierarchy, perhaps because the outcome is uncertain. The frequency of encounters between single swans and the other classes is significantly lower than expected, given the proportion of each unit present in the flock each year, suggesting that single birds either deliberately avoid aggressive encounters or that they are attacked less often because they do not pose such a threat to the other social units. When there are encounters between the different classes, the percentage of fights initiated by the dominant class increases as the overall rank of the opposing class decreases. Conversely, birds from subordinate classes make fewer unprovoked attacks upon swans from the higher-ranking groups. Yet, intriguingly, the individual that initiates the encounter is likely to be the eventual winner, irrespective of its social class change to class (Rees 1988).

The hypothesis that individual animals adjust their behaviour in dominance disputes to ensure that they obtain maximum benefits from each encounter, whilst incurring minimum costs to themselves, was proposed during the 1970s (Parker 1974; Maynard-Smith & Parker 1976; Clutton-Brock *et al.* 1979). Thus, contests between closely matched combatants may be more likely to escalate as they assess each other's fighting ability (Parker 1974). In the case of Bewick's Swans, this might be reflected in the differing intensity of displays during encounters with different opponents. However, Scott (1978) did not find any evidence to suggest that there was a higher proportion of the intense 'open wing' display in encounters between similar-ranking swans at Welney. Further observations at Slimbridge confirmed that, although Bewick's Swans tend to use low-grade threats in encounters with the low-ranking single class, there was little to suggest that they further adjust their displays according to the strength of their opponents. For instance, none of the classes showed a tendency to employ higher-grade displays in encounters with members of their own class (Rees 1988). It is possible that Bewick's Swans establish dominance relationships with those of similar class or rank through one intense dispute at the beginning of the season and thereafter have lower intensity disputes with the same opponents, but further behavioural studies would be needed to clarify this point.

6.2.3 The importance of mate proximity in aggressive encounters

The importance to members of a pair of remaining close to each other in the wintering flock was shown in studies of the Bewick's Swans wintering on the Ouse Washes (Scott 1980c). In these studies, Dafila Scott introduced the method of measuring proximity between birds as the number of swan body lengths ('swan-lengths'), since this provides a ready means of recording distances in the field. When paired females were separate from their mates (defined as more than four swan-lengths apart) while feeding on the fens, they were less successful in aggressive encounters and spent less time feeding (61.4%, compared with 80%) than when they were close to them. This pattern does not hold for female parents, however, possibly because the presence of cygnets ensures that female parents retain their status even when the male is elsewhere.

Males also suffer reduced success in encounters when their mates are at a distance, with the proportion of encounters won declining with increasing distance from the mate, but the effect is less pronounced than for females (Figure 6.4). Males also have a tendency to spend less time feeding when their mates are absent (56%, compared with 75%), although the difference is not statistically significant (Scott 1980c). The more consistent success of the male birds, irrespective of whether or not their mate is present, is perhaps due to fighting ability (measured as dominance rank) being closely associated with physical characteristics such as size and body mass. As males tend to be larger than females, they are at less of a disadvantage when separated from their mates. Nevertheless, that both members of the pair are more successful in aggressive encounters when close together differs from the situation for some other species; studies of Common Eiders *Somateria mollissima* (Ashcroft 1976) and Barnacle Geese (Owen *et al.* 1977; Black and Owen 1989) found that only the female benefited significantly from proximity.

The precise manner in which the female supports the male is still obscure, since the highest intensity of encounter (physical combat) generally involves only one bird, and that

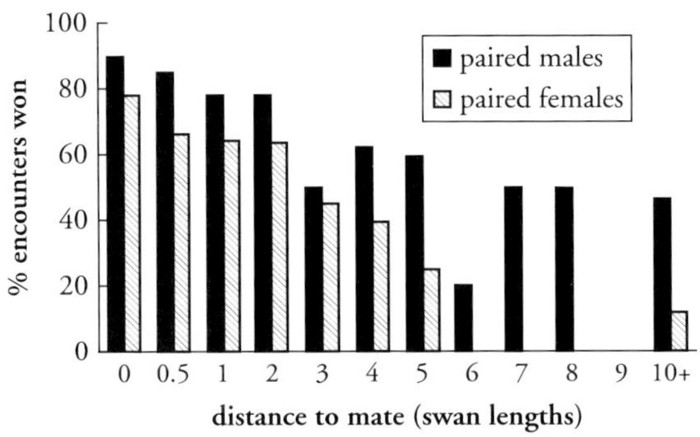

Figure 6.4 Percentage of encounters won by paired males (closed columns) and paired females (hatched columns) at varying distances to their mates. From Scott 1980c.

bird is usually the male (Scott 1980c). Very occasionally the female engages with the opposing female when their mates are grappling, but it is the outcome of the males' combat that determines the relative dominance of the pairs for the rest of the winter. It is possible that the male fights harder in his mate's presence, perhaps in order to defend resources adequate for them both (Scott 1980c), but it is unlikely that he fights harder to defend his mate from usurping males because Bewick's Swans are almost totally monogamous (see Chapter 7). By performing together, partners appear to reinforce each others' displays, which may serve to encourage the male to fight harder, and so send stronger signals of their physical condition to other swans in the vicinity.

Partners appear to assist one another by approaching their mate when either is threatened, by joining in any aggressive encounters, and also through their presence 'inhibiting' other birds from threatening their mates. When apart, paired males are threatened 1.9 times more frequently and females 3.6 times more frequently than when they are with their mates (Scott 1980c). Thus, a paired bird whose mate is absent seems to be perceived as more subordinate by other members of the flock, and in a position analogous to that held by single swans in the social hierarchy. That the female is most at risk of incurring threats may again be due to social dominance being associated mainly with the characteristics of the male.

Although members of a pair generally remain closer to each other than to other birds in the flock, the proximity of partners varies with the situation and between different pairs. For instance, partners maintain greater proximity in dense flocks than in dispersed flocks. Pairs with cygnets tend to be further apart, but this is probably due to the larger size of the unit and to cygnets frequently positioning themselves between their two parents. Paired birds, both with and without cygnets, also remain closer together when the male by himself was high ranking, but there is no association between mate proximity and the overall dominance of the pair, which may perhaps be due to there being a greater benefit to the female in remaining close to such a male (Scott 1980c). Whether the proximity maintained by members of a pair reflects the quality of the pair (measured both as dominance ranks and breeding success), the number of years that they have been together, or familiarity with the wintering site, has yet to be determined.

6.2.4 Costs and benefits of achieving social dominance

Maintaining a dominant position in the social hierarchy is physically demanding, and achieving a high rank in the wintering flock may involve costs as well as benefits. One possible immediate benefit of dominance, access to the prime feeding areas and roost sites, has not been fully investigated for Bewick's Swans, although the amount of time that females spend feeding on the grain supplied at Slimbridge increases with the dominance rank of their social unit. Females with cygnets fed for longer than paired females, which in turn fed longer than single females, giving some concrete support for this hypothesis (Bowler 1996). In Barnacle Geese, dominance has been linked with the right to feed at the leading edge of the flock. Here the birds have first choice of food plants and can spend more time in areas with more abundant vegetation (Black & Owen 1989). The fact that Bewick's Swans' offspring from previous years are more likely to associate with their parents and their new brood if the family size is large, suggests that something may be gained by the family and former offspring alike if the unit is at the upper end of the social hierarchy (Bowler 1996). On the other hand, since some pairs and families repeatedly chase away their former offspring, the optimum size for a family unit in winter may not always be the largest

(Bowler 1996), but may be determined by a balance between the resources achieved and the number of birds in the family using these resources.

The costs and benefits of achieving high dominance rank were studied in some detail by John Bowler in Gloucestershire. He spent three winters in the fields around Slimbridge, observing the effect that social dominance has on body condition (as determined by the abdominal profile indices recorded for individual birds; methods are described in Section 4.5.2), and found that the association between dominance rank and body condition differs markedly between the sexes. Perhaps surprisingly, low-ranking single males were in better condition (with higher abdominal profile scores) than higher-ranking paired males, whereas low-ranking single females were in poorer condition than the higher-ranking paired females. This suggests that females benefit from the presence of a mate, perhaps due to the male's ability to defend feeding areas or to protect her from interference when feeding; that this incurs a cost to the male is shown by a reduction in his own fat reserves (Bowler 1994, 1996). Moreover, those females which are in good body condition on leaving Slimbridge in the spring are significantly more likely to return with cygnets the following year (Bowler 1996). Both the benefit to the female and the cost to the male increases with the dominance rank of the pair. The abdominal profile scores of the cygnets also increase with the dominance rank of the family unit, confirming that they, too, benefit from the presence of a dominant male parent. Yearling condition, however, was not obviously affected by the rank achieved by their parents in that winter (Bowler 1994, 1996).

In the longer term, the dominance achieved by Bewick's Swans in winter is associated with their life-time breeding success (Scott 1988; see also Chapter 7). This suggests that measures of fighting ability in winter are relevant throughout the year, with the same birds being successful in defending prime territories in the breeding grounds. Nevertheless, there has been no indication that single swans are more dominant in the winter prior to their returning with a mate. Given the ritualised nature of the Bewick's Swans' displays, however, it seems reasonable that these might be sufficient indicators of the quality of potential partners, so this subject may warrant some further investigation.

6.2.5 Maintaining dominance from one year to the next

Once established, dominance rank is usually held throughout the winter (Section 6.2.2). There is also a consistency between winters in the dominance rank achieved by paired birds (with or without cygnets) that do not change their social class (Rees 1988; Bowler 1996). Single swans differ in that their dominance rank tends to increase from one year to the next, perhaps because increasing familiarity with the site enables them to feed more efficiently and so gain weight, which would improve their fighting ability (Bowler 1996). Bearing in mind the major effect that social class has on the outcome of aggressive encounters, birds that change class between years do not immediately achieve the average rank of the new class. Thus, although the dominance ranks of swans recorded as singles are significantly higher on returning the following winter with a mate, their rank is still lower than that of birds paired for at least one previous winter. Pairs seen without young in one year that return with cygnets the following winter similarly show an increase in dominance rank over the two years (Bowler 1996), although they are not as dominant within the family class as they were within the pair class the previous year.

A downward shift in social class has less predictable effects on social dominance than an improvement in class, suggesting that there is some retention of dominance rank. For instance, pairs which breed successfully in one year then return without cygnets the next do not show an overall decrease in rank; indeed they retain higher social status than pairs that did not bring cygnets to the wintering site in either year (Rees 1988; Bowler 1996). Thus, not only is the rank achieved by paired birds or families in different years generally stable, but the rank achieved as a family unit may be held for at least one winter if the birds return without young the following year.

The mechanism by which pairs retain their status on returning without cygnets is not clear, but may perhaps be through birds recognising each other in a relatively stable flock. Flock stability at Slimbridge is high, with some 40%–60% of the adults and yearlings recorded each year having been identified at the site in previous winters (Rees 1988), and allegiance to particular roost sites at Slimbridge is even higher (Bowler 1996; see Section 6.4.1). Another possibility is that the absence of cygnets in one winter may not necessarily reflect the physical condition (and thus fighting ability) of the pair that year, since the pair may have bred but lost their cygnets to predators or because of other adverse conditions in the breeding range (Chapter 5). The loss of a mate did result in a major drop in dominance rank for the surviving member of the pair, however (Bowler 1996), further emphasising that the presence of a mate helps to achieve dominance (Section 6.2.3).

6.2.6 Inheritance of dominance

There is no evidence to date to suggest that the dominance ranks of grown offspring are related to those of their parents, or based on the parents' rank in the years when the offspring were associating as cygnets (Bowler 1996). Although associating cygnets assume their parents' rank, an effect that continues into the second winter for offspring associating with their parents when yearlings (Scott 1980b), there is no substantive evidence for the influence of the parents' success perpetuating once male offspring aged two years or more are independent (Bowler 1996). However, sample sizes are small, so although the outcome of aggressive encounters has been recorded (and dominance ranks determined) for Bewick's Swans wintering at Slimbridge in most years since winter 1979–80, we are still some way from discovering for swans first seen as cygnets whether any inheritance of social dominance translates in due course to improved lifetime breeding success.

6.2.7 Factors affecting social dominance

As has been mentioned earlier, the dominance rank of male swans is closely linked to their social class and to their body size, even though by the end of the winter their physical condition (as reflected in abdominal profile) may be rather low. Dominance rank is also related to age, increasing for two to six-year-old males, but not for females (Bowler 1996). Bewick's Swans reach adult size within two years, so the increase in male dominance rank with age thereafter may be due to an improvement in their fighting ability through greater experience of aggressive encounters, rather than to an increase in physical strength over several years. The number of years that an individual has previously wintered at Slimbridge also contributes to the dominance rank of single males. How accrued knowledge of the wintering site might help is not known. It could be due to one or several factors such as the birds having greater confidence and focus in somewhat artificial surroundings, or to a better knowledge of feeding opportunities improving their condition and fighting ability.

For paired swans, it seems that success in aggressive encounters increases with the number of years that the pair have been together, with dominance rank increasing linearly with pair duration (Rees *et al.* 1996). The study also found that there was no such increase when only the years in which the birds were accompanied by cygnets were considered, so it was initially thought that the overall increase in dominance rank with pair duration was due to the improved breeding success of established pairs (further details in Chapter 7), since families are dominant over paired birds. Although this may still be the main reason for the correlation, further consideration of pairs without cygnets found that their dominance also increases from one year to the next, indicating that dominance rank does increase with pair duration alone (Bowler 1996). This may appear at odds with the consistency from one year to the next in the dominance ranks of paired birds and family units, but can perhaps be explained as a more gradual improvement in a pair's success over a longer (in this case four-year) period, through the increase in age, experience and fighting ability of the pair.

6.2.8 Timing of arrival and social dominance

The possibility of a relationship between social dominance and the timing of migratory movements is an interesting question. By comparing the arrival and departure dates of swans of known dominance rank, it was found that prompt arrival in autumn establishes a bird for the rest of the season. Swans that returned early to a known wintering site (in this case, Slimbridge) tended to achieve a higher dominance rank than those returning later in the season when all classes (i.e. families, pairs and single swans) are considered together. Within-class differences followed a similar pattern, but were not significant. Moreover, low-ranking birds tended to leave early, whilst dominant swans remained later in the spring. The relatively early arrival and late departure of the more dominant birds results in there being a positive correlation between dominance rank and the number of days the swans spent at Slimbridge (Rees 1988).

The question of whether a change in arrival date could lead to a corresponding change in dominance rank was assessed for 16 individuals or pairs that did not change their social class in two winters at the site. It was anticipated that early arrival would enable the birds to replace the weight lost during migration and improve their physical condition before the return of birds arriving later in the season, giving them an advantage in aggressive encounters. Observations showed, however, that the birds remain as successful in aggressive encounters in years when they return late as when they return early. Moreover, dominance during one winter (1983–84) was significantly related to early arrival the following autumn. This suggests that early arrival does not necessarily enable a swan to achieve dominance, but that dominant birds are more likely to arrive early (Rees 1988). There may be an advantage to dominant swans in returning early to a known site, as this would enable them to repossess their preferred feeding and roosting areas, whilst subordinate units may investigate alternative sites on arrival in the wintering range. Certainly, swans found to be subordinate in the winter flock are less likely than dominant birds to return in a subsequent winter, although it is not yet known whether this is due to lower survival or to them transferring their allegiance to another site in subsequent years (Bowler 1996; Section 6.5).

6.3 PARENTAL CARE IN WINTER

Although in the avian world there is a wide diversity of mating systems, ranging from polygyny (i.e. one male with more than one female) to polyandry (one female with more than one male) and polygyandry (several males mating with several females), many species are monogamous for the lifetime of their mates. It is suggested that monogamy has evolved due to each member of a pair leaving, on average, more descendants if they share in raising a brood (Lack 1968; Black 1996). Prolonged care of the brood is less common, but among the waterfowl it occurs in the whistling ducks, swans, geese and sheldgeese (Kear 1970). Among these birds, including Bewick's Swans, the family remains together within flocks during the non-breeding season, usually separating before the start of the next breeding season.

Dafila Scott studied the interactions of Bewick's Swans and their offspring in winter to see whether parental proximity is important to cygnets in protecting them, while feeding, from interference from other adults in the flock. She found that, when separated from their parents, cygnets were less successful in aggressive encounters, were frequently displaced and spent less time feeding than when close to their parents. For instance, on flooded meadows, cygnets won 75% of the encounters in which they engaged when less than four swan-lengths from their parents, but only 33% when further away. Similarly, cygnets were displaced by other swans 0.14 times per minute when away from their parents, but only 0.02 times when close to them (<4 swan-lengths away). These findings are analogous to the greater interference experienced by female swans when separated from their mates. The relatively small size of the cygnets makes them even more susceptible to interference, which underlines the importance of adult protection. Interestingly, cygnets were just as successful when close to a female parent as a male one, and they did not generally remain closer to one parent than to the other. In aggressive encounters, however, cygnets approached or followed the male more closely than the female, suggesting that they recognised that he provided more protection at this time (Scott 1980b).

The distance between parents and their cygnets decreases as distances to the nearest non-related swan decrease (i.e. with increasing flock density), although parent–offspring distances appeared to reach a plateau at 3.3–3.4 swan-lengths (about 2.2–2.3m) on the fens and at 2.2 swan-lengths (1.5m) in front of the observatory. Cygnets from the more dominant families tend to venture further from their parents than those of subordinate families, and this seems to be due neither to differences in brood size (i.e. the larger area occupied by the family unit), nor to variation in density around high-ranking and lower-ranking family groups. Feeding habitat also plays a part in determining parent–offspring distances; cygnets maintained closer proximity to their parents on waste potatoes than on winter wheat, and the frequency of aggressive encounters was higher on potatoes than on wheat, despite flock density being lower (i.e. the nearest non-related neighbour distances were greater) on potatoes. This may perhaps reflect the importance of the two food sources to the swans. Cygnets remained closer to their parents whilst the family was asleep or roosting than during feeding, both in front of the observatory and on the fens (Scott 1980b).

In addition to changing with social situation and location, parent–cygnet relationships change over time, with distances to the parents being significantly greater at the end of the winter (February–March) than earlier (October–January). The responsibility for maintaining family cohesion, measured as the frequency with which parents and offspring approach or leave each other, also changes during the winter. At the beginning of the season parents and cygnets are equally responsible for maintaining the cohesion of the family group, whereas cygnets assumed more responsibility for maintaining proximity to the male parent (less so to the female) thereafter (Scott 1980b). This trend appears to continue during the spring migration: cygnets observed at Matsalu Bay, Estonia, spent long periods and distances away from their parents in spring (Rees & Bowler 1991), suggesting that parents provide less support to their cygnets in late winter and early spring, as the young become better able to fend for themselves and the parents prepare for the next breeding season.

The behavioural studies indicate that one of the reasons for prolonged parental care in Bewick's Swans is the relatively slow maturation rate in this species, with cygnets not reaching adult weight until they are at least one year old, and thus less capable of competing for food in the flock without their parents' support. This does not preclude other possible benefits of parental care, such as showing them the migration route and traditional feeding sites, or protecting them from predation. A matter that still remains unclear is why some pairs tolerate offspring from previous years rejoining them in winter more than others. In general, parents (particularly females) do not tolerate proximity of previous years' offspring, particularly offspring in their third winter or older. There is also a slight tendency for cygnets to threaten yearling siblings more often than unrelated yearlings, but further observations are needed to determine whether this is a genuine trend. Aggression is not frequent between siblings from the same brood, but cygnets dominated all of their parents' offspring from previous years (Scott 1980b). It also remains to be seen whether parental care has any effect on the eventual size of their offspring, which would in turn enable them to achieve social dominance and higher reproductive success.

6.4 DISTRIBUTION OF INDIVIDUALS WITHIN THE FLOCK

Differences in the dominance ranks achieved by families, pairs and single swans might be expected to affect the distribution of individuals and social units at the wintering sites. Higher-ranking birds might be more likely to occupy the best feeding areas with minimal interference from other swans. This was considered by John Bowler (1996), who investigated the connection between social class and Bewick's Swan distribution at Slimbridge, both at the feeding sites and at two distinct roost sites.

6.4.1. Roost distribution

Bewick's Swans tend to roost communally in winter, usually standing in the shallows around areas of open water, with birds from one or more feeding areas converging at night on a particular roost. In the early years of the Slimbridge study the swans flew each evening to the River Severn or to scrapes on the adjacent saltmarsh (the Dumbles), but from the early 1970s they roosted primarily around the lake in the Rushy Pen in front of Sir Peter Scott's house. Although the Rushy Pen was, and still is, the main roost site for swans at Slimbridge, in the late 1970s they also started to roost within the enclosures that house WWT's collection of captive waterfowl (particularly in the Asian and European Pens) some 500 metres away, dubbed the Grounds roost.

The number of swans using each of the Rushy Pen and the Grounds roost sites at Slimbridge varies greatly both within and between winters, but generally swans of all social classes are faithful to a particular roost (Table 6.1). Although the Grounds roosting birds spend slightly fewer nights at their primary roost (i.e. they are more likely to turn up occasionally in the Rushy Pen than are Rushy Pen birds to roost in the Grounds), there is no difference in the proportion of singletons, pairs and family groups using each roost. Brood sizes recorded for families at the two roosts do not differ significantly, either (Bowler 1996). Swans that had visited Slimbridge for a number of years were more likely to roost in the Rushy Pen (probably because they had roosted there since before the Grounds roost became established), but the proportion of new arrivals joining each roost group by the late 1980s and early 1990s was proportionate to the number of swans already

Table 6.1 Percentage of nights spent at the primary roost site for Bewick's Swans wintering at Slimbridge between 1989–90 and 1992–93. From Bowler 1996.

	Rushy Pen roost		Grounds roost	
	mean	swan units seen	mean	swan units seen
Singles	94.2	482	88.1	162
Pairs	96.7	173	87.1	57
Families	92.5	102	91.9	34
All	94.8	757	87.9	253

using each roost (i.e. 24.2% joining the Grounds roost and 72.4% joining the Rushy Pen roost at this time, Bowler 1996). Mean roost counts during the swan season (when at least one swan was present at Slimbridge) were 89 birds for the Rushy Pen and 23 for the Grounds between winters 1989–90 and 1992–93, with peak roost counts exceeding 250 and 100 birds respectively.

Roost-site fidelity within a winter is generally very high, but it differs between the social classes. Despite there being no overall difference in the proportion of singletons, pairs and family groups using each roost, individual single swans are significantly more likely to be seen at a secondary roost site (8.6% of 24,767 bird observations) than paired birds (6.3% of 14,069 observations) or family parties (6.6% of 8,385 observations) (Bowler 1996). Several reasons have been suggested to explain this which, given that food is not usually provided at the swans' roost at other wintering sites, relate more to feeding competition than to roost-site selection. Firstly, being subordinate in the social hierarchy, singles may be unable to obtain access to the food distributed at their main site, and so may seek better opportunities elsewhere. Conversely, since paired birds and families are able to obtain better positions in the flock at feeding and roost sites, they have less need to make changes when conditions deteriorate. Finally, single swans may use the opportunity of being unattached to explore other roost sites and so acquire information for use should they return with a mate and cygnets. Whatever the reason, the movement of singles between roosts is a reflection of the larger number of staging and wintering sites that they use along the migratory flyway, compared with paired birds and family groups (Rees & Bacon 1996; see also Chapter 3).

Despite this insight into the structure and function of the separate roosts, why the Grounds roost developed is still not clear, although the presence of captive Bewick's Swans in the European Pen probably decoyed the first wild swans to the enclosures. One possible reason may be the reduced level of aggression between related birds; independent offspring tend to continue using the roost site favoured by their parents, and there were significantly fewer interactions in flocks that consisted mostly of swans from the smaller Grounds roost (Bowler 1996). Unfortunately, it is not yet possible to test this hypothesis because only a small number of the swans that winter at Slimbridge (and hardly any from the Grounds roost) were ringed there as cygnets, so the relative proportions of closely related birds in the two roost groups cannot be ascertained.

6.4.2 Association between dominance and feeding distribution for birds from different roost groups

Although there is little difference in the overall proportion of swans from each social class using the two main roost sites at Slimbridge, the same is not true for the swans' selection of fields for day-time feeding in the surrounding area. As described in Chapter 4, the proportion of families in flocks tends to decrease with increasing distance from the roost sites, whereas the proportion of single swans tends to increase. The dominant family groups therefore may be able to establish themselves at feeding sites closer to the roost, whereas lower-ranking single swans fly further afield either because they are displaced or less constrained by the requirements of mate and offspring. Moreover, the proportion of cygnets in the flock decreases with increasing flock size, reinforcing the view that the larger families are able to monopolise their preferred feeding areas, and that these are sites close to the roost (Bowler 1996).

There is a major difference in the distances that birds from the two roost groups fly in order to feed. The proportion of swans originating from the Rushy Pen roost in any flock increases with distance from the roost. By comparison, the Grounds-roosting birds are more likely to feed closer to 'home' during the day. Moreover, the Rushy Pen swans appear to select feeding sites where the grass sward has a relatively high protein content, whereas the protein content of the vegetation seems unimportant to the Grounds-roosting swans. John Bowler found that swans from the Rushy Pen spent less time than those in the Grounds feeding on the grain distributed by WWT staff at the site, and that they also spent more time feeding in the fields. One possible explanation for the longer distances flown by the Rushy Pen birds therefore may be that the selection of day-time feeding sites is more critical to them than to the swans roosting in the Grounds. This appears to have negative consequences, however, in that the Grounds birds also tend to have poorer breeding success, particularly in years when breeding success is low for the population as a whole (Bowler 1996).

6.5 SOCIAL BEHAVIOUR AND RETURN RATES

Whether a swan that wintered at Slimbridge returns in subsequent years is influenced by a number of factors. These include its sex, with males being more likely to return than females, and whether or not it has found a new mate (see Chapter 3). Moreover, single swans have lower return rates than paired birds, irrespective of their sex and irrespective of whether the latter are accompanied by cygnets. Cygnets and yearlings are also less likely to return than adult swans (50%, compared to 65.9%), and the dominance rank of the family does not seem to determine whether or not the cygnets return the following year (Bowler 1996). Although there is no evidence that dominance rank *per se* affects the likelihood of a unit returning to the site within the family and pair classes, these findings still suggest a low return rate for those of lower social status, since females are generally of lower status and less likely to return than males, young birds are subordinate to and less likely to return than single adults, and single adults are less likely to return than the dominant paired birds and families. Whether the lower return rates are due to lower survival or to swans using a different site in subsequent years has yet to be investigated for birds of different age and social class, but the lower return rates of females is known to be primarily due to them switching wintering sites (Rees 1987).

There are different return rates for units in the two roost groups, with swans from the Grounds being more likely to return than those from the Rushy Pen. This occurs despite the former's poorer feeding success, and is due to yearlings and single females being more likely to return to the Grounds roost. This difference is especially marked for yearlings, and suggests that both these and low-ranking females may in some circumstances benefit from returning to a known wintering site, perhaps in cases where there are related birds present and therefore less of a threat of aggressive encounters. Single females have lower dominance ranks than single males of the same age, so may find it harder to improve their food intake at new and unfamiliar sites.

6.6 OVERVIEW OF RELATIONS WITHIN THE WINTERING FLOCK

To the casual observer, the vigorous and vocal fights between Bewick's Swans on arrival at the wintering sites might seem like the simple establishment of a peck order, with the large families demonstrating their superiority to the lesser opposition. Yet we have seen here that more complex relationships exist between the birds, both between the swan units and within pairs and family groups. Dominant male swans are presumably the fittest, but achieving dominance and maintaining vigilance whilst his mate and cygnets feed has a detrimental effect on the body condition of the high-ranking breeding males during the course of the winter. There is no obvious evidence for this reducing his breeding success the following summer, but whether the running down of body condition during the winter has a longer-term effect on the survival and longevity of breeding males has yet to be studied in this species. That females gain protection from interference from other swans by remaining close to their mate, and similarly that cygnets are displaced more frequently when away from their parents, reinforces the value of monogamy and prolonged parental care for the birds. Even within a wintering site, the swans may separate into sub-flocks, as shown by the variation in feeding-site selection by swans from two groups using different roost sites at Slimbridge. Thus, life within the swan flocks seems to be fluid and ever-changing, which adds greatly to their already substantial appeal.

CHAPTER 7

Life-history studies

... of their weatherings and their marryings and their buryings and their natural selections ...

James Joyce (*Finnegans Wake*)

The arrival of the Bewick's Swans each autumn is a key time for the Wildfowl & Wetlands Trust's swan research team and for ornithologists awaiting the arrival of the swans across Europe. As groups of birds fly in to their wintering sites, calling vociferously as they establish the dominance hierarchies (Chapter 6), they provide a wonderful spectacle for rookie and veteran birdwatchers alike. For each new wave of swans, WWT staff and an enthusiastic network of ring-readers are keen to check which individuals have survived from one winter to the next, which have lost a mate or acquired a new one, and the number of young (if any) accompanying each pair. In addition to reading leg rings and neck-collars, and counting numbers at the major swan sites, the Bewick's Swan monitoring network also records the proportion of cygnets in the flocks and the number of cygnets in each family (brood sizes) at a range of wintering sites to determine whether the previous summer had been a good breeding season for the population as a whole. The information is sent to Jenny Earle at Slimbridge, who has been responsible for adding incoming updates on the swans to the Bewick's Swan database since the late 1980s. Meanwhile, the detailed monitoring of Bewick's Swans returning to Slimbridge each year adds to the information accrued over many years of each individual's longevity (lifespan). The age of the swans at first pairing, the age at which they are first known to have bred, frequency of mate change and annual variation in breeding success are also observed for birds that return to the site over several years. Life histories have thus been described for many of the individuals wintering at Slimbridge, and for many it has been possible to determine the age that they reached and their lifetime breeding success.

As we have already seen, environmental conditions (particularly weather in the high Arctic) have a major impact on Bewick's Swan breeding success. Yet at a finer level it is the annual survival and productivity of individuals that determines the total number of swans on the planet. By monitoring individuals over a number of years, it is possible to determine whether some pairs are consistently successful breeders (i.e. they bring cygnets to the wintering range even when the breeding season has been a poor one for the population as a whole), whereas others may be more erratic, and whether there is a downside to raising cygnets each year, such as having a shorter lifespan. There may also be a delay in the consequences of annual variation in the swans' breeding success and survival impinging on population trends. For instance, a poor breeding year results not only in lower recruitment to the population in that year, but also to there being fewer birds in some age cohorts, and thus potentially fewer birds joining the breeding population in 3–4 years' time. Yet in areas with high population density, the reduction in numbers may reduce competition between individuals for resources such as food and nest sites, and ultimately improve survival and breeding success for swans that were raised in the poor breeding years. The dynamics of the Bewick's Swan population as a whole therefore cannot be understood without knowledge of the behaviour of individuals within the population. Since demographic processes are regulated both by external environmental events (weather and food supply) and by individual characteristics, it is important to know how individual and environmental variables operate in relation to each other, not only at different times of year but through the swans' lifetimes.

Following the same individuals over many years at Slimbridge has not only helped to describe general patterns in the swans' life histories, but has provided a valuable opportunity to distinguish the traits that characterise the more successful swans, and thus provide insight as to why certain traits may have evolved in the species. This section therefore describes the strategies identified to date (particularly in relation to breeding success), considers why some partnerships are more fruitful than others, and notes gaps in knowledge that could be addressed in future studies.

7.1 FIRST PAIRING

Given their slow maturation rate, with birds still retaining flecks of grey juvenile plumage into their second (and occasionally their third) winter, and also being below full adult size in their second winter (Evans & Kear 1978), it is perhaps not surprising that Bewick's Swans take several years before they join the breeding population. As part of the long-term study of the swans at Slimbridge, Evans (1979a) followed known-age birds (i.e. those first recorded as cygnets or yearlings) seen each winter through to pairing or until they disappeared (i.e. paired status no longer known), and her observations have been updated here with records to the 1999–2000 winter. Of 628 birds monitored, only one male ('Post') definitely associated with a mate ('Penny Black') during his second winter (i.e. when one year old), the pair spending four winters together at Slimbridge. A second yearling male ('Lanyard') also appeared to be paired (with 'Rope'), but neither swan returned in subsequent years. By two years of age (i.e. in their third winter), 20% of the females were paired, but only 13% of the males. This apparent difference is not statistically significant, and any sex bias in the age of first pairing certainly was not maintained thereafter, as the males caught up the females, with a higher proportion of males than females pairing for the first time at age three.

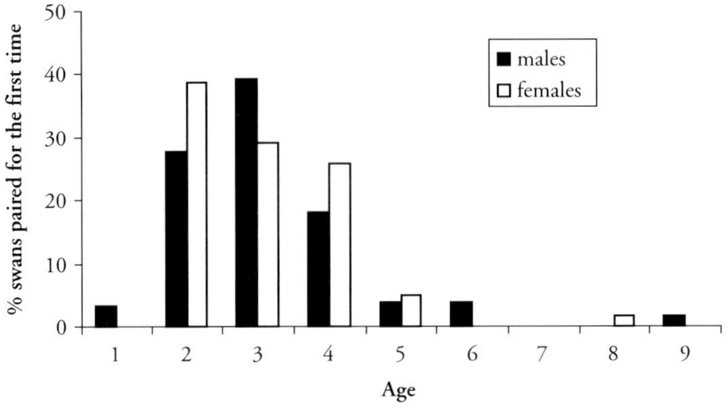

Figure 7.1 Age of first pairing for Bewick's Swans seen at Slimbridge between 1963–64 and 1999–2000 that are known to have paired. Records are for birds of known age (61 males and 62 females) that were seen each winter until they returned with a mate. From Evans 1979a; WWT unpubl. data.

Of 123 swans followed each winter that were also known to have paired, 34% were recorded with a mate for the first time at age two, 69% had paired by their third year and 91% by their fourth year (Figure 7.1). For a few slow starters, first pairing occurred as late as six and nine years. One bird, 'Saturday' spent 13 winters without a mate (between 1986–87 and 1998–99), when aged 8 to 21 years. She spent several winters away from Slimbridge whilst a young bird, however, so it is not known if she had a mate during this time. The mean age at first pairing is 3.18 (SD ±1.35) years for males and 3.05 (SD ±1.12) years for females (Figure 7.1). The reason for the slightly earlier pairing of females is unclear, and since not statistically significant could be discounted were it not that the pattern appears to be consistent over many years. If a genuine tendency, it may perhaps be due to physiological or behavioural differences between the two sexes. It seems unlikely to be due to a preponderance of males, since there is no evidence for an imbalance of the sexes in the population. For instance, of 2,335 Bewick's Swans caught at WWT centres (Slimbridge, Welney and Martin Mere) between 1967 and 2003, and 110 non-breeding swans caught in Russia in summer 2003, 50.8% and 48.2% respectively were recorded as male (WWT unpubl. data).

7.2 FIRST BREEDING

By monitoring each of the Bewick's Swans wintering at Slimbridge, we have also been able to discover the age at which the birds first bred successfully, in that the cygnets fledged and accompanied the parents to the wintering range. Between 1963–64 and 1999–2000, 40 of the 123 known-age swans seen until pairing were also seen each year until they returned with young (Evans 1979a; WWT unpubl. data; Figure 7.2). A small number of birds are known to have bred at a young age, with three individuals recorded with cygnets in their third winter (i.e. they bred at age two years), but this appears to be exceptional. Most (57.5%) breed successfully for the first time between the ages of four and six years. As with the age of first pairing, there was little evidence of a difference between the sexes in the age of first successful breeding; although 35% of males and just 20% of females known to have bred did so by age 4 years this difference was not significant. The sex ratio was closer to parity by the

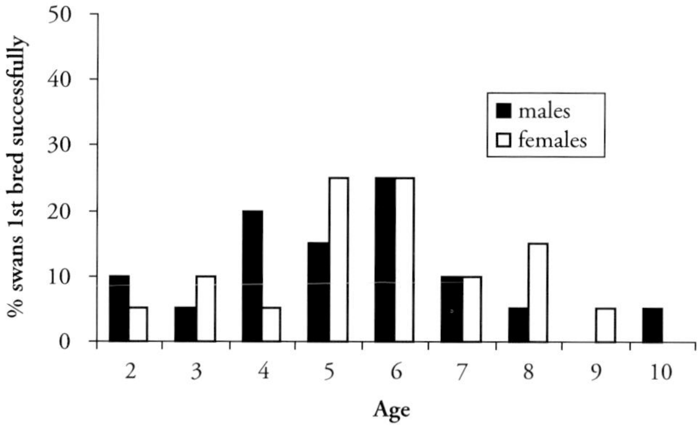

Figure 7.2 Age of first successful breeding (i.e. recorded with cygnets in winter) for Bewick's Swans seen at Slimbridge between 1963–64 and 1999–2000 that were known to have bred. Records are for birds of known age (20 males and 20 females) that were seen each winter until they returned with offspring. (From Evans 1979a; WWT unpubl. data).

age of five (50% of males and 45% of females having bred by then), but the number of birds for which this information is available remains quite small.

Successful first breeding can occur at quite a late age; two males ('Victory' and 'Money') were recorded with cygnets for the first time at age ten and 11 years respectively, and a female swan ('Kama') did not breed until age nine. It is of course possible that these older swans had established a breeding territory and hatched a clutch in previous years, but did not succeed in raising their young to the first winter. Lack of breeding experience may account for the loss of cygnets early in the breeding season. Certainly pairs that have been together for several years are more likely to bring cygnets to the wintering range than newly formed pairs (see Sections 7.6.2 and 7.6.3). Overall, mean age at first breeding was 5.55 (SD ±2.33) years for males and 5.70 (SD ±1.84) years for females (Figure 7.2). These values are likely to be underestimates because a failure to bring young to the wintering grounds does not, as mentioned previously, mean a failure to breed.

7.3 PAIR-BOND DURATION AND MATE FIDELITY

The number of years that a pair has been together ('pair-bond duration') has been investigated for those birds where there is at least five years' information for one or both members of the pair (Evans 1979a; Rees *et al.* 1996), and this information is updated here. Cases where the exact duration of the bond, and also the minimum pair-bond duration are considered. The latter includes swans already paired when first seen at Slimbridge, and also those still paired in the last winter considered (1999–2000). Pairs with non-consecutive records, having missed one or more years at Slimbridge, are assumed to have been paired to each other in the missing years since mate-switching ('divorce') is extremely rare in this species. Only one of the swans recorded at Slimbridge has changed mates whilst its previous partner was still alive, and there is also only one record of divorce amongst the swans at Welney. The Slimbridge pair ('Piotr' and 'Patch') were together in winters 2000–01 to 2002–03 inclusive, but in 2003–04 and 2004–05 'Patch' was paired to 'Peploe', whereas 'Piotr' returned alone. The situation was similar for the Welney Bewick's

Swan pair, named 'Josh' and 'Olive', who separated after seven years together, during which time they failed to rear any young. 'Olive' was recorded with a new mate and bred in three of the four years seen with her second partner. 'Josh' returned for a further two years but was not seen paired with a new mate (Rees *et al.* 1996). Only one other case has been recorded; a neck-collared swan originally seen nesting at Khabuicka (our main study site in the Russian Arctic; see Chapter 5) returned to the breeding grounds in the same year that its previous mate was seen breeding with a new mate.

Pair bond duration varies substantially, ranging from 1–19 years, (Figure 7.3). The average duration of the pair bond was 3.47 (SD ±3.05) years for 331 pairs where the actual duration was known, and 4.67 (SD ±3.47) years for the 376 pairs where the minimum number of years spent together was known. In the former group, nearly 39% of pair bonds were of just one year's duration. This may reflect the annual mortality rate of around 15% of the population dying each year (Section 7.5), although whether survival rates for paired birds are the same as for the population as a whole has yet to be determined. For the slightly more substantial number of cases where bond length was minimal, the proportion of one-year bonds was lower (22%), although it was still the most common. Pair duration is generally underestimated; nevertheless 34% of paired birds are together for at least five years, and 8% for at least ten years. Thus Bewick's Swans are not only monogamous, but some paired birds may be together for many years. A higher proportion of Bewick's Swans spend more than two years with the same mate (63.6%) than do Mute Swans on the River Thames (44.8%) or Whooper Swans from the Icelandic breeding population (42.9%) (Rees *et al.* 1996; WWT unpubl. data). The pairing and breeding of individual Whooper Swans has been followed for fewer years than it has for Bewick's Swans. Whooper Swans have been ringed and monitored at WWT centres only since 1979–80, and the breeding success of individual Whooper Swans has been studied in Iceland since 1988 (Black & Rees 1984; Einarsson 1996; Rees *et al.* 2002), but this difference in study time is unlikely to account for the major difference in pair-bond duration recorded for the two species.

Most Bewick's Swans have only one mate during their lifetime although, not surprisingly, birds that survive to old age are likely to have several mates (Table 7.1).

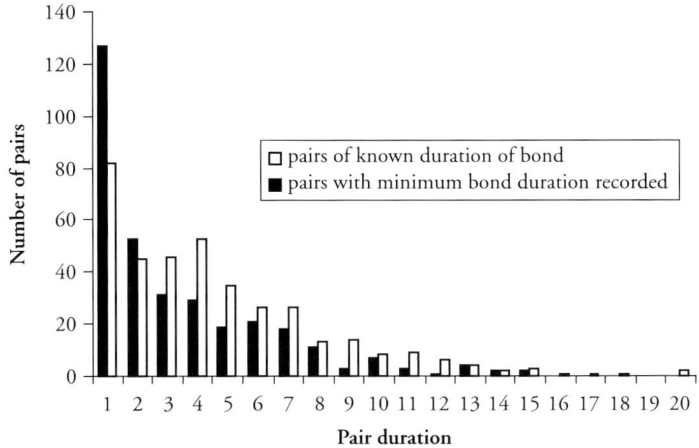

Figure 7.3 Variation in pair-bond duration for Bewick's Swans wintering at Slimbridge. Closed columns indicate pairs with minimum pair-bond duration recorded, and open columns are pairs where the length of the pair bond is known.

Table 7.1 Percentage of birds recorded with 1 to 4+ mates during their known lifetime. Data for both sexes combined. Rees *et al.* 1996 and WWT unpubl. data. Only one bird had more than 4 mates.

Years of Data	Birds (n)	Percentage of birds with:			
		1 mate	2 mates	3 mates	4+ mates
1–5	2960	97.6	2.3	0.1	0
6–10	362	71.0	24.9	3.9	0.3
11+	158	38.0	38.6	14.6	8.9
Total	3480	91.2	6.3	1.2	0.4

One exceptional bird ('Offset') had seven mates during the 21 years he was recorded at Slimbridge; other long-lived birds had at most only three or four. The mean length of the pair bond is longer with the first than with subsequent mates, although three birds had long pair bonds (of at least five years) with their fourth mates (Figure 7.4). However, some individuals breed more successfully with their second or subsequent mates (Section 7.6.3). This might be due to the experience already gained by the swans when with their first mates – improving their ability to hatch and rear young with subsequent mates – or, perhaps more debatably, to an improvement with age in a swan's ability to identify and win a good partner.

A collaborative analysis of long-term swan data sets was undertaken by the WWT and the Edward Grey Institute (EGI) at Oxford University, to compare the pair-bond characteristics of migratory Bewick's Swans and Whooper Swans with those of the more sedentary Mute Swans (the latter studied by the EGI on the River Thames), and thus assess inter-specific differences in mate-change frequency and mate fidelity. It was considered that mate change might be more costly for Bewick's Swans, in terms of lost breeding success, than for the other two species, because the close synchrony of the migratory and breeding cycles needed for Bewick's Swans nesting in the Russian Arctic allows them only limited time for courtship and pair formation before the onset of the breeding season. Although Bewick's Swans are monogamous, and there are only three known cases of divorce (see above), the swans do re-pair quite rapidly following the loss of a mate. On average, they take

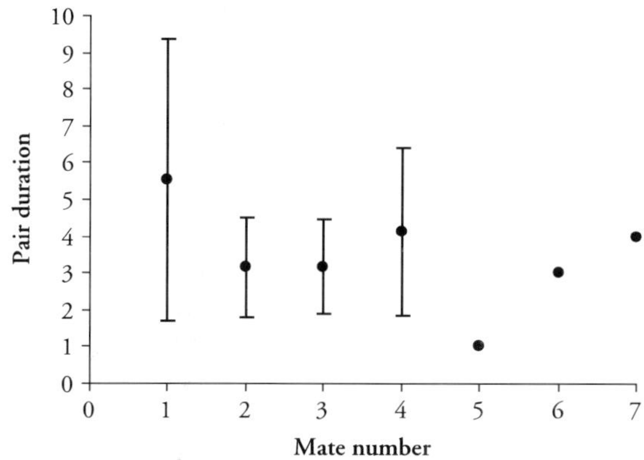

Figure 7.4 Mean pair duration (± SD) for swans with their first to seventh mates. The mate numbers 5, 6 and 7 are all for one bird, 'Offset', with his last 3 mates.

2.6 years (SD ±1.7, range 1–9 years) to return to the wintering grounds with a new partner, and there is no difference between the sexes in re-pairing rates (Rees *et al.* 1996). This is significantly longer than the time taken by the Icelandic-breeding Whooper Swans to find a new mate, however, with the former re-pairing in, on average, 1.9 years (SD ±1.2, range 1–7 years) (Rees *et al.* 1996). Data on time taken to replace a mate have not been collected for Mute Swans on the River Thames, but is considered likely to occur within weeks (C. Perrins pers. comm.). Mute Swans were found to change mates significantly more frequently than the other two species. Of swans seen for up to five years, 15.8% of Mute Swans, 7% of Whooper Swans and only 3.5% of Bewick's Swans in this category were recorded with two or more mates (Rees *et al.* 1996).

7.4 AN UNUSUAL ASSOCIATION

In addition to occasional cases of Bewick's Swans pairing with Whistling Swans in areas where there breeding grounds are adjacent or overlap (Chapters 1 and 2), there is also an incident of a Bewick's Swan ('Sligo') consorting with a Bean Goose. In December 1994, Wim Tijsen was checking a mixed Bewick's Swan and Bean Goose flock in the Wieringermeerpolder, Netherlands (looking for Bewick's Swan rings), when he saw a Bean Goose with her head low and neck stretched chasing one of the swans. The goose finally attacked the swan and pecked it on the neck. Curious at seeing such courage in the goose, Tijsen continued watching, to see what the goose was defending. After some time it became apparent that the goose was remaining close to a pair of Bewick's Swans, one of which was ringed with a leg ring YXH ('Sligo'), and three birds finally flew off as a family. A few days later Wim saw the same three birds in the Anna Paulownapolder, and saw that 'Sligo' was head-pumping (preflight signalling) about 100 metres away from his mate because he wanted to fly to another sugar beet field, but that his mate was asleep and did not react. Charmingly, the Bean Goose pecked slightly at the tail feathers of the sleeping female, as if to wake her, and after a few minutes the three birds again flew off together.

A full year later, in November 1995, Wim Tijsen was racing through the polders during his lunch break (as usual!), checking the Bewick's Swans in his area, and paused to scan a large flock of 500 Bewick's Swans for leg rings. Through his telescope he found not only the ring YXH (i.e. 'Sligo') but also noted that there was a single Bean Goose in the Bewick's Swan flock. The farmer took him out on his tractor to get closer, and he was able to confirm the Bean Goose was immediately adjacent to 'Sligo', and again attacking other swans in the vicinity. This time 'Sligo' and his mate had three cygnets as well as the brown goose in their family party. A week later, on driving home from the swans' main roosting site at Normerven on the Dutch Waddensea, he saw above a family of two white swans, three grey cygnets and a brown goose, flying with him at an average speed of 64km/h, and naturally knew which these birds were.

The family remained in the Netherlands in the early part of the winter and was seen in Anna Paulownapolder on 23 December 1995. In February 1996 Henk Schobben and Wim Tijsen made their first of several visits to study the Bewick's Swans on the Ouse Washes around Welney. On their last day, and after four hours in a field at Block Fen checking a large flock of over 1,100 Bewick's, they spied a very small group behind a hedge at Langwood Fen in the middle of nowhere. Despite a long day, persistent rain, and bad road conditions (a muddy track), they couldn't resist. The flock comprised just nine Bewick's Swans and a Bean Goose. After five minutes the last swan decided to stand up and showed its ring (YXH), making two Dutchmen very happy (W. Tijsen, pers. comm.).

7.5 LONGEVITY AND SURVIVAL

The record for the oldest age known to have been achieved by a Bewick's Swan is still held by a captive bird ('Mrs Noah'), who was taken as an adult during the 1940s and eventually died at Slimbridge at the age of at least 36 years. A few wild Bewick's Swans reach their early to mid-twenties; one bird ('Casino') was last seen at age 26 years, and another ('Brimstone') was found freshly dead in the Russian Arctic when at least 29 years old. The average lifespan of a Bewick's Swan is substantially lower, at 5.4 years (Rees *et al.* 1996). This is likely to be an underestimate because only minimum age is known for many of the swans considered, for instance for birds first seen as adults, or where both members of a pair disappear at the same time and their fate is not known. There is currently no evidence for one sex outliving the other (Scott 1988; Figure. 7.5), although the number of swans in the older age categories is still quite low. Of the 177 swans that reached at least 15 years, 88 were male and 89 female.

There have been two main investigations of Bewick's Swan survival rates during WWT's long-term study of the species (Evans 1979a; Scott 1988). Both assumed that, given the high level of site fidelity in this species and the good ring-reading effort across the swans' wintering range, ringed birds not seen in subsequent years were dead. Of course some birds are still likely to be missed, and others might lose leg rings, so these early indications of mortality levels are likely to be overestimates.

Sightings of paired birds, first seen at Slimbridge during the 1960s and monitored to winter 1970–71 put the percentage of swans surviving from one winter to the next at this time at 87.1%. A longer (14-year) run of data for swans at the same site, up to the 1977–78

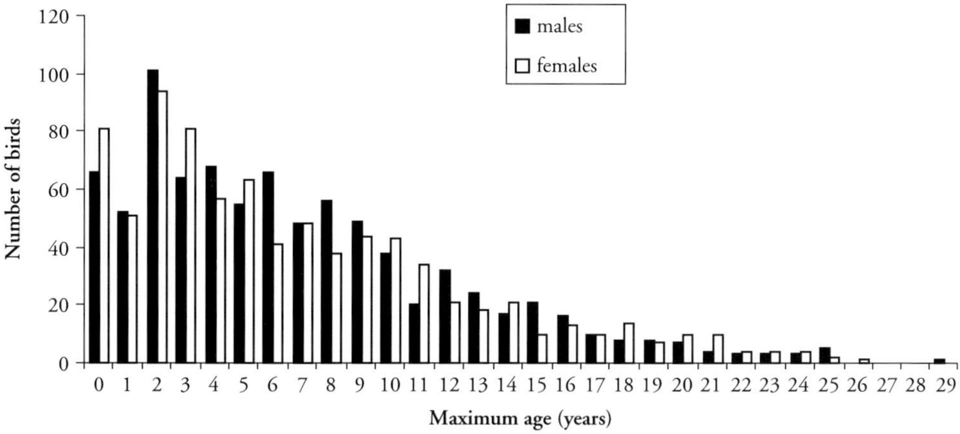

Figure 7.5 Minimum age recorded for male and female Bewick's Swans marked with colour rings at Slimbridge. Ringed birds only are included, because it was possible to continue monitoring these individuals following a change of wintering site.

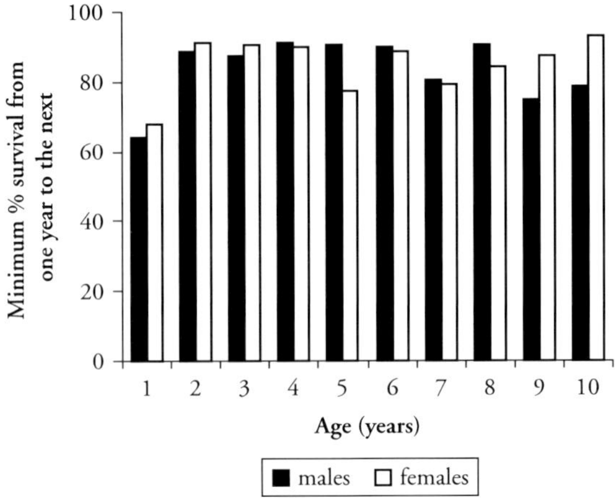

Figure 7.6 Annual survival rates for swans in different age categories. From Scott 1988.

winter, found survival rates to be slightly lower at 82.2% per annum (Evans 1979a). This is below the 85%–90% survival rates estimated for the Svalbard Barnacle Goose population during the 1970s (Owen & Norderhaug 1977), but compares favourably with a range of 74%–87% recorded for the non-migratory Mute Swans breeding in the UK (Bacon & Perrins 1991; McCeery *et al.* 2002). The survival estimates for Bewick's Swans in relation to their age gave similar results. Although survival rates were lower for young birds, with only 64% of males and 68% of females surviving from their first to their second winters, thereafter they increased to over 80% in both sexes (Figure 7.6; Scott 1988).

These figures are currently being revisited for this species using more sophisticated analytical techniques (MARK and M-SURGE software), which derive the probability of survival from the probability of an individual being sighted over several winters. Preliminary results suggest that overall annual survival will not differ substantially from the earlier estimates (WWT unpubl. data), perhaps because a high proportion of the ringed birds are re-sighted each winter.

7.6 LIFETIME REPRODUCTIVE SUCCESS

7.6.1 Overall breeding success

The breeding success of 157 pairs present at Slimbridge together for at least five consecutive winters shows that the successful raising of young can be very erratic in Bewick's Swans (Evans 1979a; WWT unpubl. data; Table 7.2). Twenty-one pairs had no young in five to eight winters, and a further 33 pairs brought young only once in five winters or more. These together represented 34% of the total birds included. Only two pairs ('Boniface' with 'Bunnygirl', and 'Pampisford' with 'Brinckley') were accompanied by young in all the years they were recorded at Slimbridge, and these pairs each spent just five years together at the site. Of 27 pairs with long pair bonds (10+ years) whose breeding success was recorded each winter, only seven were consistently successful breeders, breeding on average at least one year in every two (Table 7.2). The most successful of these ('Dougie' with 'Estralita', and 'France' with 'Valois') bred in nine of 13 winters, and a third pair ('Bionic Diz' with 'Chemical Fiz') in eight of 12 winters. Of the two most durable pairs ('Laburnum'

Table 7.2 Breeding records of Bewick's Swan pairs seen together for at least five consecutive winters. From Evans 1979a and WWT unpubl. data.

Number of years recorded	Number of years of successful breeding									
	0	1	2	3	4	5	6	7	8	9
5	9	13	9	5	2	2	–	–	–	–
6	5	9	10	8	2	1	–	–	–	–
7	5	5	8	4	4	3	1	–	–	–
8	2	2	4	–	2	2	2	–	–	–
9	–	2	3	3	2	1	–	–	–	–
10	–	2	1	4	–	2	–	–	–	–
11	–	–	–	1	1	–	–	1	–	–
12	–	–	2	2	–	–	–	–	1	–
13	–	–	–	–	–	–	1	–	1	2
14	–	–	–	–	–	2	1	–	–	–
15	–	–	–	–	1	–	1	–	–	–
16	–	–	–	–	–	–	–	–	–	–
17	–	–	–	–	–	–	–	–	–	–
18	–	–	–	–	–	–	–	–	–	–
19	–	–	–	–	–	–	–	–	–	–
20	–	–	–	–	–	–	–	1	–	–

with 'Limonia' and 'Night' with 'Day'), both together for at least 20 years, 'Laburnum' with 'Limonia' brought cygnets seven times to the wintering range. 'Night' and 'Day' skipped several winters at Slimbridge, so their full breeding history is not known.

This variation in individual breeding success reflects marked annual variation in breeding success recorded for the population as a whole. The percentage of young is generally higher among the Bewick's Swans seen at Slimbridge compared with those in other parts of the wintering range (notably the larger herds on the Ouse Washes or in the Netherlands), particularly in recent years (Figure 7.7). Nevertheless, even Slimbridge-wintering swans may have exceptionally poor breeding seasons. In winters following the 1992 and 1997 breeding seasons, only 3.8% and 4.1% cygnets were recorded amongst the swans at Slimbridge, when the percentage juveniles recorded on the Ouse Washes was also very low at

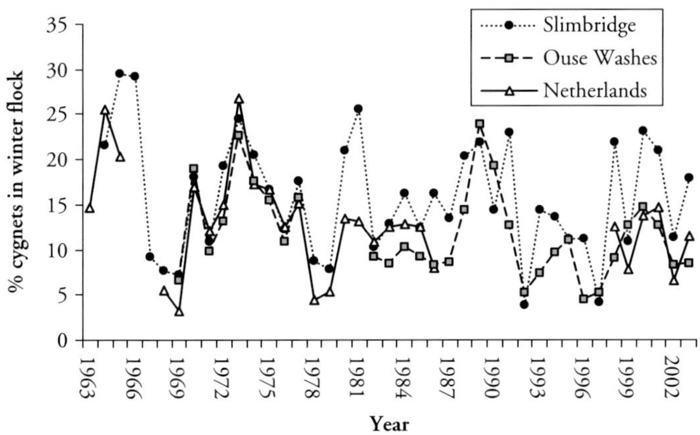

Figure 7.7 Annual variation in breeding success for Bewick's Swans at Slimbridge, the Ouse Washes and the Netherlands. Data for the Netherlands from Poorter (1991) and W. Tijsen pers. comm.

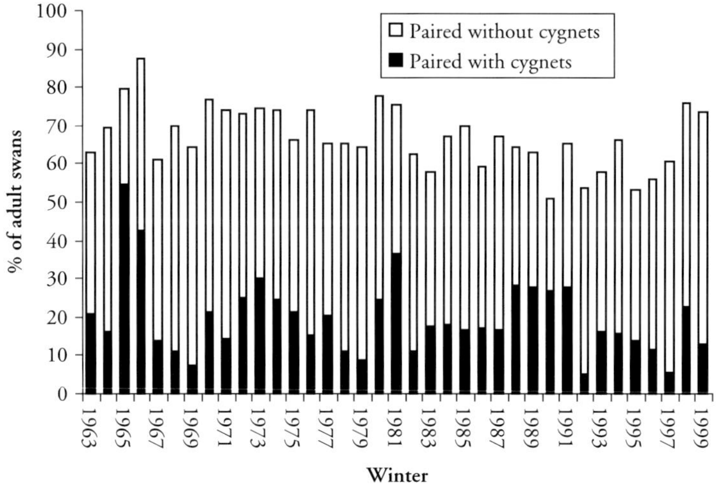

Figure 7.8 Variation in the proportion of paired birds with and without cygnets each winter at Slimbridge.

around 5%. The years 1964, 1965, 1973 and 1989, by comparison, seem to have been particularly good breeding years, with >20% cygnets recorded in different parts of the wintering range (Figure 7.7). Although 1981, 1988, 1991, 1998, 2000 and 2001 were also good breeding years for the swans wintering at Slimbridge, swans wintering on the Ouse Washes and in the Netherlands had more average (12%-15%) breeding success.

The proportion of pairs recorded with young is recorded only at Slimbridge, where detailed behavioural observations identify all paired birds, including all those without young. This also varies substantially between years, from a low of 9.5% of paired birds having cygnets in 1997–98 to 69% of pairs recorded with young in 1965–66 and 53% in 1990–91 (Figure 7.8). The 1965–66 observations may perhaps be an anomaly, because the Slimbridge flock was still increasing at this time and the study was in its infancy. Nevertheless, 1965 was also a good breeding year for swans wintering elsewhere (Figure 7.7), so a high proportion of pairs with young at Slimbridge concurs with the success of the population as a whole. In comparison with the varying ratio of pairs that breed successfully to those without young on arrival in the wintering range, the proportion of adults in the population that are paired (with or without young) has been comparatively stable over the years. On average, 67% of the Slimbridge adults are paired each year, although up to 87% were paired in the 1966–67 winter (Figure 7.8). Fluctuations in the proportion of paired birds may perhaps reflect the number of young birds in the population, which have not yet had time to find a mate.

7.6.2 Effects of mate change on individual breeding success

The collaborative study by WWT and the EGI, which compared the pair-bond characteristics of Bewick's Swans, Whooper Swans and Mute Swans, also considered the implications of mate change on individual breeding success for each of the three species. The study indicated that Bewick's Swans incur significantly higher short-term cost, in terms of reduced breeding success, in changing a mate. This was attributable not only to the longer time taken to find a new mate (described in Section 7.3), but also in the time taken to breed successfully with the new partner. At the outset, the probability of breeding in any year is low for Bewick's Swans unpaired in the previous winter. Only 3.2% of birds that returned with cygnets were single in the preceding year, and on only 1% of occasions did birds return with cygnets and a different mate from that in the previous winter (Scott 1988). For paired birds that lose a mate then re-pair, Bewick's Swans take an average of 2.3 years (SD ±4.2, range = 1–8) to breed successfully with their new mate. There was no difference between the two sexes in time taken to find a new mate or to breed. This is significantly longer than the 1.5 years recorded for both Whooper and Mute Swans (Rees *et al.* 1996). Moreover, on controlling for the annual variation in the breeding success of the population, and also for age effects, Bewick's Swan breeding success was significantly lower in the first year after pairing with a new mate (adjusted mean = 0.4 ± 0.2 cygnets) than the average for swans paired with the same partner (adjusted mean = 0.7 ± 0.4 cygnets). Whooper Swans also appeared to bring fewer cygnets to the wintering grounds in the year following re-pairing (adjusted means = 0.6 ± 0.3 and 1.3 ± 0.7 cygnets with a new and the same mate, respectively; Rees *et al.* 1996), but this was not statistically significant.

If they should lose a mate, it therefore takes Bewick's Swans about five years between losing one mate and breeding successfully with the new partner (i.e. 2.6 years to re-pair, see Section 7.3, plus 2.3 years to breed successfully with the new mate), which is quite a substantial gap in an individual's productivity. In Whooper Swans from the Icelandic breeding population, this time-lag is much shorter (1.9 plus 1.5 years), with a few individuals able to find a new mate and to breed successfully between leaving the wintering grounds in one season and returning the next. Bewick's Swans also tend to pair one year later, and to breed later, than Mute Swans (Rees *et al.* 1996). The Whooper Swans have a relatively short migration of some 800 km from their wintering quarters in Britain and Ireland to their breeding grounds in Iceland. Moreover, the more hospitable Icelandic climate enables several hundred Whooper Swans to remain in Iceland over winter, including a high proportion of family parties (Garðarsson & Skarphéðinsson 1985). Whooper Swans therefore enjoy more time than Bewick's Swans for courtship and pair formation upon arrival on the breeding grounds. The non-migratory British Mute Swans similarly have greater flexibility in the timing of the breeding programme. The very tight schedule under which the Bewick's Swans operate therefore appears to make the formation of long-term pair bonds especially advantageous for their breeding success, and consequently for the viability of the species (Rees *et al.* 1996). Certainly, although Whooper Swans have been seen mating whilst wintering at WWT centres, and Whoopers also occasionally pair with birds from the same wintering site, this has not been observed for Bewick's Swans in the UK. Observations of Bewick's Swans in non-breeding flocks in the Russian Arctic, suggest that much courtship activity occurs during the summer months. The evolution and maintenance of mate fidelity in Bewick's Swans therefore may be at least partly attributed to the energy costs of their long-distance migration requiring the swans to spend more time feeding, coupled with the short time available for breeding at high latitudes.

Although the loss of a mate results in a short-term reduction in a swan's breeding capacity, and reproductive success is positively linked with increasing pair duration (Figure 7.9), mate change does not seem to result in a longer-term reduction in breeding success. For instance, mean lifetime reproductive success does not diminish significantly with the number of mates, for individual swans seen for at least 10 years (mean number of cygnets raised per year being 0.82, 0.57, 0.71 and 0.70 for 1 to 4 mates,

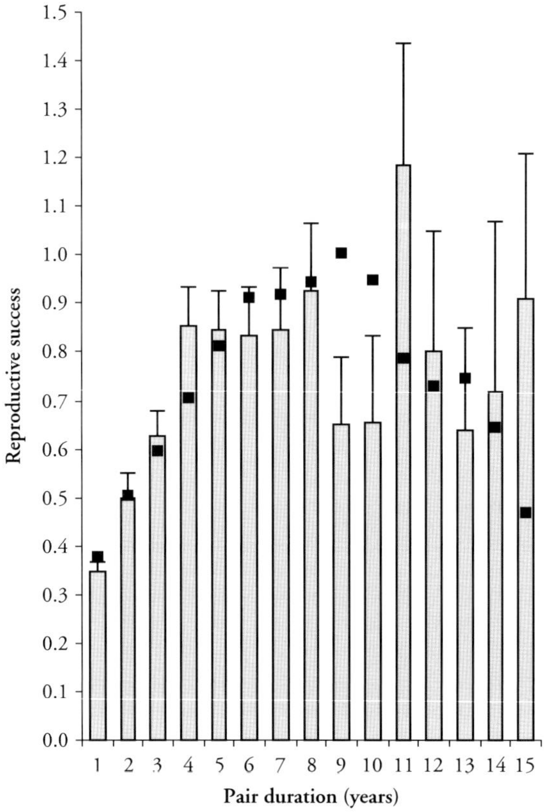

Figure 7.9 Association between pair duration and reproductive success. Histogram and SE bars shows mean reproductive success; black squares denote fitted values after controlling for year and age effects. The increase in reproductive success with pair-bond duration is statistically significant. From Rees *et al.* 1996.

respectively). Individual productivity (i.e. the total number of cygnets raised divided by the number of years seen) similarly was not affected by the number of mates recorded (Rees *et al.* 1996).

Given the apparent short-term decline in the swans' reproductive output on losing a mate, the lack of a longer-term impact of mate change on their lifetime breeding success seems counter-intuitive. Although individuals are marginally more successful with their second and/or third mates this trend was not significant, and there was substantial variation between individuals in their breeding success on re-pairing (Figure 7.10). Thus there is currently no evidence to suggest that mate selection improves with experience, which might have enabled the initial shortfall in breeding success following a change of mate to be made up subsequently if the new mate is of higher quality than the previous one, or if the newly formed pair is more compatible. The lack of any long-term costs on re-pairing may perhaps be due to the overriding effect that weather conditions and predator levels at the breeding sites can have on the annual success of individual pairs, making them somewhat erratic. The reduction in breeding success due to the time taken to find a new mate therefore would be obscured by the effects of poor breeding seasons when considering an individual's lifetime productivity (Rees *et al.* 1996).

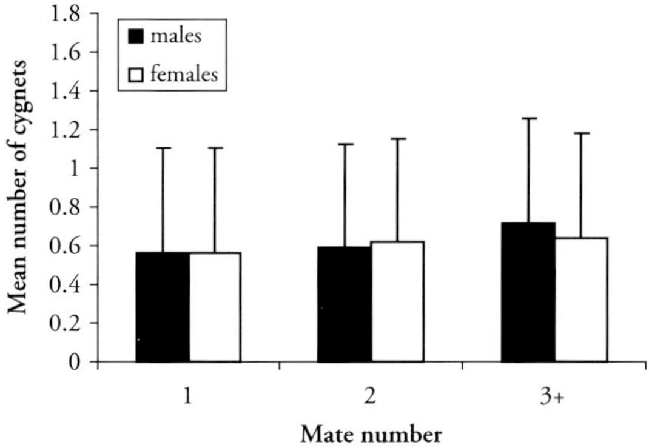

Figure 7.10 Reproductive success (mean number of cygnets) for male and female Bewick's Swans with their first, second and third mates. Differences were not statistically significant. From Rees *et al.* 1996.

7.6.3 Pair-bond duration and breeding success

Breeding success, measured as the number of cygnets associating with a pair on arrival at Slimbridge each winter, increases with the number of years that a pair has been together. This improvement in productivity is evident for at least six years (after controlling for the effects of annual variation in the breeding success of the population as a whole), and may continue for up to nine years (Figure 7.9; Scott 1988; Rees *et al.* 1996). This reflects greater consistency in the breeding success of pairs that have spent several years together, because maximum brood size is just five cygnets per pair. Dominance rank also increases with pair duration, suggesting that cooperation between the pair during aggressive encounters may improve over time (Rees *et al.* 1996). Yet whether this is the reason for the correlation between pair bond duration and breeding success (perhaps through the pair being better able to defend their territory and protect their young) is unclear. When only the years in which the birds are accompanied by offspring are considered, there is no association between their dominance ranking and the number of years paired to each other. The improved breeding success therefore may be at least partially due to greater cooperation in parental care of the young, rather than to an improvement in their fighting ability. If so, the increase in dominance rank associated with pair duration could be purely a consequence of the improved breeding success of the pair, since large families are dominant in the winter flocks. Social dominance increases over time for pairs that do not bring cygnets to Slimbridge, however, so it seems that the swans' fighting ability, if not their breeding success *per se*, does improve with pair duration (Bowler 1996).

In addition to pair duration having a significant effect on Bewick's Swans breeding success, it also appears to be more important in some years than in others in influencing the number of cygnets (if any) associating in winter (Rees *et al.* 1996). Established pairs embark more promptly on their breeding programme than pairs new to an area in cold springs (Chapter 5), which suggests that their years together and prior knowledge of the breeding site may give them an advantage. The level of cooperation by members of a pair in defending the territory and offspring also may be more important for the survival of the young in some years than in others, for instance when predator levels are high. Pair duration and breeding experience are likely to be less important in milder years, when conditions are favourable and a higher proportion of pairs nest and raise young.

7.6.4 *How reproductive success is affected by the combination of male and female characteristics, and the implications of this for mate choice*

Although mate change does not appear to have any long-term implications for an individual's breeding success (Section 7.6.2), certain characteristics both of the male and of the female do have a bearing on the overall success of a pair. Initial analyses of the effects of body size found that, within pairs aged five to ten years, the number of cygnets produced was closely linked to male size, as determined by his skull length (Scott 1988). Male weight, the dominance of the pair, and the size and weight of the female are also significantly, but less strongly, associated with the pair's breeding success, with dominance rank being closely correlated with male size (Scott 1988). It therefore seems that a combination of the male swan's size and his associated fighting ability has a major influence on the breeding success of the pair, the larger and stronger birds being more likely to bring cygnets to the wintering grounds.

Further analyses of the effects of body size have confirmed the association between male size and breeding success, this time using a combination of size measures (i.e. derived from both skull and tarsus length) to classify each individual as small, medium or large according to its size relative to the mean for its sex (Rees *et al.* 1996). Female size was also correlated with the number of cygnets accompanying the pair in winter but only when taken in combination with male characteristics. It is possible that female traits have greater influence in determining clutch size and hatching success, and male size in influencing subsequent survival of the young through protection of the brood, since measurements of parent birds on their Russian breeding grounds have found that female size (rather than male size) is significantly associated with brood size in August for pairs that succeeded in gaining breeding territories (WWT unpubl. data and J. Bowler pers. comm.).

Age may also play an important part in determining reproductive success. After controlling for the pair bond duration, it was found that reproductive success increases rapidly with age for both sexes during the first six to seven years, then reaches a plateau thereafter. The combination of male and female age groupings within the pair (with up to age five years being young, age 6–15 being the middle years, and 16+ being old) has a significant effect on the number of cygnets brought to the wintering site (Figure 7.11). In Bewick's Swans, young males paired with young females have the lowest breeding success, with middle-aged and old birds of both sexes being most successful when being paired with mates from the same age category (Rees *et al.* 1996; Figure 7.11). Moreover, young Bewick's Swans of either sex are more successful when paired with older partners. Overall, old female Bewick's Swans have a high level of reproductive success irrespective of the age categories of their mates, indicating that their breeding performance is not reduced when paired with younger males (Rees *et al.* 1996). There is little evidence to date for a reduction in breeding success in the oldest of our swans; six females and four males have returned with cygnets in at least one winter when aged at least 20 years, and one female ('Casino') bred at ages 20, 22 and 23 years. It is possible that consistently successful breeders have shorter lifespans, due to the physiological demands of defending territories and rearing young, but this has yet to be investigated in Bewick's Swans. Moreover, since male parents are more vigilant and have lower food-intake rates during the winter than their mates or males without young (see Chapter 4), any reduction that breeding success may have on longevity may be more evident in males than in females.

The importance of the characteristics of both male and female in determining the breeding success of the pair has implications for mate selection. Female swans would appear to

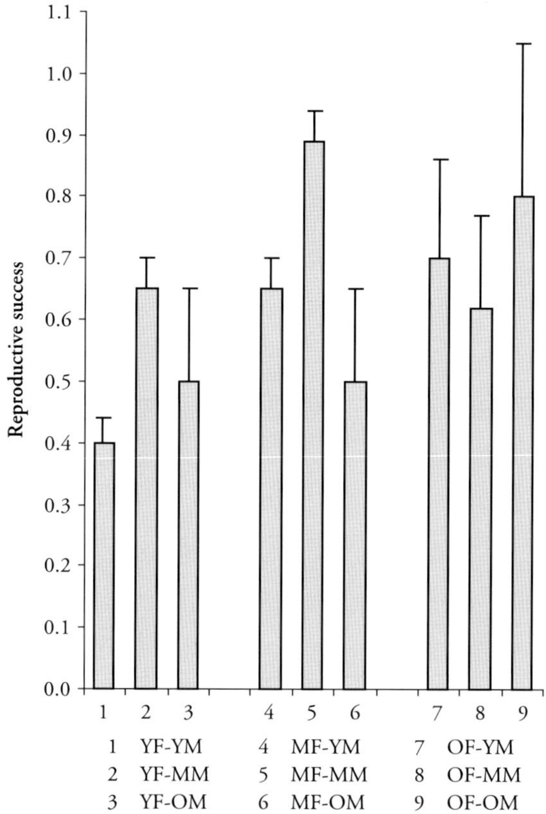

Figure 7.11 Reproductive success (number of cygnets associating in winter) of Bewick's Swan pairs in relation to the combination of male and female age categories; young (Y), middle-aged (M), and old (O) for females (F) and males (M). From Rees *et al.* 1996.

have a better chance of breeding successfully if they were to pair with large males. Similarly, while swans of either sex are more successful with mates of similar age, young swans would improve their reproductive success if they paired with older mates, suggesting that the greater breeding experience of older swans could compensate for an inexperienced partner (Rees *et al.* 1996). However, there is no evidence to date for such assertive mate selection. This may be due to limited availability, since paired Bewick's Swans remain with their mates until one or both dies. Moreover, in the case of the swans' age, this variable is not fixed and young mates gain in age and experience over time. As Scott (1988) points out, if male characteristics (notably size) are more important than female characteristics in influencing breeding success, then females that re-mate should show a greater change in breeding rate than males that re-mate. This change may be either an improvement or a decline in breeding success. She found that the mean change in the number of cygnets raised per year between partners was 0.676 for males and 0.918 for females. It therefore seems that male traits predominate in influencing the number of cygnets brought to the wintering grounds, and thus the overall success of the pair, despite female characteristics being important in determining clutch size and hatching success, in the earlier stages of the annual breeding programme.

7.7 IMPLICATIONS OF THE SWANS' LIFE-HISTORY STRATEGIES FOR SPECIES SURVIVAL

The results of the long-term Bewick's Swan study at Slimbridge have demonstrated that the Bewick's Swan is a long-lived species that is slow to mature. Birds take several years to pair and to breed and have a low reproductive output each year. Moreover, a high proportion of pairs apparently breed successfully once or twice only, and even good breeding pairs are only successful about one year in two. These results from the wintering range have recently been reinforced by observations made on the breeding grounds (Chapter 5), which found that only a low proportion of the population attempts to breed each summer, and that the success of the pairs that do attempt to breed is highly variable due to the vagaries of the Arctic summer. As with a number of other species (e.g. Mute Swans and migratory geese), recruitment to the population therefore is highly biased and seems to be dependent on the success of those relatively few birds that succeed in raising young over a number of years. The swans' low overall reproductive output is to some extent compensated for by high annual survival rates. This enables the breeding pairs to increase the number of offspring produced during their lifetime, but the proportion of these offspring which go on to join the breeding population is as yet unknown. The next generation of breeding birds could be recruited equally from the offspring of both good and poorer breeds, or perhaps the offspring of successful parents are fitter and more likely to breed themselves in due course.

Mate fidelity has proved a highly important strategy for ensuring breeding success in Bewick's Swan. Loss of a mate causes an immediate drop in reproductive output whilst the swan re-pairs, and even after re-pairing breeding success may initially be lower because success generally increases with pair duration. The apparent lack of any long-term costs of mate change may be due to the variability in a pair's success over several years; any reduced breeding success in one or more seasons due to the time taken to find a new mate could be obscured by the effects of other poor breeding seasons when considered over the individual's lifetime and its lifetime productivity. In Bewick's Swans, long-term monogamy may have evolved because of the increase in breeding success with increasing pair duration, which in turn may be linked to the very tight schedule under which they complete their annual migratory and reproductive cycles. This view is reinforced by the lower short-term costs of mate change noted for Whooper Swans and Mute Swans (Section 7.6.2), which do not have the same constraints on their breeding programmes. The value of parental care to the cygnets, and of the presence of the mate to the female during winter (Chapters 4 and 6) would also reinforce the evolution of monogamy in this species. The importance of longevity and of pair duration for enabling Bewick's Swans to breed and thus bring new birds to the population means that even small increases in annual adult mortality rates could have a major effect on numbers in the population as a whole. That population growth in long-lived, slow-breeding species is most sensitive to adult survival is a well-established concept, and has also been shown for several goose species (e.g. Trinder *et al.* 2005).

CHAPTER 8

Threats and conservation measures

'Mince finely 3 lbs of rump steak with 3 shallots, and season liberally with salt, pepper and grated nutmeg. Truss the bird like a goose, stuffing it with the rump steak, etc., sewing it up to prevent the stuffing escaping. Wrap it in well greased or buttered paper, and cook for about 2 hours, basting very liberally. When cooked, remove the coverings and froth it with a little flour and butter, and dish up with brown gravy round it, and port wine sauce.'

Old Norfolk recipe (from *Birds and their Eggs*, compiled by Andre Simon)

Although swans have long been revered for their grace and beauty, it was only in the 20th century that man began to question the effects of his exploitation of the species. Thus when Henry Seebohm visited the Bewick's Swans' breeding areas in the late 19th century, it was not considered unusual for a swan skin to be the longed-for trophy, or for the expedition party to make an omelette out of swan eggs. Seebohm (1901) reported that the skin he had acquired had been left by a peasant with his wife and that '*she, good soul, had cut off the beak and given it to her children as a plaything*'. The skin (with feet still attached) was purchased for one rouble, with the beak included in the deal, and Seebohm went on to note that '*this, undoubtedly, was the skin of a Bewick's Swan; the beak also was indisputable*'.

Seebohm's cavalier approach towards the swans seems strange today. His description of a Russian 'bath' or sauna is much more familiar to those on recent expeditions to the High Arctic. The birds have been protected from persecution for many decades, and measures have been introduced by governments and non-governmental organisations to improve the Bewick's Swan's conservation status throughout its range. Meanwhile, the work of previous generations means that we are more aware than the pioneering natural historians of the various threats facing the swans, although sadly this is offset by our increased ability with modern technology to cause them harm. This final chapter dwells on the current threats to the birds and on-going conservation effort, and considers conservation measures that may be required for the future.

8.1 DIRECT THREATS TO THE SWANS: CAUSES OF MORTALITY

The continued survival of a species may be threatened by a reduction in breeding success (possible reasons are described in Chapter 5) or by an increase in mortality levels. If both operate at the same time, the rate of decline is accentuated, and the risk of extinction becomes critical. An increase in mortality could be immediate, for instance through an open hunting season or licensed hunting removing a proportion of the population each year, as now occurs for Whistling Swans in America. It can also be more insidious, such as progressive habitat loss resulting in the birds being unable to find sufficient food for survival at key sites.

Standard post mortem examinations of wild Bewick's Swans recovered in the vicinity of Wildfowl & Wetlands Trust centres have been undertaken since 1951 to gain some insight into the main causes of death for swans wintering in Britain. Samples for bacteriology, virology, toxicology and histopathology being taken to confirm the initial diagnoses. Only a few birds are found dead each year, but a total of 182 post mortems were completed by the end of the 2003–04 winter. Flying accidents were found to be the most frequent cause of death, followed by lead poisoning and avian tuberculosis (Brown *et al.* 1992; WWT unpubl. data; Figure 8.1). Most of the predated birds (6.1% of those recovered) had been shot,

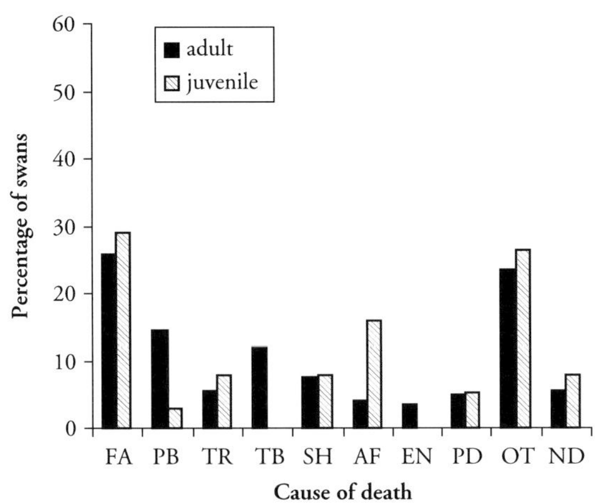

Figure 8.1 Main cause of death in Bewick's Swans, showing the percentage to die from different n = 121 adults; 29 juveniles. Causes of death: FA = flying accidents; PB = lead poisoning; TR = trauma; TB = avian tuberculosis; AF = aspergillosis; EN = enteritis; PD = predated; OT = other; ND = no diagnosis.

despite national and international legislation banning the hunting of the Bewick's Swan throughout its migratory range.

Causes of death reported to the British Trust for Ornithology (BTO) for 47 birds ringed in Britain recovered along the northwest European flyway similarly found that 30% had collided with man-made structures (mainly power lines), but that a further 32% had been 'deliberately taken by man' (Rees & Bowler 2002). It seems likely that this higher figure of 32% may be an overestimate due to the small numbers involved and, since post mortems were not usually conducted, to hunting (like power-line collisions) being relatively obvious. Nevertheless, illegal hunting does seem to be a significant cause of death for the species.

The post mortem examinations carried out routinely at Wildfowl & Wetlands Trust centres identify the primary cause of death for most Bewick's Swans examined. Since they are made mainly for birds recovered in Britain in winter, however, and generally in areas with a good food supply, they may not be representative of other parts of the wintering range, and are unlikely to reflect causes of mortality during the migratory and breeding seasons. The numbers dying of starvation or exhaustion during migration, or of predation on the breeding grounds, therefore, is still unknown, although predation is seen to be a major cause of death in newly hatched cygnets (Chapter 5). There are relatively few man-made structures in the Russian Arctic, so the frequency of flying accidents there should be low, and the swans are more likely to die of other causes during the summer months. In particular, sick swans unable to embark on or complete autumn migration would be lost, because they would be unable to survive an Arctic winter.

8.1.1 Collisions with power lines

Flying accidents (mainly collisions with power lines, but also with trees and other structures, particularly in foggy conditions) accounted for the death of 25.7% of adults and 28.9% of juveniles recovered and examined (Brown *et al.* 1992; WWT unpubl. data; Figure 8.1). This is similar to the 21.7% of deaths attributable to power-line collisions reported for Mute Swans ringed in Britain (Perrins & Sears 1991), and an overall estimate of 22%–23% of deaths for all swans (Bewick's, Whooper and Mute) checked by the Wildfowl & Wetlands Trust's pathologist (Brown *et al.* 1992). Perrins and Sears suggest that their estimate might be artificially high, since swans that hit wires may be more likely to be reported than those dying of other causes, for example because a repair engineer might visit the site following a power failure and return the ring. On the other hand, they also mention that the likelihood of power-line casualties being reported might have been affected by the introduction of auto-reclose mechanisms of the local supply lines (whereby the power supply is restored after a few seconds, thereby reducing the need for a visit by an engineer), and by the prompt removal of dead birds by Red Foxes. Most (66%) of the swans examined by the Wildfowl & Wetlands Trust died within five miles of a WWT centre, so the swans were from closely monitored flocks and the results less likely to be biased towards power-line casualties. Since swans are large birds, they have greater difficulty than some other species in changing tack at short notice in flight. Thus despite their good eyesight it is not surprising that they are vulnerable to flying accidents, particularly if (as has been noted on the Ouse Washes) their webbed feet become caked in mud after feeding on arable land, which makes it more difficult for them to manoeuvre in flight.

In order to make their overhead wires more visible to the swans, several electricity companies have marked power lines with bird diverters in 'hot-spot' areas, where the frequency of collisions on a stretch of wire has caused repeated power failure. Orange balls appear to be the most effective diverters, although coils and corks were also used on local supply lines in earlier years. Following the death of a number of swans due to collisions with the 400kV power lines on the Ouse Washes, the National Grid Company fitted brightly coloured orange balls to the earth wire (25 balls per span), in an attempt to make it more obvious to the birds. A comparison of the frequency of power-line casualties in five years before and four years after marking the wires found that the rate of collisions per year and per span was lowered considerably where the balls were attached (Butler 1999). The system was not infallible, as the total number of collisions at the Ouse Washes actually increased from ten Bewick's Swans per winter prior to marking, to 15 per winter after the markers had been fitted, but this was due to 44 birds colliding with an unmarked section of wires over a four-week period during the 1996–97 winter (Butler 1999). Overall, the balls fitted to wires on the Ouse Washes and elsewhere have proved reasonably successful in reducing collisions in key areas, but changes in the swans' flight-paths between their roost and feeding sites do result in fresh (unmarked) stretches of wires becoming a risk to the swans over time.

8.1.2 Illegal shooting

Despite being legally protected from hunting throughout its range (Section 8.4.1), 11 (7.6%) of the adult and three (7.9%) of the juvenile Bewick's Swans examined by the WWT had been killed by hunters (Brown *et al.* 1992). Since the birds were recovered in Britain, it is likely that they had all been shot in this country. Certainly one swan, named 'Brockbank', survived having been shot whilst visiting flooded meadows on the River Avon near Tewkesbury at the start of the 1998–99 winter. She was nursed back to health at an animal rescue centre, released at Slimbridge in December 1998, and remained in the area until her deparure on spring migration in February 1999. She was not resighted in Britain for several years, but was seen safe and well wintering in the Netherlands, at staging sites in Germany, and finally returned to Britain (to Welney) on 14 December 2004.

The 32% of Bewick's Swans 'taken by man' in swan deaths reported to the BTO includes nine (69%) of 13 recoveries from Russia ascribed to hunting, compared with two (11%) of 18 birds recovered in Britain and Ireland having been shot (Rees & Bowler 2002). Although there are likely to be substantial biases in these figures, nevertheless they may indicate regional variation in the causes of death. It is likely that any bias in mortality attributable to illegal hunting would be linked not only to human population density, but to the amount of time that birds remain within an area, variation in hunting ethos, and variation in policing levels within different countries along the flyway.

Since Bewick's Swans clearly were being illegally killed by hunters, the WWT made a further assessment of the level of hunting by X-raying swans caught at Slimbridge, to identify those that had survived shooting from the presence of lead pellets in their body tissues. The first phase, during the early 1970s, found that of the 272 individuals checked, 92 (34%) had been shot; 44% of adults, 25% of yearlings and 12% of cygnets contained at least one lead pellet. One individual ('Pie') had 21 pellets in its body (Evans *et al.* 1973; Figure 8.2). When X-raying was resumed from 1988–89 onwards, the situation was little better. Despite attempts to increase the public's awareness of the swans and their conservation status throughout Europe, a worrying 39% of 191 adults, 19% of 43 yearlings and 4% of 51 cygnets X-rayed in Britain between 1990 and 2000 were found to be carrying lead shot in their tissues (Rees *et al.* 1997a; WWT unpubl. data). On this occasion, one bird ('Luciano') had 30 pellets. The extent to which carrying lead shot incapacitates Bewick's

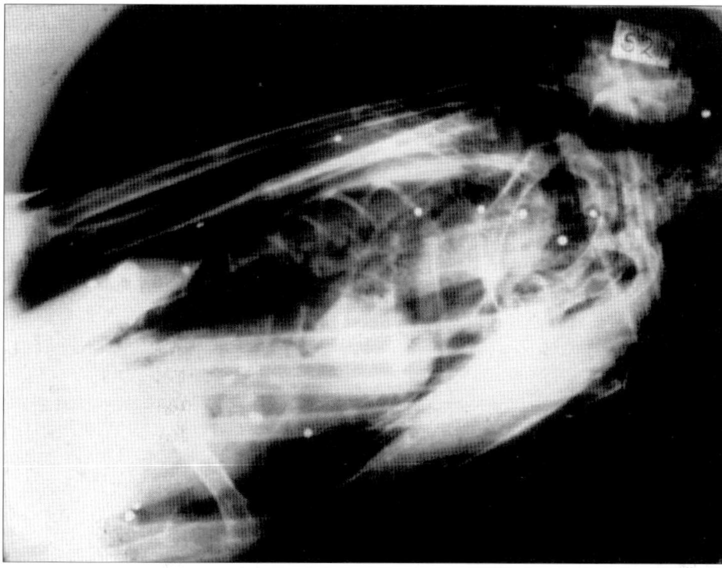

Figure 8.2 X-ray of the Bewick's Swan 'Pie', which had 21 pellets in its body tissues (WWT).

Swans has not been examined in detail, but studies of Pink-footed Geese have found that survival rates are significantly diminished in the short-term in birds that initially survived being shot (Noer & Madsen 1996; Madsen & Noer 1996).

The hunting of Bewick's Swans of the eastern population is also illegal, but again incidents are known to occur. In China, people have used cyanide baits to kill swans (Yu & Liu 1987), and both Whooper Swans and Bewick's Swans have traditionally been eaten in Shanghai, Chong Ming Island, Suzhou and Shazhou (Ma & Cai 2000). Although the hunting of swans is now illegal in China, they were hunted there for many years for medicinal purposes as well as for food (Pang 1974, 1976; Ma & Cai 2002). Ma and Cai (2000) report that every winter one can still find fresh swan meat in the markets, selling by the 'catty' (1.5lb). It seems, therefore, that mortality due to hunting is not being unduly overstated in either the east or the west, and that tighter enforcement of hunting laws is needed to improve the situation for Bewick's Swans throughout their range.

8.1.3 Disease and other causes of mortality

A number of other causes of death have been noted in Bewick's Swans, including lead poisoning (in 14.6% of adults subjected to post mortem examination at WWT), avian tuberculosis (11.8% of adults), aspergillosis (4.2% of adults and 15.8% of juveniles), enteritis (3.5% of adults) and predation. Predation by Red Foxes was found to be significant, accounting for 4.9% of adults and 5.3% of juveniles. Less frequent causes of death were renal failure, cardiac failure, visceral gout, gizzard impaction and pneumonia (Brown *et al.* 1992; WWT unpubl. data; Figure 8.1). Gizzard worm (*Amidostomum* spp.) is a regular contributory cause of death. When swans are already debilitated by another condition such as lead poisoning they become more susceptible to parasites, and gizzard worms are frequently pathogenic in these circumstances (M. J. Brown pers. comm.).

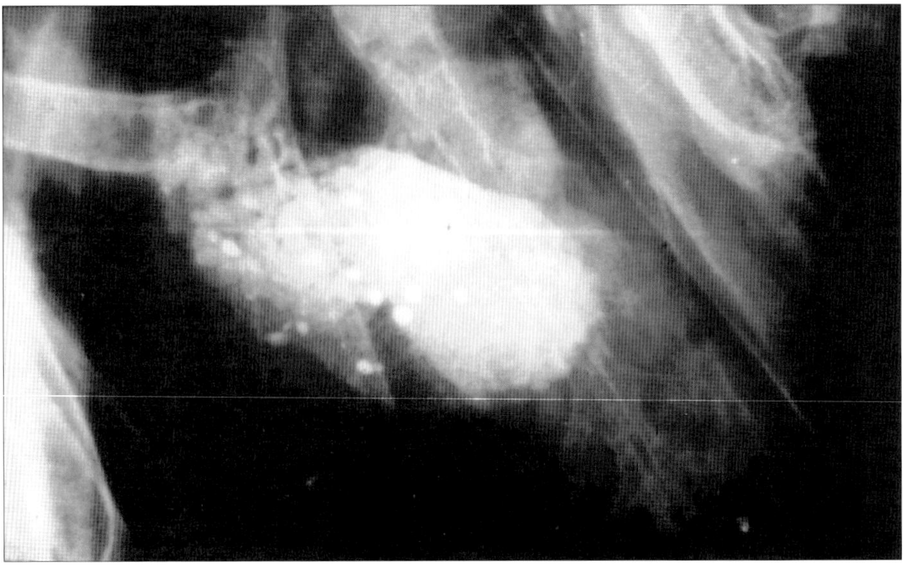

Figure 8.3 X-ray of a Bewick's Swan gizzard, showing lead pellets that had been ingested as grit (WWT).

Lead poisoning is usually due to a bird ingesting one or more lead fishing weights or shotgun pellets, which are retained as grit in the gizzard (Figure 8.3). In the cases of Bewick's and Whooper Swans, the lead ingested is almost entirely shotgun pellets, whereas lead fishing weights and split shot are more of a problem for Mute Swans. As the lead is ground down, it forms soluble lead salts, which are then absorbed into the bloodstream. Many governments and local authorities have now banned the use of lead for shooting or angling, but spent lead in the environment remains a hazard for the swans. Further efforts therefore are needed to reduce lead levels, and management measures introduced (such as the provision of non-toxic grit) to minimise the risk of lead poisoning in the birds (O'Halloran *et al.* 2002). It has been suggested that flying accidents could be associated with lead poisoning (Birkhead 1982), because high lead levels affect the neuro-muscular system, but of all the swans (Bewick's, Whooper and Mute) checked at Slimbridge only two of those killed by colliding with overhead wires were also suffering from lead poisoning (Brown *et al.* 1992; WWT unpubl. data).

Avian tuberculosis is a chronic disease that normally affects older birds, so is more likely to be evident in long-lived species. The relatively high proportion of adult Bewick's Swans recorded as having tuberculosis may not be representative of the population as a whole, because most of those diagnosed were recovered at Slimbridge where the bacterium *Myobacterium avium* may be more prevalent. Additionally, the daily observations would ensure that the few occasions when sick birds were present in the flock were noted, so swans with progressive illnesses such as tuberculosis or lead poisoning are more likely to be recovered (Brown *et al.* 1992). The higher incidence of the fungus *Aspergillus fumigatus* in juveniles compared with adults is likely to be due to their immune system being less well developed than that of adult birds. Five Bewick's Swans that died of enteritis developed the condition after ingesting oil which damaged the gut; four were recovered from the River Mersey after an oil spillage in 1973 and the other from the Manchester Ship Canal in the same year.

Blood samples are occasionally taken at the swan catches made at WWT centres to check the birds for pathogens and parasites. Of 164 Bewick's Swans surveyed for *microfilariae* parasites in the late 1970s, six (3.6%) were found to be infected with heartworm *Sarconema euryerca* and one juvenile bird with a short *microfilariae* (Seegar 1979). More recently, assessment of 147 swans at Slimbridge in winters 1999–2000 to 2004–2005 found that the birds were quite healthy. Overall, ectoparasite loads were low, although juveniles had higher burdens and higher levels of associated feather damage. Seventy blood samples tested showed no evidence of Newcastle disease virus or pathogenic avian influenza viruses (R. Cromie pers. comm.).

Pesticide poisoning is relatively infrequent, but can kill large numbers of birds in a short period of time. For instance, the use of carbophenothion in seed dressing, aimed at protecting the crops from wheat bulb fly, has caused death of both geese and swans feeding on farmland. Because the swans (and also geese) graze in flocks, a large number may die of this poisoning if treated fields are selected as feeding sites (Owen *et al.* 1986). Incidents of geese dying in large numbers (several hundred at a time) in Britain during the 1970s, and the ensuing publicity, resulted in the Ministry of Agriculture, Fisheries and Food (MAFF) and the distributors of carbophenothion agreeing that it should not be used in central Scotland, since wheat bulb fly was not a particular problem in this area, and the poison was having adverse effects on Greylag and Pink-footed Geese. About 30 Bewick's Swans were killed by the same pesticide in East Anglia in 1982 (where crops were highly vulnerable to wheat bulb fly), but the birds were scared from the field after the first deaths were reported, and this seems to have prevented more birds from being affected.

The radiation leak following damage to the nuclear reactor at Chernobyl, in the Ukraine, on 26 April 1986 caused concern for the welfare of swans migrating through the Baltic Sea and Gulf of Finland at this time. Although it was thought that the birds would have passed through the Baltic well before the reactor incident, and there was no indication that the plume of radioactivity released reached the swans' breeding grounds in the Russian Arctic, there remained a slight possibility that the swans might have been affected by the incident if their migration had been delayed. When the swans returned the following winter, scientists from the Berkeley Nuclear Laboratories therefore attended the Slimbridge swan catch to measure the radiation levels in individual swans, excluding cygnets since these would have hatched after the Chernobyl incident. A gamma-ray spectrophotometer, which is capable of detecting minute amounts of radioactivity, was brought to Slimbridge for this purpose. It did not detect radioactivity above background levels in any individual swan, and the caesium isotopic ratio determined from data for all 46 birds checked was indicative of nuclear weapons tests rather than Chernobyl. Two swans that died during the winter (one on the Ouse Washes and one at Caerlaverock) were also checked for radioactivity, including samples taken of their muscle and internal organs. Again radioactivity levels were very low; the ratio of caesium-134 to caesium-137 could have been due to Chernobyl in one bird, but for the other the isotope ratios were indicative of weapons testing in the 1960s (Hancock & Woollam 1987).

Although mass mortality has not been recorded in Bewick's Swans, three diseases in particular pose this type of risk to the birds. Of these, the one most likely to be seen in the wintering range is necrotic enteritis which is unusual but becoming increasingly frequent. It was the cause of death for around 20 Whooper Swans on the Ouse Washes in 1999 and for a further 30 Whooper Swans at the same location in November 2003, with occasional cases also occurring in the Whooper Swan flock at Martin Mere (Pritchard *et al.* 2004,

M. J. Brown pers. comm.). Necrotic enteritis was also found in several Mute Swans recovered in Scotland in 2003 (Pennycott *et al.* 2004), and has been blamed for earlier sporadic losses of swans elsewhere in Britain (MacDonald *et al.* 1987). The condition is associated with the birds eating a high carbohydrate diet (e.g. spilt grain and waste potatoes or sugar beet) when they are not used to it, for instance on arrival at the wintering sites in autumn. The change of diet raises the acidity in the gut, which enables the bacteria *Clostridium perfringens* to thrive. The provision of lime or limestone, put out for the birds to take as grit, helps to neutralise the acid, and may have alleviated the situation on the Ouse Washes; there were no further cases that winter. In the case of the Mute Swans, the outbreak of necrotic enteritis was attributed to the birds ingesting toxins produced by cyanobacterial blue-green algae, during an algal bloom on their pond (Pennycott *et al.* 2004).

The two other possible causes of an outbreak of mortality, duck viral enteritis (DVE, or 'duck plague') and botulism could also occur at migratory sites or in the breeding grounds. Captive Bewick's Swans are susceptible to DVE, which is contracted through contact with infected birds. An increase in temperatures through global warming could make botulism more prevalent, because epidemics are associated with hot weather, eutrophication and lowered water levels. Mortality due to a pathogenic influenza virus (e.g. the H5N1 strain) has not been recorded in Bewick's Swans to date, and there was no evidence for the disease in Bewick's Swans checked (by droppings or cloacal swabs) across Europe in winter 2005–06.

8.2 THREATS TO SWAN SITES: HABITAT DEGRADATION AND POLLUTION

Despite efforts to reduce and reverse the rate of loss of wetlands, the northwest European Bewick's Swan population continues to be threatened by the drainage of their traditional wetland habitats. The availability of areas of open water for roosting, and for providing safe havens when the swans are disturbed in the fields, is essential for their welfare. The loss of wetland habitats also reduces the feeding opportunities for the birds, and reinforces their feeding on arable land and pastures where they may come into conflict with farmers (Section 8.3). The swans' wintering distribution is closely linked to areas where secure roosts are located within areas of intensive farming, so a major change in farming activity would have major implications for the swans' food supply. Management programmes for the restoration of aquatic ecosystems are underway, particularly in the Netherlands, and this may allow birds access to more of their traditional foods (Bowler 2005b). Many important wintering sites for the swans in northwest Europe receive some level of legal protection, particularly their roosting areas (Section 8.4.2). Further protection of feeding areas is required, however, to allow for the movement of birds between different fields and crop types during the course of the winter.

Swans in the eastern population similarly face habitat loss and degradation, through the conversion of floodplain wetlands to agriculture, and through developments such as the construction of dykes, dams and other infrastructure. Moreover, human activity has greatly increased the disturbance to birds at remaining wetland sites. In the Repbublic of Korea, a large part of the Nakdong Estuary is being reclaimed despite national legislation protecting the site, and recreation and fishing result in high disturbance levels at Joonam (N. Moores pers. comm.). In the D.P.R. Korea, developments on the coast and the damming of rivers have affected wetland sites used not only by Bewick's Swans but many other waterbirds

(Perennou *et al.* 1990). The development of heavy industry and agriculture started later in China than in western Europe, but pollution and drainage of wetlands, and increased disturbance levels, has caused major habitat loss since the 1960s (Ma & Cai 2000). In the lower Yangtze floodplain, the total area of lakes is reported to have declined by 62% between the 1950s and 1980s, with more than 1,100 lakes totally reclaimed (Barter *et al.* 2004). The construction of the Three Gorges Dam, which started to fill in mid-2003, and associated changes in hydrology in the region, may have a further negative impact on the wetlands downstream. Even in Japan, where Bewick's Swans have been protected since the early 20th century (apart from a resurgence in hunting during and immediately after the Second World War), the loss of many wetlands means that the migratory swans are now concentrated at relatively few sites (Brazil 2003).

As for the other half of the year, the swans' breeding habitat is increasingly threatened by oil and gas exploration throughout the Russian Arctic (Beekman *et al.* 1994a; Bowler 2005b). The increase in disturbance associated with developing and servicing oil and gas installations, particularly in previously remote areas, could have a major impact on the swans' breeding distribution and productivity. The increase in human activity on the tundra could impinge on the swans through one or more mechanisms, including temporary or long-term habitat loss, disturbance levels influencing the swans' site selection and behaviour, collisions with powerlines and oilfield structures, and increased mortality through spills of hydrocarbons and other compounds in the oil or gas fields (Ritchie & King 2000; Ritchie *et al.* 2002).

Bewick's Swans are also vulnerable to oil spills at their moulting and pre-migratory fattening areas, where they congregate in large numbers in August and September. For instance, a major oil spill in the Pechora Delta in the mid-1990s could have affected some 15,000 birds (Beekman *et al.* 1994a), and even in years when smaller numbers are present, thousands of swans would still be at risk. The swans are flightless at this time, because they are undergoing their annual moult, so would be less able to escape from a major environmental incident. Moreover, if they were unable to put on the body fat needed to fuel migration, they would not have sufficient energy reserves to complete the long journey to the wintering sites. The leak from the pipeline at Usinsk, near Ukhta, in the Komi Republic of northern Russia, in February 1994 could have had a devastating effect on the northwest European population, because the oil seeped into the Kolva River (a tributary of the Pechora), and thousands of tons of oil had been spilt in a case considered to be worse than the *Exxon Valdez* disaster. The rate of spread was relatively slow because many of the waterways were frozen in the Arctic winter, and a clean-up operation was initiated. Despite this effort, there was much concern about the impact of the oil spill not only on Bewick's Swans but on the high densitities of other wildfowl and shorebirds that breed in the extensive network of channels, pools and marshes of the Pechora Delta. On visiting the delta region in summers 1994 and 1995, we were therefore relieved to find no evidence of a slick; it seemed that the huge outflow from the Pechora had pushed the oil well out into the Barents Sea. Nevertheless, oil and gas exploration has been underway much closer to home, in the immediate vicinity of the delta, where a spillage would have a much more rapid and direct effect.

Given that a single incident could have catastrophic consequences for the species, environmental impact assessments of proposed developments including oil and gas installations should be undertaken routinely by independent international organisations. Preliminary investigations of the effects of oilfield infrastructures on Whistling Swans breeding in the vicinity of the Kuparuk and Milne Point oilfields in northern Alaska found that some mitigation can be introduced to reduce the disturbance to the birds. Examples include locating the main structures away from prime habitat and locating access routes in drier upland areas several kilometres from the nest sites (Stickney *et al.* 2002). However, such measures would not ameliorate a major oil spill on the tundra.

A new potential threat to the swans' summer haunts is habitat loss due to global climate change. The sea level rises in the Russian Arctic associated with the melting of the polar ice-caps might be viewed as a threat, particularly since much of the swans' breeding range is just a few metres above sea level. The rises in arctic sea levels in the second half of the 20th century were measured in centimetres per year rather than in metres, and it is possible that future increases could be offset by isostatic rebound as the weight of the polar icepack on the earth's crust diminishes, and land rises in some (but not all) parts of the globe. Of much greater concern are the habitat changes predicted to follow a warming of the Arctic regions, in particular the northward extension of the boreal zone. The maritime tundra, with its rush of spring vegetation and open landscape, which currently provides the best breeding habitat for the birds, is already limited to a narrow strip along the northern coast of Russia. Encroachment by shrubs and trees could have devastating consequences for the swans' breeding success, since there is no evidence for them using such habitats, despite these being as little as 100km from their nest sites in some areas. Continued monitoring of the Bewick's Swans and timely introduction of conservation measures (e.g. site protection) in the breeding range therefore could prove crucial for the future welfare of the species. The conservation of key breeding, moulting and pre-migratory areas is of critical importance, since the swans are vulnerable to environmental changes in the summer months.

8.3 POTENTIAL CONFLICT: SWANS AND AGRICULTURE

Before the drainage of wetland habitats and the intensification of farming, the Bewick's Swans frequented the shallow margins of lakes and estuaries where they fed mainly on aquatic vegetation (Owen *et al.* 1986; Chapter 4). The move onto farmland has not necessarily been to the swans' disadvantage, however, because some foods may be more nutritious and easier to collect in the farmland monocultures, and fertilisation of the sward provides palatable and digestible food for grazing species – livestock and swans alike. It cannot be said for certain that the usual farmland pasture of mixed ryegrass and Timothy Grass provides better food for the birds than the more traditional meadow grasses, although no doubt this could be tested. However, the increase in population size during the second half of the 20th century suggests that it is not having a negative impact on population trends.

Bewick's Swans feeding on arable land caused some conflict with farmers in the late 20th century, particularly where the fields used were wet, because trampling by webbed feet exacerbates the loss of crops already damaged by water lying on the land for several weeks. In some areas farmers are keen to ensure that the pasture is not depleted by swan grazing once the grass begins to grow, because the fresh growth provides an early 'spring bite' for their livestock. Bewick's Swans are present for relatively short periods each winter, however, and leave the wintering grounds in late February to mid-March before the cattle are ready to be returned to the land. Bewick's Swans are therefore tolerated more readily by farmers than are flocks of geese (van Roomen & Madsen 1991) or Mute Swans

(Birkhead & Perrins 1986), which remain until later in the spring, or throughout the year, respectively.

Following an increase in the number of complaints about swans in Denmark during the 1980s, Bjarke Laubek circulated a questionnaire to the game consultants of the Danish Ministry of the Environment. The questionnaire asked how many reports of crop damage due to swans the game consultants had received annually from farmers between 1975 and 1992. Laubek found that the first damage reported (for any swan species) in Denmark was in 1974, and confirmed that there was indeed a substantial increase in complaints during the 1980s. Of the 62 incidents registered, however, most were due to Whooper Swans, which are more numerous in Denmark and spend more time than Bewick's Swans on arable land (Laubek 1995b). Although incidents of Bewick's Swans causing crop damage have not been recorded officially in Britain, there is similar potential for conflict with agricultural interests because large numbers of swans do congregate at a small number of arable sites. This is particularly true on the Ouse Washes where the birds are most numerous (Rees *et al.* 1997b; Chapter 4). In areas where crop damage is a concern, swans can be discouraged from using particular fields by being persistently disturbed by humans, by scaring devices, or by stringing tapes across key fields (Owen *et al.* 1986). Compensation for damage to crops is also paid by government agencies in some countries, although this currently is mainly for damage attributed to geese rather than to swans.

8.4 CONSERVATION MEASURES

8.4.1 National and international legislation

Bewick's Swans and their habitats have long received protection through both national and international legislation. A ban on hunting was introduced at different times in different countries (for instance, 1954 in UK, 1964 in Russia and 1989 in China), but the extent to which these laws are implemented remains variable, particularly in remote areas. A more integrated approach to conservation has been developed under international conventions. The Bewick's Swan is included under Appendix I of the EU Birds Directive, in Appendix II of the Convention on the Conservation of European Wildlife and Natural Habitats (the 1979 'Bern Convention'), and in Appendix II of the Convention on Migratory Species (the 1979 'CMS', also known as the 'Bonn Convention'). The Bewick's Swan is also listed as a SPEC 3 species in BirdLife International's species of conservation concern, and is recognised as being vulnerable in Europe (BirdLife International 2004a), due to a high proportion of the population congregating at relatively few sites during migration and in winter (Beekman *et al.* 1994a), and to the overall decline in the northwest European population (particularly in Ireland and the Netherlands) in recent years (BirdLife International 2004a; see Chapter 2). The pioneering Bonn Convention is an inter-governmental agreement and was the first to aim at conserving migratory species across political boundaries. All countries throughout the Bewick's Swans' range are either parties to the Convention (e.g. those in northwest Europe) or non-parties that participate in Convention agreements (e.g. Russia and China), with the notable exceptions of Japan and Korea. Appendix II of the Bonn Convention (which includes the Bewick's Swan) lists species that need or would benefit significantly from international cooperation. For this reason, the Bonn Convention encourages the range states for migratory species to conclude global or regional management agreements.

With the most recent (1995) estimate of population size putting numbers at around 29,000 birds (Beekman 1997), the northwest European Bewick's Swan population is listed under Category A(3)c in the Agreement on the Conservation of African and Eurasian Waterbirds (the 'AEWA Agreement'). Further east, the Caspian-wintering population is also listed under AEWA, in Category A(1)c because of its smaller population size. The 'AEWA Agreement', which was developed under the Bonn Convention, requires that signatories to the Agreement prepare and implement national action plans to improve the conservation status of a population or species. As for the Bonn Convention, countries with key migratory and wintering sites for the northwest European population are mostly signatories of AEWA: the United Kingdom, Ireland, the Netherlands, Denmark, Germany, Lithuania and Estonia are all parties to the Agreement. Although Russia is not yet a signatory of AEWA, swan hunting has been illegal across Russia for the last 40 years except for an open season on Whooper Swans in Yakutia for one year in the mid-1990s. Whether there should again be an open hunting season for Whooper Swans in Yakutia is still being debated. The Yakutian Hunting Society and some representatives of the Yakutian Ministry of Nature Conservation argue that Whooper Swans are very aggressive during the breeding season, and that this accounts for a decrease in the numbers of game waterfowl, mainly ducks. The hunting lobby is particularly strong in the Kolyma and Indigirka taiga, and in forest-tundra lowlands, which are thought to have high Whooper Swan densities in summer (Syroechkovski 2002). Although Whooper Swans have a more southerly breeding distribution than Bewick's Swans, it seems likely that any permission to hunt Whoopers would increase the risk of Bewick's Swans being shot during migration through misidentification in the field.

Swans in the eastern Bewick's Swan population are also protected at their migratory and wintering sites, mainly through national legislation. From 1975 to 1992 the eastern birds were included as an Appendix II species in the Convention on International Trade in Endangered Species of Wild Flora and Fauna (CITES), where the species was listed as *Cygnus bewickii jankowskii*. In 1992 the swan was deleted from CITES Appendix II (which lists species that are not necessarily threatened with extinction, but which may become so unless trade is closely controlled) on the basis that there was little scientific support for separating the western and eastern populations, and because numbers globally had increased since the 1970s. The birds are included in the list of protected birds under China's 1989 Wildlife Protection Law (Li 1996), which gives them some protection from habitat loss and hunting (although, as in other areas, illegal persecution still occurs; Section 8.1.2), and reserves have been established at some wintering sites in China (Ma & Cai 2000; Section 8.4.2, Appendix 3).

There was a major decline and degradation of wintering habitat in Korea during the 1960s following reclamation of coastal and inland wetlands for farm expansion programmes, and for industrial development (Won 1981). This has continued more recently, for instance with the reclamation of wetlands along the west coast (Moores *et al.* 2001) and continued development along the Nakdong Valley. The latter includes construction work (roads, dykes and a pumping station) at the Upo Ramsar site, hinterland conservation at Joonam, and continued reclamation and road-building works at the estuary itself. This has contributed to the decline of several waterbird species (Moores 2002). The swans' current status in Korea is unclear, although wintering numbers have recently been estimated at just 20 individuals in D.P.R. Korea (Pak 1995; Miyabayashi & Mundkur 1999). It is probable that significantly fewer than the 500–1,000 Bewick's Swans estimated for the Republic of

Korea by Miyabayashi & Mundkur (1999) now winter in the country (N. Moores *in litt.*). Migratory swans (both Bewick's and Whooper) have long been revered in Japan, and artificial feeding of the swans in winter is widespread. Drainage of important wintering sites remains a threat, however, and changes in agricultural practice may cause future conflict with rice farmers (Ohmori 1981, Albertsen & Kanazawa 2002).

8.4.2 Site protection

A key measure for ensuring the Bewick's Swans' future viability is to protect key sites in summer and winter, and at staging areas along the flyway. In western Europe, many of the main sites used by the birds are protected or partially protected, although the concentration of the swans into relatively few areas puts them at risk to any changes at these sites or to diseases within the population. Also, not all the habitats currently used by the swans in winter are protected, because there has been a general (possibly erroneous) perception that the farmland that provides much of the food for the birds in winter will continue to be managed without major changes by the farming community.

A review of the sites of international and national importance for Bewick's Swans in Britain and Ireland found that, of the 20 sites identified, only River Lagan at Flatfield had no site protection (Robinson *et al.* 2004). The other 19 have all been classified as Special Protection Areas (SPAs) under the EC Birds Directive, and the same 19 sites are given further international protection by being classified as Ramsar sites under the Convention on Wetlands of International Importance especially as Waterfowl Habitat (the Ramsar Convention). A number of Bewick's Swan sites in the Netherlands, Germany and Denmark are also protected. The Netherlands currently supports 79 SPAs and is continuing to implement the Natura 2000 network (van Roomen *et al.* 2003), but of the 63 sites identified as being important for the species in the Netherlands in recent years, only 20 are recorded as being totally or partially protected (Appendix 2). Further north along the flyway, the swans use a wide range of wetlands in Estonia, and several of these are protected for the swans and other migrating waterbirds including Matsalu Bay (Appendix 2). The important spring staging area at Dvina Bay on the White Sea has only partial national protection, however, and there are no international agreements covering the site (Sviridova 2000).

Efforts to conserve waterbirds at a flyway level are more recent in the Asia–Pacific region, but an atlas of key sites for waterbirds in the eastern Asian flyway has been developed (Miyabayashi & Mundkur 1999). There are gaps in the information provided in the atlas due to a lack of information in many areas, but it still provides a very useful list of sites currently known to be important for the swans and is a useful basis for future monitoring. Most of the key Bewick's Swan sites listed are in Japan, where the species has been monitored for a number of years. Here 26 sites were identified, with a further two sites from the Republic of Korea. Of the two Korean sites (the upper Guem River, previously known as the Kum River, and the Nakdong Estuary), the upper Guem River is no longer internationally important for the species, although more than 900 birds were recorded in the area in 1994 (Miyabayashi & Mundkur 1999; Appendix 3). Just 45 birds were counted there in 1998 (Park 2002), and there have not been any significant numbers reported in more recent winters (N. Moores pers. comm.). Just three sites in China are listed in the atlas (Qinghai Hu, Chongming Island and Poyang Lake NNR), but these seem to be among the key areas for the swans in the country (Ma & Cai 2000). Although Poyang Lake and Bird Island at Qinghai Lake are nature reserves, the important site of Chonming Island in Shanghai is not protected, and there are calls from Chinese environmentalists to save this piece of habitat for the birds (Ma & Cai 2000). Several more areas have been recognised as being important for Bewick's Swans in China since the atlas was published, and a number of these sites are protected, notably in the Shangjin Lake region (BirdLife International 2004b; Appendix 3). It is likely that there are still gaps in information, however, particularly for key staging sites used during migration.

In Japan, supplementary food is provided at nine of the 26 sites of international importance for Bewick's Swans (Albertsen & Kanazawa 2002). Supplementary feeding of swans started in the middle of the 20th century in Japan (the first record was at Lake Hyoko in 1955), to help the birds survive particularly harsh winters (Hatakeyama 1981; Ohmori 1981), and this practice has now spread considerably to many man-made and traditional wintering areas. Watching the swans being fed, mainly on rice and bread, has become a major recreational activity in Japan, and the practice may be sufficiently widespread to have increased the swans' winter survival rates in recent years. This may partly explain the rapid growth in numbers in Japan during the 1980s, although conditions on the breeding grounds and in other parts of the wintering range will also have influenced population trends and winter distribution (Albertsen & Kanazawa 2002).

To date there is only one site protected for the swans on the breeding grounds in the Russian Arctic. This, the Nenetskiy State Nature Reserve (the Nenetskiy *zapovednik* in Russian) was designated as a state nature reserve by the Russian Government in December 1997 following several years of international interest and collaborative research projects undertaken in the region (Chapters 1 and 5). It covers an area of 313,400ha, including 181,900ha of water within the Nenets Autonomous District, in the northeastern part of the Malozemelskaya tundra and the lower reaches of the Pechora River Delta, north of the city of Nar'Yan Mar. Within the Nenetskiy State Nature Reserve, Zakharin Bereg has one of the highest breeding densities of Bewick's Swans in European Russia. Whilst the designation of the Nenetskiy *zapovednik* and the increase in protection of the flora and fauna at the site are major steps towards conserving the swans in their breeding range, other important breeding areas currently remain unprotected or receive only lower levels of protection. Thus the conservation status of Bewick's Swans in the eastern population would be much enhanced if their key breeding grounds, such as on the Lena Delta, the Kolyma lowlands and in Chuckotka, were similarly designated as state nature reserves. Similarly, protection of the swans' migratory site at the mouth of the Amur River would be important for ensuring that the network of sites used by the birds between the wintering grounds and the breeding range are maintained.

8.5 A VIEW OF THE FUTURE

There have been major increases in the Bewick's Swan populations in both the western and eastern parts of the range since the 1970s, which indicates that their conservation status has improved over this time. This is partly attributable to reduced levels of persecution in the second half of the 20th century, the conservation of key sites, and possibly also to more intensive agriculture providing food for the birds on the wintering grounds. Yet it should be noted that, as a long-lived species with low annual recruitment to the population, the swans are likely be vulnerable to population declines if there were to be an increase in adult mortality levels. Moreover, any major changes in climate or habitat in the swans' Arctic breeding areas are likely to impinge on the viability of both the eastern and western populations.

Further monitoring and research programmes will help to identify problems facing the migratory swans and their conservation needs, and to identify reasons for the decline in the northwest European population in recent years. Meanwhile the substantial improvement in international communication about the swans and the mechanisms provided within the international conventions should be used to take forward the protection of key areas and to develop flyway management plans for the species. These beautiful and still mysterious birds must be allowed to continue their annual flights across continents, linking people from the densely populated wintering grounds with those from their remote summer haunts. Hopefully, with so much human goodwill towards the Bewick's Swans, people will work together to ensure their future welfare.

APPENDIX I

Winter weights and measurements of adult Bewick's Swans and Whistling Swans

			BEWICK'S SWAN		WHISTLING SWAN	
			Mean (range)	Sample size	Mean (range)	Sample size
Weight (kg)	Male	Adult	6.6 (4.2–8.6)	1201	7.2 (3.8–10.5)	1447
		Yearling	6.3 (4.3–7.9)	206	6.6 (5.1–8.6)	87
		Cygnet	5.3 (3.7–7.8)	325	6.1 (3.4–8.9)	299
	Female	Adult	5.9 (4.1–8.3)	946	6.3 (4.1–9.0)	1290
		Yearling	5.7 (3.9–8.0)	170	5.8 (4.1–7.0)	57
		Cygnet	5.1 (3.3–7.0)	351	5.6 (3.2–8.0)	403
Wing (mm)	Male	Adult	533 (480–574)	512	538 (501–569)	8
		Yearling	520 (480–563)	128	–	–
		Cygnet	507 (455–557)	143	–	–
	Female	Adult	513 (474–578)	414	532 (505–561)	15
		Yearling	504 (468–563)	93	–	–
		Cygnet	491 (445–558)	168	–	–
Skull (mm)	Male	Adult	163 (148–187)	485	–	–
		Yearling	162 (146–177)	135	–	–
		Cygnet	162 (135–177)	168	–	–
	Female	Adult	158 (142–179)	417	–	–
		Yearling	158 (144–173)	100	–	–
		Cygnet	157 (141–172)	188	–	–
Bill (mm)	Male	Adult	95 (81–110)	112	104 (90–118)	305
		Yearling	91 (80–100)	63	98 (87–107)	27
		Cygnet	76 (65–93)	69	88 (79–104)	34
	Female	Adult	91 (79–102)	112	101 (89–117)	164
		Yearling	89 (76–99)	46	95 (87–105)	13
		Cygnet	76 (64–82)	65	89 (73–104)	38
Tarsus (mm)	Male	Adult	107 (92–120)	386	116 (94–146)	290
		Yearling	108 (87–117)	122	113 (104–127)	27
		Cygnet	107 (94–118)	140	110 (100–122)	33
	Female	Adult	104 (87–116)	325	110 (92–134)	160
		Yearling	103 (86–113)	86	105 (96–113)	10
		Cygnet	104 (92–115)	147	111 (98–126)	37

Notes: Bewick's Swan measurements from Evans & Kear 1978 and Rees *et al.* 1997a. Whistling Swan measurements from Limpert *et al.* 1987 and Limpert & Earnst 1994.

APPENDIX 2

Key staging and wintering sites for the northwest European Bewick's Swan population

Site	Protected status	Season	Max count	Year of max	Most recent average count	Years considered for average count	Source
BELGIUM							
Ijzervallei COORDINATES 51:00N 02:50E	Protected	Winter	195	2004	124	1999–2004	K Devos pers. comm.
Krekengebied Noord-Oost-Vlaanderen COORDINATES 51:16N 03:40E	Protected	Winter	553	2000	260	1999–2004	Anselin *et al.* 2000; K Devos pers. comm.
Oostkustpolders COORDINATES 51:15N 03:13E	Protected	Winter	200	2003	108	1999–2004	K. Devos, E. Kuijken & C. Verscheure pers. comm.
DENMARK							
Bolle and Try meadows COORDINATES 57:07N 10:12E	Protected	Passage	2,475	1994	–	–	Rasmussen *et al.* 2000
Droninglund COORDINATES 57:08N 10:15E	Unknown	Passage	–	–	–	–	Scott & Rose 1996
Fiilsø COORDINATES 55:42N 08:14E	Protected	Passage Winter	– 479	– 1995	–	–	Rasmussen *et al.* 2000; Scott & Rose 1996
Lønnerup Fjord COORDINATES 57:00N 08:47E	Protected	Passage	316	1994	–	–	Rasmussen *et al.* 2000
Nissum Fjord COORDINATES 56:21N 08:14E	Protected	Passage	320	1989	–	–	Rasmussen *et al.* 2000
Nors So and Vendlet So COORDINATES 57:02N 08:35E	Unknown	Passage	–	–	–	–	Scott & Rose 1996

Appendix 2. (*continued*)

Ringkøbing Fjord COORDINATES 56:00N 08:15E	Protected	Passage	1,091	1994	–	–	Rasmussen *et al.* 2000
Roskilde Fjord, Selsø and Kattinge Søerne COORDINATES 55:45N 12:05E	Protected	Passage	300	1993	–	–	Rasmussen *et al.* 2000
Skjern Å COORDINATES 56:56N 08:30E	Protected	Passage	700	–	–	–	Rasmussen *et al.* 2000; Scott & Rose 1996
Stadil Fjord and Vest Stadil Fjord COORDINATES 56:11N 08:09E	Protected	Passage	1,000	1995	–	–	Rasmussen *et al.* 2000; Scott & Rose 1996
Store Vildmose, Ryå and Stavad Enge COORDINATES 57:13N 09:50E	Not protected	Passage	1,179	1994	–	–	Rasmussen *et al.* 2000
Tipperne COORDINATES 55:53N 08:14E	Protected	Passage Winter	– 387	– 1992	– 188	– 1984–93	Scott & Rose 1996
Tissø, Lille Åmose and Hallenslev Mose COORDINATES 55:35N 11:20E	Protected	Passage	475	1996	–	–	Rasmussen *et al.* 2000
Tøndermarsken, Magisterkog and Rudbøl Sø COORDINATES 54:54N 08:43E	Protected	Passage Winter	332 304	1993 1981	– 6	– 1984–93	Rasmussen *et al.* 2000; Scott & Rose 1996
Tylstrup COORDINATES 57:10N 09:58E	Unknown	Passage	–	–	–	–	Scott & Rose 1996
Ulvedybet and Nibe Bredning COORDINATES 57:02N 09:35E	Protected	Passage	4,320	1994	–	–	Rasmussen *et al.* 2000; Scott & Rose 1996
Vejlerne (east) COORDINATES 57:03N 09:00E	Protected	Winter	377	1994	–	–	Rasmussen *et al.* 2000
Vejlerne (west), Arup Holm and Hovsør Røn COORDINATES 56:58N 08:52E	Protected	Passage	402	1994	–	–	Rasmussen *et al.* 2000; Scott & Rose 1996

Appendix 2. (*continued*)

ESTONIA

Alam-Pedja wetland complex Coordinates 58:30N 26:10E	Protected	Passage	600	1996	–	1996	Kalamees 2000
Assamalla Coordinates 59:14N 26:16E	Not protected	Spring	200	1993	200	1984–93	Scott & Rose 1996
Audru Polder Coordinates 58:24N 24:21E	Not protected	Spring	17,500	1996	–	1984–96	Scott & Rose 1996; Kalamees 2000
Emajõe Suursoo & Piirissaar Island (Lake Peipsi) Coordinates 58:24N 27:15E	Partially protected	Autumn	800	1994	<200	1984–99	Scott & Rose 1996; Kalamees 2000
Eru Bay Coordinates 59:35N 25:49E	Protected	Passage	<400	1990s	>200	1990–99	Kalamees 2000
Häädemeeste-Võiste coast Coordinates 58:08N 24:29E	Not protected	Spring	>700	1990s	–	1984–99	Scott & Rose 1996; Kalamees 2000
Haapsalu-Noaroosti Bays Coordinates 58:59N 23:35E	Not protected	Passage	<4,300	1990s	>300	1984–99	Scott & Rose 1996; Kalamees 2000
Halvati Coordinates 58:41N 23:53E	Not protected	Passage	460	1993	–	1984–93	Scott & Rose 1996
Hullo and Sviby Bays (Vormsi Island) Coordinates 58:58N 23:16E	Unknown	Passage	<400	–	–	1991–99	Kalamees 2000
Irbe Strait (between Saaremaa Island & Latvia) Coordinates 58:48N 22:47E	Unknown	Passage	300	1990s	>200	1991–99	Kalamees 2000
Ihamaa Coordinates 58:25N 26:28E	Protected	Spring	600	1993	–	1984–93	Scott & Rose 1996
Kahtla-Kübassaare coast (Saaremaa Island) Coordinates 58:26N 23:10E	Not protected	Passage	<500	1990s	>200	1991–99	Kalamees 2000

Appendix 2. (*continued*)

Käina Bay (Hiiumaa Island) COORDINATES 58:48N 22:47E	Protected	Passage	230	1990s	–	1991–99	Kalamees 2000
Kallaste-Kodavere coast (Lake Peipsi) COORDINATES 58:40N 27:10E	Not protected	Autumn	1,000	1996	>200	1984–96	Scott & Rose 1996; Kalamees 2000
Kalli COORDINATES 58:32N 24:01E	Protected	Spring	1,750	1995	200	1984–93	Scott & Rose 1996
Kidise COORDINATES 58:29N 23:57E	Not protected	Spring	700	1995	–	1984–93	Scott & Rose 1996
Kilksama COORDINATES 58:27N 24:33E	Not protected	Spring	800	1994	300	1984–93	Scott & Rose 1996
Küdema Bay COORDINATES 58:32N 22:16E	Partly protected	Passage	400	1990s	–	1995–99	Kalamees 2000
Kuresoo/Kaansoo (Soomaa mire complex) COORDINATES 58:25N 25:05E	Protected	Passage	2,000	1996	200	1984–96	Scott & Rose 1996; Kalamees 2000
Lahepera Lake (Lake Peipsi) COORDINATES 58:34N 27:12E	Protected	Autumn	1,000	1996	>250	1984–96	Scott & Rose 1996; Kalamees 2000
Lao-Lui coast/Pärnu Bay COORDINATES 58:12N 24:10E	Partially protected	Passage	<10,000	1996	1,000	1984–99	Scott & Rose 1996; Kalamees 2000
Lemmetsa COORDINATES 58:27N 24:25E	Not protected	Autumn	260	1993	–	1984–93	Scott & Rose 1996
Matsalu Bay COORDINATES 58:45N 23:40E	Protected	Passage	14,500	1994	8,500	1984–99	Scott & Rose 1996; Luigujõe *et al.* 1996; Kalamees 2000
Mustvee-Omedu coast (Lake Peipsi) COORDINATES 58:48N 26:58E	Not protected	Spring	6,000	1990s	>2,000	1984–98	Scott & Rose 1996; Kalamees 2000

Appendix 2. (*continued*)

Nästi-Võllur Bog COORDINATES 58:30N 24:00E	Partially protected	Spring	600	1996	–	1996	Scott & Rose 1996; Kalamees 2000
Omedu (Lake Peipsi) COORDINATES 58:46N 27:02E	Not protected	Autumn	420	1993	200	1984–93	Scott & Rose 1996
Paldiski Bay COORDINATES 58:20N 24:00E	Not protected	Spring	>300	1990s	200	1984–99	Scott & Rose 1996; Kalamees 2000
Petaaluse COORDINATES 58:40N 23:43E	Not protected	Spring	260	1993	–	1984–93	Scott & Rose 1996
Rajaküla COORDINATES 59:14N 26:16E	Not protected	Autumn	210	1993	150	1984–93	Scott & Rose 1996
Tagamõisa Peninsula COORDINATES 58:29N 21:56E	Protected	Passage	240	1996	–	1996	Kalamees 2000
Vaida COORDINATES 59:16N 24:58E	Not protected	Spring	400	1994	200	1984–93	Scott & Rose 1996
Väike Väin Strait COORDINATES 58:32N 23:11E	Protected	Passage	<3,000	1990s	>500	1984–99	Scott & Rose 1996; Kalamees 2000
Varnja-Kolkja (Lake Peipsi) COORDINATES 58:30N 27:14E	Not protected	Autumn	1,000	1993	500	1984–93	Scott & Rose 1996
Vihtra COORDINATES 58:34N 25:02E	Not protected	Spring	1,200	1994	500	1984–93	Scott & Rose 1996
FINLAND							
Kirkon-Vilkkilantura COORDINATES 60:32N 27:45E	Unknown	Spring	200	–	–	1984–93	Scott & Rose 1996
GERMANY							
Aland–Elbe lowlands COORDINATES 53:00N 11:37E	Protected	Spring	600	1995	–	–	Unselt *et al.* 2000

Appendix 2. (*continued*)

Aller lowlands (from Buchholz to Etelsen) COORDINATES 52:47N 09:24E	Partially protected	Winter Spring	417 400	– –	– –	1984–93 1990–99	Scott & Rose 1996 Melter & Schreiber 2000
Lowlands near Bramel COORDINATES 53:34N 08:43E	Partially protected	Spring	410	–	–	1990–99	Melter & Schreiber 2000
Lowlands of Dosse and Jäglitz COORDINATES 52:49N 12:20E	Partially protected	Spring	–	–	300	1990–99	ABBO 2003
Lowlands of the Rivers Eider, Treene and Sorge COORDINATES 54:23N 09:20E	Partially protected	Spring	2,083	2001	–	1977–2004	O. Ekelöf pers. comm.
River Ems lowlands near Laten COORDINATES 52:52N 07:18E	Not protected	Winter	716	1997	–	–	Unselt *et al.* 2000
Ems lowlands (from Lingen to Papenburg) COORDINATES 53:02N 07:16E	Partially protected	Winter Spring	180 2,310	– 1999	– –	1984–93 1990–2004	Scott & Rose 1996 A. Degen pers. comm.
Elbe from Stade to Otterndorf (lower tidal reaches of the River Elbe) COORDINATES 53:51N 09:10E	Partially protected	Winter Spring	1,627 1,870 (4,460)	1990 –	– –	– 1990–99 (1970–89)	Scott & Rose 1996; Unselt *et al.* 2000 Melter & Schreiber 2000
Elbe, Bishorst to River Krückau COORDINATES 53:38N 09:32E	Partially protected	Spring	723	–	–	1984–93	Scott & Rose 1996
Elbe, Haseldorfer Marsh COORDINATES 53:36N 09:35E	Partially protected	Spring	348	–	–	1984–93	Scott & Rose 1996
Elbe, Wedeler Marsh COORDINATES 53:37N 09:40E	Partially protected	Spring	954	–	–	1984–93	Scott & Rose 1996
Elbe from Schnackenburg to Hohnstorf (Niedersachsen) COORDINATES 53:10N 11:09E	Partially protected	Spring	1,476	–	–	1990–2004	A. Degen pers. comm.

Appendix 2. (*continued*)

Elbe from Polz to Boizenburg (Mecklenburg-Vorpommern) COORDINATES 53:09N 11:15E	Partially protected	Spring	–	–	950	1990–99	Scheller *et al.* 2002
Elbe from Werben to Gaarz (Brandenburg) COORDINATES 52:45N 12:53E	Partially protected	Spring	–	–	>2,000	1990–99	ABBO 2003
Elbniederung (Elbe floodplain) from Schnackenburg to Lauenburg including Amt Neuhaus COORDINATES 53:18N 10:45E	Partially protected	Winter	2,155	1990	–	–	Scott & Rose 1996; Unselt *et al.* 2000
Greifswalder Bodden COORDINATES 53:13N 13:31E	Partially protected	Spring	–	–	3,500	1990–99	Scheller *et al.* 2002
Groß Fullener marsh COORDINATES 52:40N 07:08E	Not protected	Winter	441	1997	–	–	Unselt *et al.* 2000
Haaler-Au COORDINATES 54:10N 09:30E	Partially protected	Spring	2,572	2001	–	1977–2004	O. Ekelöf pers. comm.
Hamme lowlands COORDINATES 53:16N 08:55E	Partially protected	Spring	273	–	–	1990–99	Melter & Schreiber 2000
Lowlands of Havel, Lake Schollene and Lake Gülpe COORDINATES 52:45N 12:16E	Partially protected	Winter Spring	450 –	1997 –	– 450	1984–97 1990–99	Scott & Rose 1996; Unselt *et al.* 2000 ABBO 2003
Hunte lowlands COORDINATES 53:10N 08:18E	Partially protected	Spring	255	–	–	1990–99	Melter & Schreiber 2000
Huvenhoopsmoor and Breddorfer Wiesen COORDINATES 53:20N 09:05E	Partially protected	Spring	390	–	–	1990–99	Melter & Schreiber 2000
Island Usedom COORDINATES 53:56N 14:01E	Partially protected	Spring	1,300	2001	–		D. Sellin pers. comm

Appendix 2. (*continued*)

Kudensee Coordinates 53:57N 09:12E	Partially protected	Spring	1,046	1993	–	1977–2004	O. Ekelöf pers. comm.
Kummerower See Coordinates 53:49N 12:50E	Partially protected	Spring	220	–	–	1984–93	Scott & Rose 1996
Lewitz fish-ponds Coordinates 53:25N 11:37E	Partially protected	Spring	–	–	1,600	1990–99	Scheller *et al.* 2002
Mildenitz, Lake Gägelow and Lake Upahl Coordinates 53:41N 11:57E	Partially protected	Spring	–	–	600	1990–99	Scheller *et al.* 2002
North of Nordfriesland Coordinates 54:48N 08:44E	Partially protected	Spring	1,000	1998		1977–2004	O. Ekelöf pers. comm.
Obere and Mittlere Warnow with Glöwe and Mildenitz Coordinates 53:41N 11:39E	Partially protected	Spring	–	–	245	1990–99	Scheller *et al.* 2002
Ostküste Oldenburgs Coordinates 54:20N 11:10E	Partially protected	Spring	402	–	–	1984–93	Scott & Rose 1996
Lowlands of the River Peene Coordinates 53:51N 13:47E	Partially protected	Spring	–	–	400	1990–99	Scheller *et al.* 2002
Lake Putzar and Lake Galenbeck Coordinates 53:39N 13:45E	Partially protected	Spring	–	–	170	1990–99	Scheller *et al.* 2002
Lowlands of Recknitz and Trebel Coordinates 54:03N 12:42E	Partially protected	Spring	–	–	800	1990–99	Scheller *et al.* 2002
Rheiderland Coordinates 53:14N 07:21E	Partially protected	Spring	176	–	–	1990–99	Melter & Schreiber 2000
Lake Schwerin, Lake Dambeck and Lake Neukloster-Warin Coordinates 53:47N 11:37E	Partially protected	Spring	–	–	400	1990–99	Scheller *et al.* 2002
Unterer Niederrhein (Lower Rhine area) Coordinates 51:46N 06:24E	Partially protected	Spring	400	–	–	1984–97	Scott & Rose 1996; Unselt *et al.* 2000

Appendix 2. (*continued*)

Unteres Rhinluch, Lake Dreetz COORDINATES 52:46N 12:28E	Partially protected	Spring	–	–	250	1990–99	ABBO 2003
Unterweser (Lower Weser) COORDINATES 53:21N 08:30E	Partially protected	Winter Spring	360 520	1992 –	– –	1984–93 1990–99	Scott & Rose 1996; Unselt *et al.* 2000 Melter & Schreiber 2000
Vorpommersche Bodden COORDINATES 54:26N 12:54E	Partially protected	Spring	–	–	600	1990–99	Scheller *et al.* 2002
Wadden Sea (Niedersachsen) COORDINATES 53:41N 07:31E	Protected	Winter Spring	317 2,587	– –	– –	1984–93 1990–99	Scott & Rose 1996 Melter & Schreiber 2000
Wesuwer Brook COORDINATES 52:46N 07:14E	Not protected	Winter	264	1997	–	–	Unselt *et al.* 2000
Warder See COORDINATES 53:58N 10:25E	Partially protected	Passage	300	1997	–	–	Unselt *et al.* 2000
Wismarbucht and Salzhaff COORDINATES 54:01N 11:26E	Partially protected	Winter Spring	500 –	– –	– 300	1984–93 1990–99	Scott & Rose 1996 Scheller *et al.* 2002
Wümme lowlands/St. Jürgensland COORDINATES 53:10N 08:50E	Partially protected	Spring	240	–	–	1990–99	Melter & Schreiber 2000
Wümme lowlands/Fischerhude COORDINATES 53:07N 09:00E	Partially protected	Spring	265	–	–	1990–99	Melter & Schreiber 2000
REPUBLIC OF IRELAND							
Ballycotton to Shanagarry COORDINATES 51:49N 08:02W	Partially protected	Winter	194	1991	28	1994–97	Scott & Rose 1996; Colhoun 2000
Barrow upper floodplain: Monasterevin to Athy COORDINATES 53:05N 07:03W	Not protected	Winter	191	1986	191	1984–93	Hunt *et al.* 2000; Scott & Rose 1996
The Cull & Killag COORDINATES 52:12N 06:39W	Protected	Winter	555	1996	195	1995–2000	Colhoun 2000; Robinson*et al.* 2004; Scott & Rose 1996

Appendix 2. (*continued*)

Lady's Island Lake COORDINATES 52:11N 06:24W	Partially protected	Winter	268	1972	12	1984–93	Scott & Rose 1996
Rahasane Turlough COORDINATES 53:13N 08:47W	Protected	Winter	292	1977	39	1984–93	Scott & Rose 1996
River Little Brosna callows: New Bridge to River Shannon COORDINATES 53:08N 08:03W	Protected	Winter	250	–	–	–	Hunt *et al.* 2000
River Suck callows: Shannon Bridge to Castlecoote COORDINATES 53:24N 08:10W	Partially protected	Winter	180	1982	–	–	Hunt *et al.* 2000
Tacumshin Lake COORDINATES 52:11N 06:29W	Protected	Winter	472	1997	235	1995–2000	Colhoun 2000; Robinson *et al.* 2004; Scott & Rose 1996
Wexford Harbour & Slobs COORDINATES 52:19N 06:26W	Protected	Winter	550	1976	191	1995–2000	Colhoun 2000; Robinson *et al.* 2004; Scott & Rose 1996
LATVIA							
Babite Lake COORDINATES 56:55N 23:45E	Partially protected	Passage	–	–	200	1984–93	Scott & Rose 1996
Lubana Lowlands COORDINATES 56:50N 26:52E	Partially protected	Passage	–	–	250	1984–93	Scott & Rose 1996
LITHUANIA							
Kursiu Lagoon and Nemunas River Delta COORDINATES 55:21N 21:12E	Partially protected	Passage	3,300	1992	570	1984–93	Scott & Rose 1996; Švažas *et al.* 1999; Švažas 2001; Raudonikis 2004
NETHERLANDS							
Alblasserwaard COORDINATES 51:53N 04:50E	Not protected	Winter	979	1996	410	1985–2001	Scott & Rose 1996; SOVON 1997; van Roomen *et al.* 2003

Appendix 2. (*continued*)

Ankeveense plassen COORDINATES 52:18N 05:06E	Partially protected	Winter	420	1991	279	1984–93	Scott & Rose 1996
Arkemheen, Putterpolder en randmeren COORDINATES 52:14'N 5:26'E	Partially protected	Winter	2,808	1991	1,200	1985–2001	Osieck 2000; van Roomen *et al.* 2003
Biebosch COORDINATES 51:45N 04:46E	Protected	Winter	487	1992	230	1986–2000	Scott & Rose 1996; Osieck 2000; van Roomen *et al.* 2002
Brabantse-/Gelderse Maas COORDINATES 51:49N 05:34E	Not protected	Winter	2,486	1994	916	1984–93	Scott & Rose 1996; Osieck 2000
Drontermeer e.o.: Kamperveen, Polder Oosterwolde COORDINATES 52:30N 05:51E	Not protected	Winter	1,829	1994	640	1985–2001	Scott & Rose 1996; Osieck 2000; van Roomen *et al.* 2003
Eemmeer, Nijkerkernauw, Nuldernauw e.o. COORDINATES 52:01N 05:20E	Partially protected	Winter	5,156	1993	1,800	1985–2001	Scott & Rose 1996; Osieck 2000; van Roomen *et al.* 2003
Friese noordkust COORDINATES 53:22N 05:46E	Reserve	Autumn	1,800	1993	826	1984–93	Scott & Rose 1996
Gaasterland COORDINATES 52:30N 05:51E	Not protected	Winter	247	1990	116	1984–93	Scott & Rose 1996
Gelderse Poort COORDINATES 51:53N 06:02E	Partially protected	Winter	181	1991	91	1984–93	Scott & Rose 1996; Osieck 2000
Gooimeer COORDINATES 52:19N 05:14E	Not protected	Autumn	812	1989	284	1984–93	Scott & Rose 1996
Grevelingen COORDINATES 51:46N 03:59E	Not protected	Winter	472	1990	314	1984–93	Scott & Rose 1996; Osieck 2000
Gronings-Drentse Veekoloniën COORDINATES 53:03N 06:28E	Not protected	Winter	851	1996	430	1985–2001	Scott & Rose 1996; SOVON 1997; van Roomen *et al.* 2003
Groningse Noordkust COORDINATES 53:27N 06:31E	Protected	Autumn	1,942	1993	450	1985–2001	Scott & Rose 1996; van Roomen *et al.* 2003

Appendix 2. (*continued*)

Het Bildt COORDINATES 53:11N 05:24E	Not protected	Autumn	1,410	900	900	1985–2001	van Roomen *et al.* 2003; K. Koffijberg pers. comm.
Het Binnenveld COORDINATES 52:02N 05:36E	Not protected	Winter	488	1991	170	1985–2000	Scott & Rose 1996; van Roomen *et al.* 2002
IJssel: Zwolle-Deventer COORDINATES 52:18N 06:06E	Partially protected	Winter	1,966	1992	480	1985–2001	Scott & Rose 1996; Osieck 2000; van Roomen *et al.* 2003
Krimpenerwaard COORDINATES 51:57N 04:42E	Partially protected	Winter	496	1996	240	1985–2001	Scott & Rose 1996; SOVON 1998; van Roomen *et al.* 2003
Lauwersemeer e.o. COORDINATES 53:21N 06:13E	Partially protected	Autumn Winter	6,461 514	1988 1999	1,600 356	1985–2001 1984–93	Scott & Rose 1996; Osieck 2000; SOVON 2000; Koffijberg *et al.* 1997; van Roomen *et al.* 2003
Lek COORDINATES 51:58N 05:00E	Not protected	Winter	898	1994	490	1984–93	Scott & Rose 1996
Lepelaarplassen e.o. COORDINATES 52:23N 05:12E	Partially protected	Autumn Winter	709 516	1992 1992	274 358	1984–93	Scott & Rose 1996; Osieck 2000
Limburgse Maas COORDINATES 51:22N 06:10E	Not protected	Winter	453	1994	260	1984–93	Scott & Rose 1996
Lopikerwaard COORDINATES 51:35N 04:33E	Not protected	Winter	541	1985–94	190	1985–2001	Koffijberg *et al.* 1997; van Roomen *et al.* 2003; K. Koffijberg pers. comm.
Maasland Den Bosch-Oss COORDINATES 51:47N 05:32E	Not protected	Winter	1,160	1996	540	1985–2001	Scott & Rose 1996; SOVON 1997; van Roomen *et al.* 2002, 2003
Maas Well-Mook COORDINATES 51:26N 05:34E	Not protected	Winter	232	1985–94	260	1985–2001	Koffijberg *et al.* 1997; van Roomen *et al.* 2003; K. Koffijberg pers. comm.
Midden Delfland en Oude Leede COORDINATES 51:58N 04:21E	Partially protected	Winter	515	1992	200	1985–2000	Scott & Rose 1996; van Roomen *et al.* 2002
Nederrijn COORDINATES 51:56N 05:38E	Not protected	Winter	458	1993	170	1985–2000	Scott & Rose 1996; van Roomen *et al.* 2002

Appendix 2. (*continued*)

Noord-Beveland COORDINATES 51:35N 03:47E	Not protected	Autumn	1,306	1998	730	1985–2001	Scott & Rose 1996; SOVON 2000; van Roomen *et al.* 2003
Noord-Friesland buitendijks COORDINATES 53:12N 05:27E	Partially protected	Winter	167	2001	230	1985–2001	van Roomen *et al.* 2002, 2003; K. Koffijberg pers. comm.
Noordoostpolder-west COORDINATES 52:44N 05:51E	Not protected	Autumn Winter	2,218 4,733	1996 1991	640 1,000	1985–2001 1985–2000	Scott & Rose 1996; Koffijberg *et al.* 1997; SOVON 1998, 1999; van Roomen *et al.* 2002, 2003
NW Overijssel COORDINATES 52:23N 06:02E	Partially protected	Winter	933	1985–94	410	1985–2001	Koffijberg *et al.* 1997; van Roomen *et al.* 2002, 2003; K. Koffijberg pers. comm.
Oost-en-Westsdongerdeel COORDINATES 53:21N 05:56E	Not protected	Autumn	1,191	2001	510	1984–93	Scott & Rose 1996; van Roomen *et al.* 2003
Oosterschelde COORDINATES 51:35N 03:56E	Reserve	Autumn	653	1991	149	1984–93	Scott & Rose 1996
Oost-Flevoland-noord COORDINATES 52:32N 05:44E	Not protected	Autumn Winter	2,448 2,658	2001 1990	1,400 1,758	1985–2001 1984–93	Scott & Rose 1996; van Roomen *et al.* 2003
Oost-Flevoland-zuid COORDINATES 52:28N 05:36E	Not protected	Autumn Winter	1,749 1,236	1992 1993	475 170	1984–93 1985–2000	Scott & Rose 1996; van Roomen *et al.* 2002
Oudenbosch-Made Polders COORDINATES 51:23N 04:20E	Not protected	Winter	561	1996	290	1985–2001	SOVON 1998; van Roomen *et al.* 2002, 2003; K. Koffijberg pers. comm.
Overflakkee COORDINATES 51:44N 04:12E	Not protected	Winter	793	1993	170	1985–2001	Scott & Rose 1996; van Roomen *et al.* 2003
Polder Mastenbroek COORDINATES 52:35N 06:02E	Not protected	Winter	638	2001	190	1985–2000	Scott & Rose 1996; van Roomen *et al.* 2002
Polder Zeevang COORDINATES 52:33N 05:01E	Not protected	Winter	175	1992	122	1984–93	Scott & Rose 1996
Polders Zoetermeer-Alphen aan de Rijn COORDINATES 52:06N 04:30E	Not protected	Winter	798	1994	280	1985–2000	Scott & Rose 1996; van Roomen *et al.* 2002

Appendix 2. (*continued*)

Polders rond Steenbergen Coordinates 51:50N 04:19E	Not protected	Winter	404	1993	180	1985–2000	Scott & Rose 1996; van Roomen *et al.* 2002
Raamsdonk, Heusden e.o. Coordinates 51:42N 05:04E	Not protected	Winter	880	1991	160	1985–2000	Scott & Rose 1996; van Roomen *et al.* 2002
Reeuwijkse Plassen e.o. Coordinates 52:02N 04:46E	Not protected	Winter	1,079	1992	380	1985–2001	Scott & Rose 1996; van Roomen *et al.* 2003
Reitdiepdal Coordinates 53:20N 06:23E	Not protected	Winter	423	2000	180	1988–2000	Scott & Rose 1996; van Roomen *et al.* 2003
Rottige Meenthe Coordinates 52:51N 05:53E	Partially protected	Winter	219	1991	77	1984–93	Scott & Rose 1996
Schouwen-Duiveland Coordinates 51:42N 03:55E	Not protected	Autumn Winter	1,159 1,044	1992 1998	714 770	1984–93 1984–2000	Scott & Rose 1996; SOVON 2000; van Roomen *et al.* 2002, 2003
Staphorsterveld en Haerster- en Gennerbroek Coordinates 52:36N 06:08E	Not protected	Winter	991	1994	380	1985–2001	Scott & Rose 1996; van Roomen *et al.* 2003
Texel Coordinates 53:06N 04:51E	Partially protected	Winter	1,039	1996	740	1985–2001	Scott & Rose 1996; SOVON 1998; Osieck 2000; van Roomen *et al.* 2003
Tjeukemeer e.o. Coordinates 52:54N 05:49E	Not protected	Winter	204	1993	54	1984–93	Scott & Rose 1996
Veerse Meer Coordinates 51:32N 03:43E	Partially protected	Winter	395	1993	156	1984–93	Scott & Rose 1996; Osieck 2000
Veluwemeer e.o.: Polders Harderwijk-Elburg Coordinates 52:24N 05:43E	Not protected	Autumn Winter	3,184 3,232	1990 1997	1,940 1,700	1984–93 1986–2001	Scott & Rose 1996; SOVON 1999; Osieck 2000; van Roomen *et al.* 2003
Vughtse Gement Coordinates 51:42N 05:14E	Not protected	Winter	1,160	1995	610	1985–2001	Scott & Rose 1996; SOVON 1998; van Roomen *et al.* 2002, 2003

Appendix 2. (*continued*)

Waal COORDINATES 51:50N 05:23E	Not protected	Winter	682	1992	338	1984–93	Scott & Rose 1996
Walcheren COORDINATES 51:32N 03:36E	Not protected	Winter	334	1995	145	1984–93	Scott & Rose 1996; SOVON 1998
Weeribben e.o. COORDINATES 52:47N 05:56E	Partially protected	Autumn Winter	447 933	1993 1989	248 563	1984–93	Scott & Rose 1996
West-Zeeuwsch Vlaanderen COORDINATES 51:20N 03:49E	Not protected	Winter	770	1992	270	1985–2000	Scott & Rose 1996; van Roomen *et al.* 2002
Wieden COORDINATES 52:40N 06:03E	Partially protected	Autumn Winter	531 343	1990 1994	285 218	1984–93	Scott & Rose 1996
Wieringermeer COORDINATES 52:51N 04:55E	Not protected	Autumn	2,115	2001	1,505	1976–2003	Scott & Rose 1996; W. Tijsen pers. comm.
Wonseradeel en Workum COORDINATES 52:35N 05:16E	Not protected	Winter	1,557	1997/98	380	1985–2001	Scott & Rose 1996; SOVON 2001; van Roomen *et al.* 2003
Zuidlaardermeer e.o. COORDINATES 53:09N 06:41E	Partially protected	Autumn Winter	326 498	1990 1991	174 120	1984–93 1985–2000	Scott & Rose 1996; van Roomen *et al.* 2002
Zuid-Beveland-west COORDINATES 51:29N 03:51E	Not protected	Winter	200	1993	100	1985–2000	Scott & Rose 1996; van Roomen *et al.* 2003
Zwarte Meer e.o. COORDINATES 52:37N 05:56E	Partially protected	Winter	285	1991	172	1984–93	Scott & Rose 1996; Osieck 2000
POLAND							
Ostrówek and Smolgulec Ponds COORDINATES 53:05N 17:20E	Not protected	Passage	190	1993	–	–	Gromadzki & Wieloch 2000
Ślesin and Występ Ponds COORDINATES 53:08N 17:42E	Not protected	Passage	429	1995	–	–	Gromadzki & Wieloch 2000

Appendix 2. (*continued*)

RUSSIA							
Gulf of Finland, south coast, between Bolshaya Izhora and Sosnovy Bor COORDINATES 60:00N 29:15E	Protected	Spring	4,000	1996	–	1990s	Krivenko 1999; Sviridova 2000
Gulf of Finland: Koporski Bay COORDINATES 59:47N 28:55E	Not protected	Passage	500	1997	–	–	Sviridova 2000
Gulf of Finland: Vyborg Bay (Berezovye Islands) COORDINATES 60:18N 29:00E	Protected	Spring	5,000	1996	–	1980s–1990s	Krivenko 1999; Scott & Rose 1996; Sviridova 2000
Kurgalski Peninsula COORDINATES 59:38N 28:09E	Protected	Passage	1,000s	1993	–	–	Sviridova 2000
Mshinskaya wetland system COORDINATES 58:51-59:15N 30:01-28E	Partially protected	Passage	–	–	–	1990s	Krivenko 1999
Pechora Delta/Korovinskaiya Bay COORDINATES 68:15N 53:45E	Partially protected	Autumn	15,000	1992	–	–	Beekman *et al.* 1994a, 1996b; Scott & Rose 1996
Pskov-Chudskoye Lake COORDINATES 58:48N 27:48E	Protected	Autumn	–	–	–	1956–88	Borisov *et al.* 1999; Krivenko 1999
Svir Delta, Lake Ladoga COORDINATES 60:23-47N 32:15-33:37E	Partially protected	Spring	1,000	–	–	1990s	Krivenko 1999
White Sea, Dvina Bay COORDINATES 64:51N 40:17E	Partially protected	Spring	17,972* (5,000)	1996 (1996)	–	1995–96	Nolet & Drent 1998; Nolet *et al.* 2001b
SWEDEN							
Gotland Island: coastal areas COORDINATES 57:24N 18:42E	Protected	Passage	1,000	–	–	–	Lindell *et al.* 2000
UNITED KINGDOM							
Arun Valley COORDINATES 50:55N 00:31W	Partially protected	Winter	98	1998	73	1997–2001	Pollitt *et al.* 2003; Robinson *et al.* 2004

Appendix 2. (*continued*)

Avon Valley COORDINATES 50:47N 01:48W	Protected	Winter	310	1987	85	1996–2000	Scott & Rose 1996; Robinson *et al.* 2004
Breydon Water & Berney Marshes COORDINATES 52:35N 01:39E	Protected	Winter	752	1996	360	1995–2000	Pollitt *et al.* 2003; Robinson *et al.* 2004; Scott & Rose 1996
Coombe Hill Canal COORDINATES 51:46N 02:10W	Protected	Winter	260	1986	36	1994–99	Pollitt *et al.* 2003; Rees 1988; Robinson *et al.* 2004
Derwent Ings COORDINATES 53:51N 00:56W	Protected	Winter	187	1980	59	1984–93	Scott & Rose 1996
Durleigh Reservoir COORDINATES 51:07N 03:02W	Protected	Winter	252	1982	–	1984–93	Scott & Rose 1996
Hickling Broad COORDINATES 52:44N 01:34E	Protected	Winter	202	1986	–	1984–93	Scott & Rose 1996
Lough Foyle COORDINATES 55:06N 07:07W	Protected	Winter	470	1983	57	1995–2000	Pollitt *et al.* 2003; Robinson *et al.* 2004; Scott & Rose 1996
Loughs Neagh/Beg COORDINATES 54:37N 06:24W	Protected	Winter	523	1991	69	1995–2000	Pollitt *et al.* 2003; Robinson *et al.* 2004; Scott & Rose 1996
River Lagan, Flatfield COORDINATES 54:29N 06:10W	Not protected	Winter	115	1990	53	1995–2000	Robinson *et al.* 2004
Ludham How Hill COORDINATES 52:40N 00:18W	Unknown	Winter	185	1989	72	1984–93	Scott & Rose 1996
Martin Mere/Ribble Estuary COORDINATES 53:37N 02:52W	Protected	Winter	1,046	1990	346	1995–2000	Pollitt *et al.* 2003; Robinson *et al.* 2004; Scott & Rose 1996
Medway Estuary and Marshes COORDINATES 51:23N 00:37W	Partially protected	Winter	302	1999	172	1999–2000	Robinson *et al.* 2004; Cranswick *et al.* 2005

Appendix 2. (*continued*)

Nene Washes Coordinates 52:33N 00:13W	Protected	Winter	2,585	1998	1,327	1995–2000	Pollitt *et al.* 2003; Robinson *et al.* 2004; Scott & Rose 1996
Ouse Washes Coordinates 52:21N 00:02E	Protected	Winter	7,491	2005	4,988	1995–2000	Pollitt *et al.* 2003; Robinson *et al.* 2004; Scott & Rose 1996; D. Stevens pers. comm.
Severn Estuary Coordinates 51:32N 02:44W	Protected	Winter	610	1979	364	1995–2000	Pollitt *et al.* 2003; Robinson *et al.* 2004; Scott & Rose 1996
Somerset Levels Coordinates 51:01N 02:55W	Protected	Winter	452	1982	187	1995–2000	Pollitt *et al.* 2003; Robinson *et al.* 2004; Scott & Rose 1996
St Benets Levels, Ludham Coordinates 52:42N 01:31E	Protected	Winter	550	1997	303	1995–2000	Pollitt *et al.* 2003; Robinson *et al.* 2004; Scott & Rose 1996
Walland Marsh & Dungeness Gravel Pits Coordinates 50:58N 00:50W	Protected	Winter	325	1997	255	1995–2000	Pollitt *et al.* 2003; Robinson *et al.* 2004; Scott & Rose 1996
The Wash Coordinates 52:50N 00:15W	Protected	Winter	272	–	–	1984–2000	Scott & Rose 1996; Pollitt *et al.* 2003
Walmore Common Coordinates 51:35N 02:45W	Protected	Winter	380	1982	78	1995–2000	Rees 1988; Pollitt *et al.* 2003; Robinson *et al.* 2004

Notes: * indicates number of swans flying into the White Sea area, Russia, rather than a ground count.

APPENDIX 3

Key staging and wintering sites for Bewick's Swans in the eastern population

Site	Protected status	Season	Max count	Year of max	Average count	Years considered	Source
CHINA							
Baidang Lake, Anhui Province COORDINATES 30:47.3N 117:22.8E	Protected	Winter	4,400	2004	–	–	Barter *et al.* 2004
Baishazhou, Shangrao City, Jiangxi COORDINATES 29:09.6N 116:36.9E	Not protected	Winter	2,600	2005	–	–	M. Barter *in litt.*
Bei Shen Lake, Nanchang City, Jiangxi COORDINATES 28:57N 116:20.7E	Not protected	Winter	5,000	2005	–	–	M. Barter *in litt.*
Can Dou Lake, Poyang Lake NNR, Jiangxi COORDINATES 29:04.6N 116:03.7E	Protected	Winter	1,620	2005	–	–	M. Barter *in litt.*
Chang Lake, Nanchang City, Jiangxi COORDINATES 28:59.2N 116:17.2E	Not protected	Winter	1,050	2005	–	–	M. Barter *in litt.*
Chang Hu Chi, Poyang Lake NNR, Jiangxi COORDINATES 29:08N 115:59E	Protected	Winter	680	2005	–	–	M. Barter *in litt.*
Chi'an-Xinshengyu, Shengjin Lake region, Anhui Province COORDINATES 30:23.1N 117:01.5E	Protected	Winter	807	2004	–	–	Barter *et al.* 2004

Appendix 3. (*continued*)

Chongming Island, Shanghai COORDINATES 31:31N 121:50E	Protected	Winter	3,500	1990	–	1987–96	Yu 1991, 1992; Ma & Cai 2002; Miyabayashi & Mundkur 1999
Dahuchi, Poyang Lake NNR, Jiangxi COORDINATES 28:55–29:09N 115:58–116:17E	Protected	Winter	31,900 (in November)	2004	1,161	2003–2004 (Jan/Feb)	Barter *et al.* 2004
Dahuchi west, Poyang Lake NNR, Jiangxi COORDINATES 29:08N 116:02E	Protected	Winter	1,022	2005	–	–	M. Barter *in litt.*
Dongting Lake wetlands, Hunan Province COORDINATES 29:18N 112:59E	Partially protected	Winter	–	–	–	–	Ma & Cai 2002; BirdLife International 2004b
Ergu Fan, Daguan Lake, Anhui Province COORDINATES 30:03N 116:15.2E	Protected	Winter	2,305	2005	–	–	M. Barter *in litt.*
Fenlincun, Fengsha Lake, Anhui Province COORDINATES 30:56N 117:36.5E	Protected	Winter	967	2004	–	–	Barter *et al.* 2004
Gooyou Lake, Jiangsu Province COORDINATES 32:50N 119:15E	Not protected	Winter	–	–	–	–	Ma & Cai 2002; BirdLife International 2004b
Jishan Lake, Jiujiang City, Jiangxi COORDINATES 29:18.4N 116:07.7E	Not protected	Winter	1,302	2004	–	–	Barter *et al.* 2004
Kangshan Dyke, Shangrao City, Jiangxi COORDINATES 28:48.8N 116:22.7E	Not protected	Winter	4,400	2005	–	–	M. Barter *in litt.*
Nanjiang Lake – Dalian Lake, Shangrao City, Jiangxi COORDINATES 28:49N 116:22.5E	Not protected	Winter	480	2005	–	–	M. Barter *in litt.*
Poyang Lake – Candou Hu Xixian 1, Nanchang City, Jiangxi COORDINATES 29:05.5N 116:04.9E	Partially protected	Winter	766	2005	–	1987–2005	Lu 1990; Miyabayashi & Mundkur 1999; M. Barter *in litt.*

Appendix 3. (*continued*)

Poyang Lake – Candou Hu Xidongan, Nanchang City, Jiangxi COORDINATES 29:07.2N 116:05.5E	Partially protected	Winter	732	2005	–	1987–2005	Lu 1990; Miyabayashi & Mundkur 1999; M. Barter *in litt.*
Raopoyanglian Hujiaoxi, Shangrao City, Jiangxi COORDINATES 28:50N 116:16E	Not protected	Winter	770	2004	–	–	Barter *et al.* 2004
Raopoyangdalianzi Lake, Shangrao City, Jiangxi COORDINATES 28:54.4N 116:16.4E	Not protected	Winter	770	2004	–	–	Barter *et al.* 2004
Raopoyangzhu Lake – Hanqi Lake – Qi Lake, Shangrao City, Jiangxi COORDINATES 29:07.5N 116:33E	Not protected	Winter	5,754	2004	–	–	Barter *et al.* 2004
Saicheng Lake, Jiujiang City, Jiangxi COORDINATES 29:42.4N 115:52.4E	Not protected	Winter	7,113	2003	4,269	2003–2004	Barter *et al.* 2004
San Lake, Nanchang City, Jiangxi COORDINATES 28:57N 116:20.7E	Not protected	Winter	16,000	2005	–	–	M. Barter *in litt.*
Shang Lake Bailianwei-Dayangou, Shengjin Lake region, Anhui Province COORDINATES 30:27.2N 117:08.7E	Protected	Winter	1,050	2005	–	–	M. Barter *in litt.*
Shang Lake Da Zhou, Shengjin Lake region, Anhui Province COORDINATES 30:23N 116:59.7E	Protected	Winter	755	2005	–	–	M. Barter *in litt.*
Shengjin Lake (central area), Anhui Province COORDINATES 30:23.3N 117:03.3E	Protected	Winter	520	2003	–	–	Ma & Cai 2002, Barter *et al.* 2004
Shi Lake, Shangrao City, Jiangxi COORDINATES 28:52N 116:13E	Not protected	Winter	1,537	2005	–	–	M. Barter *in litt.*
Qinghai Lake COORDINATES 36:50N 100:10E	Partially protected	Winter	403	1988	–	1987–96	Miyabayashi & Mundkur 1999

Appendix 3. (*continued*)

Wang Lake, Hubei Province Coordinates 29:52.7N 115:21–115:22E	Not protected	Winter	2,516	2004	–	–	Barter *et al.* 2004
Wuchang Lake, Anhui Province Coordinates 30:15N 116:40E	Protected	Winter	600	2003	–	–	Barter *et al.* 2004
Wu Jin Du, Baidang Lake region, Anhui Province Coordinates 30:50.6N 117:24.7E	Protected	Winter	8,760	2005	–	–	M. Barter *in litt.*
Xia Lake, Shengjin Lake region, Anhui Province Coordinates 30:27.1N 117:10.8E	Protected	Winter	1,450	2005	–	–	M. Barter *in litt.*
Xia Lake Guo He Cun, Wuchang Lake region, Anhui Province Coordinates 30:15.8N 116:46E	Protected	Winter	2,040	2005	–	–	M. Barter *in litt.*
Xia Lake Tan Bu Cha, Shengjin Lake region, Anhui Province Coordinates 30:26.8N 117:08.6E	Protected	Winter	724	2005	–	–	M. Barter *in litt.*
Xia Lake Tang Tian Cha, Shengjin Lake region, Anhui Province Coordinates 30:23N 117:11.1E	Protected	Winter	452	2005	–	–	M. Barter *in litt.*
Xiahuliuhu, Wuchang Lake region, Anhui Province Coordinates 30:18.7N 116:48.9E	Protected	Winter	3,980	2004	–	–	Barter *et al.* 2004
Xiaoluzui-chi'an, Shengjin Lake region, Anhui Province Coordinates 30:24.3N 117:02E	Protected	Winter	2,200	2004	–	–	Barter *et al.* 2004
Xiaoxihu, Shengjin Lake region, Anhui Province Coordinates 30:21.9N 117:02.1E	Protected	Winter	1,300	2004	–	–	Barter *et al.* 2004
Xin Miao Lake, Jiujiang City, Jiangxi Coordinates 29:21.4N 116:08.7E	Not protected	Winter	14,200	2005	–	–	M. Barter *in litt.*

Appendix 3. (*continued*)

Yancheng, Jiangsu Province COORDINATES 33:35N 118:25E	Protected	Winter	–	–	–	–	Ma & Cai 2002; BirdLife International 2004b
Yugankangshanei Lake, Shangrao City, Jiangxi COORDINATES 28:53N 116:30E	Not protected	Winter	541	2004	–	–	Barter *et al.* 2004
Zhou Tan Zhen, Fengsha Lake, Anhui Province COORDINATES 30:57.7N 117:38.9E	Protected	Winter	2,056	2005	–	–	M. Barter *in litt.*
JAPAN							
Abukuma-gawa River COORDINATES 37:30N 140:26E	Partially protected	Winter	938	1997	–	1987–97	Miyabayashi & Mundkur 1999; BirdLife International 2004b
Ezogatate Reservoir COORDINATES 40:45N 140:20E	Not protected	Winter	500	1990–94	–	1987–97	Abe *et al.* 1995; Miyabayashi & Mundkur 1999; BirdLife International 2004b
Lake Fukurojinuma COORDINATES 43:30N 141:45E		Passage	1,200	1990–94	–	1987–97	Abe *et al.* 1995; Miyabayashi & Mundkur 1999
Lake Fukushimagata COORDINATES 37:55N 139:16E	Protected	Winter	1,420	1997	–	1987–97	Miyabayashi & Mundkur 1999; BirdLife International 2004b
Hachirogata Rice Fields COORDINATES 40:00N 140:00E	Partially protected	Winter	3,568	1990–94	–	1987–97	Abe *et al.* 1995; Miyabayashi & Mundkur 1999
Hasama-gawa River: Toyosato Town COORDINATES 38:34N 141:13E	Not protected	Winter	425	1997	–	1987–97	Miyabayashi & Mundkur 1999; BirdLife International 2004b
Hasama-gawa River: Wakayanagi Town COORDINATES 38:34N 141:13E	Not protected	Winter	1,227	1997	–	1987–98	Miyabayashi & Mundkur 1999; BirdLife International 2004b
Lake Hyoko COORDINATES 37:50N 139:14E	Protected	Winter	2,660	1997	–	1987–97	Miyabayashi & Mundkur 1999; BirdLife International 2004b

Appendix 3. (*continued*)

Lake Inawashiro-ko Coordinates 37:28N 140:05E	Protected	Winter	1,501	1997	–	1987–97	Miyabayashi & Mundkur 1999; BirdLife International 2004b
Kamishigeoka Reservoir Coordinates 37:19N 140:59E		Winter	460	1997	–	1987–97	Miyabayashi & Mundkur 1999
Kitashinbo-ooike Reservoir Coordinates 38:10N 139:26E		Winter	425	1997	–	1987–97	Miyabayashi & Mundkur 1999
Lake Kuccharo-ko Coordinates 45:10N 142:20E	Protected	Passage	20,000	–	–	1987–97	Hamatonbetsu Town in Miyabayashi & Mundkur 1999; BirdLife International 2004b
Mawarizeki Reservoir Coordinates 40:45N 140:20E	Partially protected	Winter	2,000	1990–94	–	1987–97	Abe *et al.* 1995; Miyabayashi & Mundkur 1999; BirdLife International 2004b
Lake Miyajimanuma Coordinates 43:20N 141:43E	Protected	Passage	3,216	1990–94	–	1987–97	Abe *et al.* 1995; Miyabayashi & Mundkur 1999;
Mogami-gawa River Coordinates 38:53N 139:52E	Protected	Winter	2,100	1997	–	1987–97	Miyabayashi & Mundkur 1999; BirdLife International 2004b
Lake Nakaumi Coordinates 35:27N 133:16E	Partially protected	Winter	1,003	1997	–	1987–97	Miyabayashi & Mundkur 1999; BirdLife International 2004b
Natsui-gawa River Coordinates 37:05N 140:52E		Winter	587	1997	–	1987–97	Miyabayashi & Mundkur 1999
Nougi Rice Fields Coordinates 35:23N 133:14E		Winter	502	1997	–	1987–97	Miyabayashi & Mundkur 1999
Lake Ogawara-ko Coordinates 40:47N 141:20E	Not protected	Winter	441	1990–94	–	1987–97	Abe *et al.* 1995; Miyabayashi & Mundkur 1999; BirdLife International 2004b
Otomonuma Reservoir Coordinates 40:10N 140:00E	Protected	Winter	3,000	1990–94	–	1987–97	Abe *et al.* 1995; Miyabayashi & Mundkur 1999; BirdLife International 2004b

Appendix 3. (*continued*)

Ouchigata Rice Fields COORDINATES 36:55N 136:50E		Winter	823	1997	–	1987–97	T. Sawada pers. comm. in Miyabayashi & Mundkur 1999
Lake Sakata COORDINATES 37:49N 138:53E	Protected	Winter	2,585	1997	–	1987–97	Miyabayashi & Mundkur 1999; BirdLife International 2004b
River Shinano-gawa COORDINATES 37:37N 138:57E		Winter	949	1997	–	1987–97	Miyabayashi & Mundkur 1999
River Shiroishi-gawa COORDINATES 38:03N 140:44E		Winter	466	1997	–	1987–97	Miyabayashi & Mundkur 1999
Lake Toyanogata COORDINATES 37:53N 139:04E	Protected	Winter	1,609	1997	–	1987–97	Miyabayashi & Mundkur 1999; BirdLife International 2004b
Lake Utonai-to COORDINATES 42:42N 141:43E		Winter	839	1990–94	–	1987–97	Abe *et al.* 1995; Miyabayashi & Mundkur 1999
REPUBLIC OF KOREA							
Joonam Reservoir COORDINATES 35:18N 128:41E	Protected	Winter	166	1992	–	–	Park 2002; BirdLife International 2004b
Nakdong Estuary COORDINATES 35:08N 128:54E	Protected	Winter	1,130	1992	724	1992, 1993 & 1998	Kim *et al.* 1996; Miyabayashi & Mundkur 1999; Park 2002
Upper Kum River Estuary COORDINATES 35:45N 126:32E		Winter	905	1994	–	1987–96	Miyabayashi & Mundkur 1999
RUSSIA							
Amur River Estuary, between Berkha and Uarke COORDINATES 53:15N 141:35E	Not protected	Spring	8,150	1991	–	–	Babenko 1995; Babenko & Poyarkov 1998
Khakassiya: Lakes Ulug-Kol, Bele and Shira COORDINATES 54:36N 90:07E	Partially protected	Spring	6,100	–	–	–	Syroechkovski *et al.* 2000; Syroechkovski 2002

APPENDIX 4

Key staging and wintering sites for Bewick's Swans wintering in the Caspian region

Site	Protected status	Season	Max count	Year of max	Average count	Years considered	Source
IRAN							
Amirkelayeh Lake COORDINATES 37:18N 50:10E	Protected	Winter	16	1977	1	1984–93	Scott & Rose 1996
Anzali Mordab COORDINATES 37:25N 49:28E	Protected	Winter	51	–	–	–	Scott & Rose 1996
Bandar Krashahr Lagoon and Sefid Rud COORDINATES 37:20N 49:55E	Protected	Winter	24	1972	24	–	Scott & Rose 1996
Chaf Bala Ab-Bandan COORDINATES 37:16N 50:12E	Unknown	Winter	5	1994	3	1984–93	Scott & Rose 1996
Changiz Marsh COORDINATES 38:45N 45:20E	Unknown	Winter	46	1994	46	1984–93	Scott & Rose 1996
Ghara Gheshlaq Marshes COORDINATES 37:10N 45:50E	Partially protected	Winter	39	1991	8	1984–93	Scott & Rose 1996
Gharaso (Boralan) Marsh COORDINATES 39:45N 44:36E	Unknown	Winter	8	1993	3	1984–93	Scott & Rose 1996
Goorous Marsh COORDINATES 37:02N 45:46E	Unknown	Winter	22	1994	7	1984–93	Scott & Rose 1996
Lapoo-Zargmarz Ab-Bandan COORDINATES 36:50N 53:17E	Unknown	Winter	5	1971	–	1984–93	Scott & Rose 1996
Miankaleh and Gorgan Bay COORDINATES 36:50N 53:45E	Protected	Winter	16	–	–	1984–93	Scott & Rose 1996
Nowruzlu Dam COORDINATES 36:55N 46:10E	Not protected	Winter	27	1990	17	1984–93	Scott & Rose 1996

Appendix 4. (*continued*)

Selke Protected Region COORDINATES 37:20N 49:28E	Unknown	Winter	51	1992	10	1984–93	Scott & Rose 1996
Shur Gol, Yadegarlu and Dorgeh Sangi COORDINATES 37:00N 45:30E	Protected	Winter	68	1994	14	1984–93	Scott & Rose 1996
Uromieh Lake COORDINATES 37:50N 45:30E	Protected	Winter	63	1989	13	1984–93	Scott & Rose 1996
RUSSIA							
Dvuobje (Middle Ob River) COORDINATES 65:00N 65:30E	Protected	Passage	–	–	–	1984–96	Braude 1987; Scott & Rose 1996; Golovatin & Paskhalny 1997; Krivenko 1999; Syroechkovski 2002
Kustanav region COORDINATES 53:13N 63:21E	Unknown	Autumn	230*	1999	–	1999	Gurtovaya *et al.* 1999; Syroechkovski 2002
Ob Estuary, Kara Sea COORDINATES 66:40N 70:58E	Partially protected	Passage	–	–	–	1990s	Krivenko 1999
Turgray Depression South COORDINATES 48:00N 62:30E	Unknown	Autumn	150	1986	–	1986	Khrokov *et al.* 1990; Syroechkovski 2002
Ustje Obi (Mouth of Ob River) COORDINATES 66:38N 69:00E	Protected	Autumn	–	–	–	1984–96	Scott & Rose 1996; Syroechkovski 2002
TURKMENISTAN							
Caspian Coast: Gasankuli-Kuidzhuk COORDINATES 37:40N 53:55E	Unknown	Winter	21	1991	4	1984–93	Scott & Rose 1996; E. Rustamov pers. comm.
Caspian Coast: Khazarskiy State Nature Reserve (*zapovednik*) COORDINATES 39:52N 53:12E	Protected	Winter	103	1977	18	1979–2002	V.I. Vassiliev and E. Rustamov pers. comm.

Notes: * indicates number of swans seen in flight, rather than a ground count.

APPENDIX 5

Scientific names of plants and animals

PLANTS

Bent, Creeping *Agrostis stolonifera*
Birch, Dwarf *Betula nana*
Cloudberry *Rubus chamaemorus*
Clover, White *Trifolium repens*
Cowberry *Vaccinium vitis-idaea*
Crowberry *Empetrum nigrum*
Eelgrass *Zostera* spp.
Foxtail, Marsh *Alopecurus geniculatus*
Grass, Tundra *Dupontia fisheri*
Hornwort
  Ridged *Ceratophyllum demersum*
  Soft *Ceratophyllum submersum*
Marsh Grass, Arctic (also known as Pendent Grass) *Arctophila fulva*
Meadow-Grass
  Smooth *Poa pratensis*
  Rough *Poa trivialis*
Pondweed
  Fennel *Potamogeton pectinatus*
  Horned *Zannichellia palustris*
  Perfoliate *Potamogeton perfoliatus*
  Slender-leaved *Potamogeton filiformis*
Purslane, Sea *Atriplex portulacoides*
Rape, Oilseed *Brassica napus*
Reed, Common *Phragmites australis*
Reedmace, Common *Typha latifolia*
Ryegrass, Common *Lolium perenne*
Sedge
  Water *Carex aquatalis*
  Haresfoot *Carex lachenalii*
Small-Reed, Narrow *Calamagrostis neglecta*
Stitchwort, Low (also known as Saltmarsh Starwort) *Stellaria humifusa*
Stoneworts *Chara globularis, C. contraria* and *C. aspera*
Sweet Grass
  Floating *Glyceria fluitans*
  Reed *Glyceria maxima*
Tasselweed *Ruppia* spp.
Timothy Grass, Large-leaved *Phleum pratense*
Water Milfoil species *Myriophyllum* spp.
Yellow-Cress, Marsh *Rorippa palustris*
Yorkshire Fog *Holcus lanatus*

INVERTEBRATES

Clam, Soft-shelled *Mya arenaria*
Krill, Antarctic *Euphausia superba*
Macoma *Macoma baltica*
Ragworm *Nereis* spp.
Shrimp
  Mud *Corophium* spp.
  Tadpole *Notostraca* spp.

BIRDS

Buzzard, Rough-legged *Buteo lagopus*
Coot *Fulica atra*
Duck, Long-tailed *Clangula hyemalis*
Eagle, White-tailed Sea *Haliaeetus albicilla*
Gadwall *Anas strepera*
Goose
  Barnacle *Branta leucopsis*
  Bean *Anser fabalis*
  Brent *Branta bernicla*
  Canada *Branta canadensis*
  Cape Barren *Cereopsis novaehollandiae*
  (Eurasian) White-fronted *Anser albifrons albifrons*
  Greylag *Anser anser*
  (Lesser) Snow *Anser caerulescens caerulescens*
  Pink-footed *Anser brachyrhynchus*
  Tundra Bean *Anser fabalis rossicus*
Gull
  Glaucous *Larus hyperboreus*
  Herring *Larus argentatus*

Owl, Snowy *Nyctea scandiaca*
Skua, Arctic *Stercorarius parasiticus*
Swan
  Bewick's *Cygnus columbianus bewickii*
  Black *Cygnus atratus*
  Black-necked *Cygnus melanocoryphus*
  Coscoroba *Coscoroba coscoroba*
  Mute *Cygnus olor*
  Trumpeter *Cygnus buccinator*
  Tundra *Cygnus columbianus*
  Whistling *Cygnus columbianus columbianus*
  Whooper *Cygnus Cygnus*
Wigeon, European *Anas penelope*

MAMMALS

Bear
  Brown *Ursus arctos*
  Polar *Thalassarctos maritimus*
Fox
  Arctic *Alopex lagopus*
  Red *Vulpes vulpes*
Hare, Brown *Lepus europaeus*
Lemming *Lemmus* and *Dicrostonyx* spp.
Reindeer *Rangifer tarandus*
Wolverine *Gulo gulo*

APPENDIX 6

List of abbreviations used in the text

BTO – British Trust for Ornithology
CDNA – Commissie Dwaalgasten Nederlandse Avifauna [Dutch Rarities Committee]
EGI – Edward Grey Institute for Field Ornithology, University of Oxford
IBAs – Important birds areas (as listed in Heath & Evans 2000)
IWC – International Waterfowl Census
I-WeBS – Irish Wetland Bird Survey
IWRB – International Waterfowl & Wetlands Research Bureau
NERI – National Environment Research Institute, Denmark
NIOO – Nederlands Instituut voor Oecologisch Onderzoek [Netherlands Institute of Ecology]
PRISM – Pechora Basin Integrated System Management programme
RIZA – Institute for Inland Water Management and Waste Water Treatment, the Netherlands
RSPB – Royal Society for the Protection of Birds
WeBS – Wetland Bird Survey
WWF – World Wide Fund for Nature
WWT – Wildfowl & Wetlands Trust

References

Arbeitsgemeinschaft Berlin-Brandenburger Ornithologen (ABBO) (2003) *Important Bird Areas (IBA) in Brandenburg and Berlin.* Natur & Text, Rangsdorf.

Abe, M., Fujimaki, Y., Iijima, H., Kurechi, M., Miyabayashi, Y., Sato, F. & Sugimori, F. (1995) Status and conservation needs of Anatidae and their habitats in Japan. Unpublished report, *Workshop on Action Plan for Anatidae,* Northeast Asia and North Pacific Environmental Forum, Kushiro, Japan, 1995.

Acland, C.N. (1923) Notes on Bewick's Swans in Glamorganshire. *British Birds* 17: 63–64.

Albertsen, J.O. & Kanazawa, Y. (2002) Numbers and ecology of swans wintering in Japan. *In* E.C. Rees, S.L. Earnst & J. Coulson (eds), Proceedings of the Fourth International Swan Symposium, 2001. *Waterbirds,* 25 (Special Publication 1): 74–85.

Alphéraky, S.N. (1904) *Gussi Rossyi.* Kuschnerew, Moscow. [In Russian.]

Alison, R.M. (1977) Homing of subadult oldsquaws. *Auk,* 94: 383–384.

Anderson, P. (1913) The birds of the island of Tiree. *Scottish Naturalist,* 21: 193–200.

Ankney, C.D. (1975) Neckbands contribute to starvation in female Lesser Snow Geese. *Journal of Wildlife Management,* 39: 825–826.

Anselin, A., Jacob, J-P. & Van den Bossche, W. (2000) Belgium. *In* M.F. Heath & M.I. Evans (eds), *Important Bird Areas in Europe: Priority Sites for Conservation. Volume 1: Northern Europe,* 103–122. BirdLife International (BirdLife Conservation Series No. 8), Cambridge, U.K.

Antipin, V.M. (1938) Vertebrate fauna of northeast Novaya Zemlya. *Problems of the Arctic,* 2: 153–171.

Ardamatskaya, T.B. & Korzyukov, A.I. (1991) Numbers and distribution of Mute Swans *Cygnus olor,* Whooper Swans *C. cygnus* and Bewick's Swans *C. bewickii* in the Black Sea area of the Ukraine, USSR. *In* J. Sears & P.J. Bacon (eds), Proceedings of the Third International Swan Symposium, Oxford 1989. *Wildfowl Special Supplement No. 1*: 53–55.

Ashcroft, R.E. (1976) A function of the pairbond in the Common Eider. *Wildfowl* 27: 101–106.

Atkinson-Willes, G. (1976) The numerical distribution of ducks, swans and coots as a guide to assessing the importance of wetlands. *In* M. Smart (ed.), *Proceedings International Conference on the Conservation of Wetlands and Waterfowl, Heiligenhafen 1974*, 199–254. International Waterfowl Research Bureau, Slimbridge.

Atkinson-Willes, G. (1981) The numerical distribution and the conservation requirements of *Cygnus olor, Cygnus cygnus* and *Cygnus columbianus bewickii* in north-west Europe. *In* G.V.T. Matthews & M. Smart (eds), *Proceedings of the Second International Swan Symposium, Sapporo 1980*, 40–48. International Waterfowl Research Bureau, Slimbridge.

Austin, O. L. (1949) *Waterfowl in Japan.* Natural Resources Section Report No. 18. General Headquarters of the Supreme Commander for the Allied Forces, Tokyo.

Austin, O. L. & Kuroda, N. (1953) The birds of Japan: their status and distribution. *Bulletin of Miscellaneous Comparative Zoology* (Harvard), 109: 279–613.

Babenko, V.G. (1995) On spring migration of swans in the Amur River mouth. *Bulletin of the Goose Study Group of Eastern Europe and Northern Asia,* 1: 180. [In Russian with English summary.]

Babenko, V.G. & Poyarkov, N.D. (1998) Geese and swans in the low Amur area. *Casarca,* 6: 297–313. [In Russian with English summary.]

Bacon, P.J. & Perrins, C.M. (1991) Long-term populations studies: the Mute Swan. *In* B.D. Bell (ed.), *Twentieth International Ornithological Congress,* 1500–1513. New Zealand Ornithological Trust Board, Wellington, New Zealand.

Baker, R.R. (1978) *The Evolutionary Ecology of Animal Migration.* Hodder & Stoughton, London.

Bannerman, D.H. (1957) *The Birds of the British Isles. Vol. 6.* Oliver & Boyd, Edinburgh and London.

Bart, J., Limpert, R., Earnst, S., Sladen, W., Hines, J. & Rothe, T. (1991) Demography of Eastern Population Tundra Swans *Cygnus columbinanus columbinanus. In* J. Sears & P.J. Bacon (eds), Proceedings of the Third International Swan Symposium, Oxford 1989. *Wildfowl Special Supplement No. 1*: 178–184.

Barter, M., Chen, L., Cao, L. & Lei, G. (2004) *Waterbird Survey of the Middle and Lower Yangtze River Floodplain in late January and early February 2004.* China Forestry Publishing House, Beijing.

Bateson, P.P.G. (1977) Testing an observer's ability to identify individual animals. *Animal Behaviour* 25: 247–248.

Bateson, P. (1978) Sexual imprinting and optimal outbreeding. *Nature,* 273: 659–660.

Bateson, P. (1983). Optimal outbreeding. *In* P. Bateson (ed.), *Mate Choice.* Cambridge University Press, Cambridge.

Bateson, P., Lotwick., W. & Scott, D.K. (1980). Similarity between the faces of parents and offspring in Bewick's Swans and the differences between mates. *Journal of Zoology* (London), 191: 61–74.

Baxter, E.V. & Rintoul, L.J. (1953) *The Birds of Scotland.* Oliver & Boyd, Edinburgh.

Beekman, J.H. (1997) Censuses of the NW European Bewick's Swan population, January 1990 and 1995. *Swan Specialist Group Newsletter,* 6: 7–9.

Beekman, J.H. & Laubek, B. (1997) Satellite tracking Bewick's Swans on spring migration. *Swan Specialist Group Newsletter,* 6: 16–18.

BEEKMAN, J.H., DIRKSEN, S. & SLAGBOOM, T.H. (1985) Population size and breeding success of Bewick's Swans wintering in Europe 1983–84. *Wildfowl* 36: 5–12.

BEEKMAN, J.H., VAN EERDEN, M.R. & DIRKSEN, S. (1991) Bewick's Swans *Cygnus columbianus bewickii* utilising the changing resource of *Potamogeton pectinatus* during autumn in the Netherlands. *In* J. Sears & P.J. Bacon (eds), Proceedings of the Third International Swan Symposium, Oxford 1989. *Wildfowl Special Supplement No. 1*: 238–248.

BEEKMAN, J.H., REES, E.C. & BACON P.J. (1994a) Bewick's Swan *Cygnus columbianus bewickii*. *In* G.M. Tucker & M.F. Heath (eds), *Birds in Europe: their Conservation Status.* Birdlife International (Birdlife Conservation Series No. 3), Cambridge, U.K.

BEEKMAN, J.H., LAUBEK, B. & LUIGUJÕE, L. (1994b) Report on Bewick's Swan stop-over ecology studies, Estonia, spring 1993. Unpublished report, Zoological Laboratory, University of Groningen. 16 pp.

BEEKMAN, J.H., BERTHOLD, P., NOWAK, E. & QUERNER, U. (1996a) Implementation of satellite tracking in studying migration of *Anatidae*: an overview and a case study. *In* M. Birkan, J. van Vessem, P. Havet, J. Madsen, B. Trolliet & M. Moser (eds), Proceedings Anatidae 2000 Conference, Strasbourg, France, 5–9 December 1994. *Gibier Faune Sauvage, Game Wildlife* 13: 157–176.

BEEKMAN, J.H., VAN EERDEN, M.R., MINEYEV, Y.N.,LUIGUJÕE, L. & DEN HOLLANDER, H.J. (1996b). Landsat satellite images for detection of submerged macrophytes: in search of potential stop-over feeding sites for Bewick's Swans (*Cygnus columbianus bewickii*) along their migratory route. *In* M. Birkan, J. van Vessem, P. Havet, J. Madsen, B. Trolliet & M. Moser (eds), Proceedings Anatidae 2000 Conference, Strasbourg, France, 5–9 December 1994. *Gibier Faune Sauvage, Game Wildlife* 13: 421–450.

BEEKMAN J.H., NOLET, B.A. & M. KLAASSEN, M. (2002) Skipping swans: fuelling rates and wind conditions determine differential use of migratory stopover sites of Bewick's Swans *Cygnus bewickii*. *In* C. Both & T. Piersma (eds) *The avian calendar: exploring biological hurdles in the annual cycle.* Proceedings 3rd Conference European Ornithologists' Union, Groningen, August 2001. *Ardea* 90 (Special Issue), 437–460.

BELIK, V.P., DEBELO, P.V., MOROZOV, V.V. & SHEVCHENKO, V.L. (1997) Bewick's Swans (*Cygnus bewickii*) in the Volga-Ural interfluve. *Casarca*, 3: 280–285. [In Russian with English summary.]

BELL, R.H.V. (1971) A grazing ecosystem in the Serengeti. *Scientific American*, 224: 86–93.

BELOUSOVA, A.V. (1999) The breeding strategy of Bewick's Swans *Cygnus columbianus bewickii.* Candidate of Science Degree, Central Scientific Research Laboratory, Ministry of Agriculture, Russia.

BERTHOLD, P. (1996) *Control of Bird Migration.* Chapman & Hall, London.

BIANKI V.V. & SHUTOVA E.V. (1987) The distribution and numbers of swans in the European north of the USSR. *In* E.V. Syroechkovsky (ed.), *Ecology and migration of swans in the USSR*, 20–28. Nauka, Moscow.

BICKART, K.J. (1990) The birds of the late Miocene/early Pliocene Big Sandy Formation, Mohave County, Arizona. *Ornithological Monograph*, 44: 1–72.

BIRDLIFE INTERNATIONAL (2004a) *Birds in Europe: Population estimates, trends and conservation status.* BirdLife International (BirdLife Conservation Series No. 12), Cambridge, U.K.

BIRDLIFE INTERNATIONAL (2004b) *Important Bird Areas in Asia: key sites for conservation.* BirdLife International (BirdLife Conservation Series No. 13), Cambridge, U.K.

Bird Migration Research Centre, Yamashina Institute for Ornithology (1987) Swan collar-marking. *Report of the Migration Bird Research Centre (1986)*, 93–103. Abiko, Chiba Prefecture. [In Japanese.]

Birkhead, M. (1982) Cause of mortality in the Mute Swan *Cygnus olor* on the River Thames. *Journal of Zoology* (London), 198: 15–25.

Birkhead, M. & Perrins, C. (1986) *The Mute Swan.* Croom Helm, London.

Birkhead, M., Bacon, P.J. & Walter, P. (1983) Factors affecting the breeding success of the Mute Swan *Cygnus olor. Journal of Animal Ecology* 52: 727–41.

Blaauw, F.E. (1904) On the breeding of some of the waterfowl at Gooilust in the year 1903. *Ibis* 46: 67–75.

Black, J.M. (ed.) (1996) *Partnerships in Birds: The study of monogamy.* Oxford University Press, Oxford

Black, J.M. & Owen, M. (1989) Agonistic behaviour in goose flocks: assessment, investment and reproductive success. *Animal Behaviour* 37: 199–209.

Black, J.M. & Rees, E.C. (1984) The structure and behaviour of the Whooper Swan population wintering at Caerlaverock, Dumfries and Galloway, Scotland: an introductory study. *Wildfowl* 35: 21–36.

Black, J.M., Choudury, S. & Owen, M. (1996). Do Barnacle Geese benefit from life-long monogamy? *In* J.M. Black (ed.), *Partnerships in birds: the study of monogamy*, 91–117. Oxford University Press, Oxford.

Borisov, V.V., Urjadova, L.P. & Scheblikina, L.S. (1999) Autumn migration of swans in the Pskov-Chudskoye Lake region between 1956 and 1998. *Swan Specialist Group Newsletter*, 8: 15–23.

Bowler, J.M. (1994) The condition of Bewick's Swans *Cygnus columbianus bewickii* in winter as assessed by their abdominal profiles. *Ardea*, 82: 241–248.

Bowler, J.M. (1996) Feeding strategies of Bewick's Swans *(Cygnus columbianus bewickii)* in winter. Unpublished Ph.D. thesis, University of Bristol.

Bowler, J.M. (2005a) Black Swan *Cygnus atratus. In* J. Kear (ed.), *Bird Families of the World: Ducks, Geese and Swans.* Oxford University Press, Oxford.

Bowler, J.M. (2005b) Bewick's Swan *Cygnus columbianus bewickii. In* J. Kear (ed.), *Bird Families of the World: Ducks, Geese and Swans.* Oxford University Press, Oxford.

Bowler, J.M., Butler, L., Hesketh, R., Liggett, C. & Rees, E.C. (1995) Bewick's and Whooper Swans *Cygnus columbianus bewickii* and *C. cygnus*: the 1994–95 season. *Wildfowl* 46: 176–187.

Boyd, H. (1972). Classification. In: P. Scott & The Wildfowl Trust (eds), *The Swans.* Michael Joseph, London.

Boyd, J.M. (1958) The birds of Coll and Tiree. *British Birds* 51: 41–56.

Braude, M.I. (1987) Migration of Whooper Swans and Bewick's Swans in the lower Ob'. *In* E.V.Syroechkovski (ed.), *Ecology and migrations of swans in the USSR*, 97–99. Nauka, Moscow. [In Russian.]

Brazil, M.A. (1981) Geographical variation in the bill patterns of Whooper Swans. *Wildfowl* 32: 129–131.

Brazil, M.A. (2003) *The Whooper Swan.* T. & A.D. Poyser, London.

Brodkorb, P. (1964) Catalogue of fossil birds: Part 2 (Anseriformes through Galliformes). *Bulletin of the Florida State Museum of Biological Sciences* 8: 195–335.

Brouwer, G.A. & Tinbergen, L. (1939) De Verspreiding der Kleine Zwanen *Cygnus bewickii* Yarr. in de Zuiderzee, voor en na de verzoeting. *Limosa*, 12: 1–18.

Brown, M., Linton, E. & Rees, E.C. (1992) Causes of mortality among wild swans in Britain. *Wildfowl* 43: 70–79.

Butler, D.S. (1999) Rate of Bewick's and Whooper Swan collisions with power-lines: a review of a marking programme by the National Grid Company plc. Unpublished B.Sc. (final year) report, University of the West of England, Bristol, U.K.

Buzun, V.A. & Grigoryan, S.B. (2004) Distribution, numbers and breeding biology of Bewick's Swans on Zakharyin Bereg (shore) of the Pechora Bay in 1996. *Casarca*, 10: 215–238.

Cadbury, C.J. (1975) Population of swans on the Ouse Washes, England. *Wildfowl* 26: 148–159.

Callaghan, D., Rees, E. & Harshman, J. (2005) The Swans: Taxonomy. *In* J. Kear (ed.), *Bird Families of the World: Ducks, Geese and Swans.* Oxford University Press, Oxford.

Charman, K. (1979) Feeding ecology and energetics of the dark-bellied Brent Goose (*Branta bernicla bernicla*) in Essex and Kent. *In* R.L. Jefferies & A.J. Davy (eds), *Ecological Processes in Coastal Environments,* 451–465. Blackwell Scientific Publications, Oxford.

Chisholm, H. & Spray, C. (2002) Habitat usage and field choice by wintering Mute and Whooper Swans in the Tweed Valley, Scotland. *In* E.C. Rees, S.L. Earnst & J. Coulson (eds), Proceedings of the Fourth International Swan Symposium, 2001. *Waterbirds,* 25 (Special Publication 1): 177–182.

Choudhury, S. & Black, J.M. (1994) Barnacle Geese choose familiar mates from early-life. *Animal Behaviour* 48: 81–88.

Clutton-Brock, T.H., Albon, S.D., Gibson, R.M. & Guiness, F.E. (1979) The fighting strategies of red deer (*Cervus elaphus* L.). *Animal Behaviour* 27: 211–225.

Coleman, A.E. & Minton, C.D.T. (1979) Pairing and breeding of Mute Swans in relation to natal area. *Wildfowl* 30: 27–30.

Colhoun, K. (1998) *IWeBS Report 1996–97: Results of the third winter of the Irish Wetland Bird Survey.* Birdwatch, Dublin, Ireland.

Colhoun, K. (2000) *IWeBS Report 1997–98: Results of the fourth winter of the Irish Wetland Bird Survey.* Birdwatch, Dublin, Ireland.

Cooke, F., MacInnes, C.D. & Prevett, J.P. (1975) Gene flow between breeding populations of the Lesser Snow Goose. *Auk,* 92: 493–510.

Cramp, S. & Simmons, K.E.L. (1977) *The Handbook of the Birds of Europe, the Middle East and North Africa. The Birds of the Western Palearctic. Volume 1.* Oxford University Press, Oxford.

Cranswick, P.A., Bowler, J.M., Delany, S.N., Einarsson, Ó., Garðarsson, A., McElwaine, J.G., Merne, O.J., Rees, E.C. & Wells, J.H. (1996) Numbers of Whooper Swans *Cygnus cygnus* in Iceland, Ireland and Britain in January 1995: results of the international Whooper Swan census. *Wildfowl* 47: 17–30.

Cranswick, P.A., Waters, R.J., Musgrove, A.J. & Pollitt, M.S. (1997) *The Wetland Bird Survey 1995–96: Wildfowl and Wader Counts.* BTO/WWT/RSPB/JNCC, Slimbridge.

Cranswick, P., Pollit, M., Musgrove, A. & Hughes, B. (1999) *The Wetland Bird Survey 1997–98. Wildfowl and Wader Counts.* BTO/WWT/RSPB/JNCC, Slimbridge.

Cranswick, P.A., Worden, J., Ward, R.M., Rowell, H.E., Hall, C., Musgrove, A.J., Hearn, R.D., Holloway, S.J., Banks, A.N., Austin, G.E., Griffin, L.R., Hughes, B., Kershaw, M., O'Connell, M.J., Pollitt, M.S., Rees, E.C. & Smith, L.E. (2005) *The Wetland Bird Survey 2001–03: Wildfowl and Wader Counts.* BTO/WWT/RSPB/JNCC, Slimbridge

Dahms, G. & Grosskopf, G. (1978) Zum Vorkommen des Zwegschwans (*Cygnus bewickii*) im Raum Stade. *Vogelkundliche Berichte aus Niedersachsen*, 10: 1–21.

Davies, C. (2001) The European Bird Report: non-passerines. *British Birds*, 94: 125–142.

Deceuninck, B., Rufray, X. & Jourde, P. (2000) France. *In* M.F. Heath & M.I. Evans (eds), *Important Bird Areas in Europe: Priority Sites for Conservation. Volume 2: Southern Europe*, 157–245. BirdLife International (BirdLife Conservation Series No. 8), Cambridge, U.K.

Degen, A., Flore, B-O., Ludwig, J. & Südbeck, P. (1996) Rastbestände von Höcker-, Zwerg- und Singschwan (*Cygnus olor, C. columbianus bewickii, C. c. cygnus*) in Niedersachsen: Ergebnisse landesweiter Synchronzählungen im Januar und März 1995. *Vogelkundliche Berichte aus Niedersachsen*, 28: 3–18.

Delacour, J. (1954). *The Waterfowl of the World, Volume 1*. Country Life, London.

Delacour, J. & Mayr, E. (1945) The family Anatidae. *Wilson Bulletin*, 57: 3–55.

Delany, S. & Scott, D. (2002) *Waterbird Population Estimates – Third Edition*. Wetlands International Global Series No. 12, Wetlands International, Wageningen, The Netherlands.

Delany, S., Reyes, C., Hubert, E., Pihl, S., Rees, E., Haanstra, L. & van Strien, A. (1999) *Results from the International Waterbird Census in the Western Palearctic and Southwest Asia 1995 and 1996*. Wetlands International Publication No. 54, Wetlands International, Wageningen, The Netherlands.

Dementiev, G.P. & Gladkov, N.A. (1952) *The Birds of the Soviet Union, Volume 4*. Sovetskaya Nauka, Moscow.

Dirksen, S. & Beekman, J.H. (1991) Population size, breeding success and distribution of Bewick's Swans *Cygnus columbianus bewickii* wintering in Europe in 1986–87. *In* J. Sears & P.J. Bacon (eds), Proceedings of the Third International Swan Symposium, Oxford 1989. *Wildfowl Special Supplement No. 1*: 120–124.

Dirksen, S., Beekman, J.H. & Slagboom, T.H. (1991) Bewick's Swans *Cygnus columbianus bewickii* in the Netherlands: numbers, distribution and food choice during the wintering season. *In* J. Sears & P.J. Bacon (eds), Proceedings of the Third International Swan Symposium, Oxford 1989. *Wildfowl Special Supplement No. 1*: 228–237.

Dirksen, S., Noordhuis, R., Poot, M., Tulp, I. & van der Winden, J. (1998) Bewick's Swans exploiting stoneworts: recovering aquatic food sources increasingly important in autumn. *Swan Specialist Group Newsletter*, 7: 12–13.

Dorogostaiskiy, V.C. (1913) Chinese Swan (*Cygnus dividi* Swinh.) in Siberia. *Ornitologicheskiy Vestnik*, 2: 110–112. [In Russian.]

Drent, R.H. & Daan, S. (1980) The prudent parent: energetic adjustments in avian breeding. *Ardea*, 68: 225–252.

Drent, R.H. & Prins, H.H.T. (1987) The herbivore as a prisoner of its food supply. *In* J. van Andel *et al.* (eds), *Disturbance in grasslands*, 131–147. Dr W. Junk, Dordrecht.

Drent, R.H., Ebbinge, B. & Weijand, B. (1981) Balancing the energy budgets of arctic breeding geese throughout the annual cycle: a progress report. *In* M. Smart & J. Reichhalf (eds) IWRB Symposium, Feeding Ecology of Waterfowl, München 1978/79. *Verhandlungen der Ornithologischen Gesellschaft in Bayern*, 23: 239–264.

Earnst, S.L. & Bart, J. (1991) Costs and benefits of extended parental care in Tundra Swans *Cygnus columbianus columbianus*. *In* J. Sears & P.J. Bacon (eds), Proceedings of the Third International Swan Symposium, Oxford 1989. *Wildfowl Special Supplement No. 1*: 260–67.

Ebbinge, B.S. & Ebbinge-Dallmeijer, D. (1977) Barnacle Geese (*Branta leucopsis*) in the Arctic summer – a reconnaissance trip to Svalbard. *Norsk Polar Årbok*, 1975: 119–38.

Ebbinge, B., Canters, K. & Drent, R. (1975) Foraging routines and estimated food intake in Barnacle Geese wintering in the northern Netherlands. *Wildfowl* 26: 5–19.

Eber, G. & Niemeyer, H. (1982) Dokumentation der Schwimmvogelzählungen in der Bundesrepublik Deutschland von 1966/67–1975/76. Bundesministerium für Ernährung, Landwirtschaft und Forsten, Bonn.

Egorov, O.V. (1965) Status of waterfowl numbers and some other birds in the Lena Delta and in Yana-Indigirka tundra, from aerial survey data. *In* V.G. Krivosheev (ed.), *Nature of Yakutia and its protection*, 124–127. Yakut Publishing House, Yakutsk. [In Russian.]

Einarsson, Ó. (1996) Breeding biology of the Whooper Swan and factors affecting its breeding success, with notes on its social dynamics and life cycle in the wintering range. Unpublished Ph.D. thesis, University of Bristol, U.K.

Ely, C.R. (1990) Effects of neck bands on the behaviour of wintering Greater White-fronted Geese. *Journal of Field Ornithology* 61: 249–253.

Environment Agency of Japan. (1998) *Report on the 29th Annual Census of Waterfowl (Anatidae) in January, 1998*. Environment Agency of Japan, Tokyo. [In Japanese with English summary.]

Environment Agency of Japan. (1999) *Report on the 30th Annual Census of Waterfowl (Anatidae) in January, 1999*. Environment Agency of Japan, Tokyo. [In Japanese with English summary.]

Evans, M.E. (1972) Slimbridge: the wild swans 1971–72. *Wildfowl* 23: 132–135.

Evans, M.E. (1975) Breeding behaviour of captive Bewick's Swans. *Wildfowl* 26: 117–130.

Evans, M.E. (1977a) Recognising individual Bewick's Swans by bill pattern. *Wildfowl* 28: 153–158.

Evans, M.E. (1977b) Notes on the breeding behaviour of captive Whistling Swans by bill pattern. *Wildfowl* 28: 107–112.

Evans, M.E. (1979a) Aspects of the life cycle of the Bewick's Swan based on recognition of individuals at a wintering site. *Bird Study* 26: 149–162.

Evans, M.E. (1979b) The effect of weather on the wintering Bewick's Swans *Cygnus columbianus bewickii* at Slimbridge, England. *Ornis Scandinavica*, 10: 124–132.

Evans, M.E. (1980) The effects of experience and breeding status on the use of a wintering site by Bewick's Swans *Cygnus columbianus bewickii*. *Ibis*, 122: 287–297.

Evans, M.E. (1982) Movements of Bewick's Swans *Cygnus columbianus bewickii* marked at Slimbridge, England from 1960 to 1979. *Ardea*, 70: 59–75.

Evans, M.E. & Kear, J. (1975) A jacket for holding large birds. *International Zoo Yearbook*, 15: 191–193.

Evans, M.E. & Kear, J. (1978) Weights and measurements of Bewick's Swans during winter. *Wildfowl* 29: 118–122.

Evans, M.E. & Lebret, T. (1973) Leucistic Bewick's Swans. *Wildfowl* 24: 61–62.

Evans, M.E. & Rees, E. C. (1978) Slimbridge: the wild swans, 1977–1978. *Wildfowl* 29: 166–167.

Evans, M.E. & Sladen, W.J.L. (1980) A comparative analysis of the bill markings of Whistling and Bewick's Swans and out-of-range occurrences of the two taxa. *Auk*, 97: 697–703.

Evans, M.E., Wood, N.A. & Kear, J. (1973) Lead shot in Bewick's Swans. *Wildfowl* 24: 56–60.

FINSH, O. & BREM, A. (1882) *Journey to Western Siberia.* Moscow.
GARÐARSSON, A. & SKARPHÉÐINSSON, K.H. (1984) A census of the Icelandic Whooper Swan population. *Wildfowl* 35: 37–47.
GARÐARSSON, A. & SKARPHÉÐINSSON, K.H. (1985) Veturseta álfta á Íslandi [Wintering Whooper Swans in Iceland]. *Bliki,* 4: 45–56.
GÉROUDET, P. (1962) L'hivernage des *Cygnus bewickii* sur le Léman. *Nos Oiseaux,* 26: 317–318.
GÉROUDET, P. (1963) Retour des cygnes de Bewick sur le Léman. *Nos Oiseaux,* 27: 181–182.
GIROUX, J.F. & PATTERSON, I.J. (1995) Daily movements and habitat use by radio-tagged Pink-footed Geese *Anser brachyrhynchus* wintering in northeast Scotland. *Wildfowl* 46: 31–44.
GIZENKO, A.I. (1955) *The Birds of the Sakhalin Region.* USSR Academy of Sciences, Moscow. [In Russian.]
GOLOVATIN, M.G. & PASKHALNY, S.P. (1997) Spring migration of Bewick's Swans (*Cygnus bewickii*) on the lower Ob River. *Casarca,* 3: 286–297.
GORSKI, W. & JESIONOWSKI, J. (1983) Migration of the Bewick's Swan *Cygnus bewickii* in Poland. *Ornis Fennica* (Supplement), 3: 51–53.
GREENWOOD, P.J. (1980) Mating systems, philopatry and dispersal in birds and mammals. *Animal Behaviour* 28: 1140–1162.
GROMADZKI, M. & WIELOCH. M. (2000) Poland. *In* M.F. Heath & M.I. Evans (eds), *Important Bird Areas in Europe: Priority Sites for Conservation. Volume 1: Northern Europe,* 543–579. BirdLife International (BirdLife Conservation Series No. 8), Cambridge, U.K.
GURTOVAYA, E., TOLVANEN, P., ESKELIN, T., QIEN, I., BRAGINA, T., AARVAK, T., EICHHORN, G., ARKIOMAA, A. & TIMONEN, S. (1999) Preliminary results of the Lesser White-fronted Goose monitoring in Kazakhstan in October 1999. *Casarca,* 5: 145–154.
HANBY, A.M. (1986) Bewick's Swan with yellow legs. *British Birds* 79: 206.
HANCOCK, R. & WOOLLAM, P.B. (1987) Radioactivity measurements on live Bewick's Swans. Unpublished Report, Central Electricity Generating Board, Berkeley, U.K.
HARSHMAN, J. (1996) Phylogeny, evolutionary rates, and ducks. Unpublished Ph.D. thesis, University of Chicago, U.S.A.
HARTERT, E. (1920) *Die Vogel der palaarktischen Fauna.* Friedlander & Sohn, Berlin.
HARVEY, N.G. (1999) A hierarchical genetic analysis of swan relationships. Unpublished Ph.D. thesis, University of Nottingham, U.K.
HARVIE-BROWN, J.A. & BUCKLEY, T.E. (1892) *A Vertebrate Fauna of Argyll and the Inner Hebrides.* David Douglas, Edinburgh.
HATAKEYAMA, M. (1981) The feeding of swans and its results. *In* G.V.T. Matthews & M. Smart (eds), *Proceedings of the Second International Swan Symposium, Sapporo 1980,* 247–248. International Waterfowl Research Bureau, Slimbridge.
HEARN, R.D. (2004) *Bean Goose* Anser fabilis *in Britain and Ireland 1960/61–1999/2000.* Waterbird Review Series, The Wildfowl & Wetlands Trust/Joint Nature Conservation Committee, Slimbridge.
HEATH, M.F. & EVANS, M.I. (eds) (2000) *Important Bird Areas in Europe: Priority sites for conservation.* BirdLife International (BirdLife Conservation Series No. 8), Cambridge, U.K.
HEDENSTROM, A. & ALERSTAM, T. (1992) Climbing performance of migrating birds as a basis for estimating limits for fuel-carrying capacity and muscle work. *Journal of Experimental Biology* 164: 19–38.

HIGUCHI, H., SATO, F., MATSUI, S., SOMA, M. & KOHIYAMA, M. (1990) Satellite tracking of the migration routes of Whistling Swans. *Biotelemetry*, 11: 100–103.

HIGUCHI, H., SATO, F., MATSUI, M., SOMA, M. & KANMURI, N. (1991) Satellite tracking of the migration routes of Whistling Swans *Cygnus columbianus*. *Journal of the Yamashina Institute for Ornithology*, 23: 6–12.

HORIUCHI, M. (1981) Ten years of swan counts in Japan. *In* G.V.T. Matthews & M. Smart (eds), *Proceedings of the Second International Swan Symposium, Sapporo 1980*, 14–15. International Waterfowl Research Bureau, Slimbridge.

HUNT, J., DERWIN, J., COVENEY, J. & NEWTON, S. (2000) Republic of Ireland. *In* M.F. Heath & M.I. Evans (eds), *Important Bird Areas in Europe: Priority Sites for Conservation. Volume 1: Northern Europe*, 365–416. BirdLife International (BirdLife Conservation Series No. 8), Cambridge, U.K

IDESTAM-ALMQUIST, J. (1996) Waterfowl herbivory on *Potomageton pectinatus* in the Baltic Sea. *Oikos*, 81: 323–328.

IRISH RARE BIRDS COMMITTEE. (1998) *Checklist of the Birds of Ireland.* BirdWatch Ireland, Dublin.

JI, W., WU, J., YI, W., HUANG, Z. & WU, X. (2001) Aerial survey of waterbirds at Poyang Lake N.R. *China Crane News*, 5 (1): 16–17.

JOENSEN, A.H. (1974) Waterfowl populations in Denmark 1965–1973. *Danish Review of Game Biology*. 9: 1–206.

JOHNSTONE, S.T. (1957) Breeding of Bewick's Swans. *Avicultural Magazine* 63: 27–28.

JONZÉN, N., NOLET, B.A., SANTAMARÍA, L. & SVENSSON, M.G.E. (2002) *Ecological Modelling*, 147: 209–219.

KALAMEES, A. (ed.) (2000) *Important Bird Areas of Estonia.* Eesti Loodusfoto, Tartu.

KALYAKIN, V.N. (1987) Swans of the northwestern Yugor Peninsula and Vaigach Island. *In* E. V. Syroechkovski (ed.), *Swan Ecology and Migration in the USSR*, 90–93. Nauka, Moscow.

KALYAKIN, V. N. (1995) Notes on distribution of goose species in coastal regions of the Barents Sea and in the north of West Siberia. *Bulletin of the Goose Study Group of Eastern Europe and Northern Asia* 1: 150–157.

KALYAKIN, V. N. (1999) Birds of the Novaya Zemlya region and Franz-Josef Land. *In* V. K. Ryabitsev (ed.), *Materials on the distribution of birds of the Urals, Priuralye and West Siberia*, 109–137. "Ekaterinburg", Ekaterinburg.

KALYAKIN, V.N. & MOLOCHAEV, A.V. (1990) Results of aerial counts of swans in tundra of Western Siberia. *In* A. I. Koshelev (ed.) *Ecology and Conservation of Swans in the USSR Volume 2*, 47–52. Proceedings of the Second All-Union Conference on Swans of the USSR. Melitopol State Pedagogical Institute, Melitopol. [In Russian.]

KALYAKIN, V.N. & VINOGRADOV, V.G. (1987) Swans in the southern Yamal. *In* E. V. Syroechkovski (ed.), *Swan Ecology and Migration in the USSR*, 95–96. Nauka, Moscow.

KAMIYA, K. & OZAKI, K. (2002) Satellite tracking of Bewick's Swan migration from Lake Nakaumi, Japan. *In* E.C. Rees, S.L. Earnst & J. Coulson (eds), Proceedings of the Fourth International Swan Symposium, 2001. *Waterbirds*, 25 (Special Publication 1): 128–131.

KARPOVICH, V.N. & KOKHANOV, V.D. (1967) Birds of Vaigach Island and the northeastern Yugor Peninsula. *Proceedings Kandalaksha State Reserve, Forestry Industry* (Moscow), 5: 268–335.

Keane, E.M. & O'Halloran, J. (1992) The behaviour of a wintering flock of Mute Swans *Cygnus olor* in southern Ireland. *Wildfowl* 43: 12–19.

Kear, J. (1963) The history of potato-eating by wildfowl in Britain. *Wildfowl Trust Annual Report*, 14: 54–65.

Kear, J. (1970) The adaptive radiation of parental care in waterfowl. *In* J.H. Crook (ed.), *Social Behaviour in Birds and Mammals*, 357–392. Academic Press, London.

Kear, J. (1972) Reproduction and family life. *In* P. Scott & The Wildfowl Trust (eds) *The Swans.* Michael Joseph, London.

Kear, J. (1977) Aviculture, 1976. *Wildfowl* 28: 170–175.

Kear, J. (1978) Aviculture, 1977. *Wildfowl* 29: 170–175.

Kear, J. (1993) Duck decoys, with particular reference to the history of bird ringing. *Archives of Natural History*, 20: 229–40.

Kear, J. (ed.) (2005) *Bird Families of the World: Ducks, Geese and Swans.* Oxford University Press, Oxford.

Kennedy, P.G., Ruttledge, R.F. & Scroope, C.F. (1954) *The Birds of Ireland.* Oliver & Boyd, Edinburgh.

Khrokov, V.V., Auezov, E.M., Buketov, M.E. & Grachev, A.V. (1990) New and rare species of the Turgay depression. *Ornitologia*, 24: 164–165. [In Russian.]

Kim, J-H., Kim, S-W., Park, J-Y. & Yi, J-Y. (1996) Wintering status of waterbirds on major wetlands in Korea. *Journal of Korean Biota* 1: 127–168. [In Korean with English summary.]

Kishchinski, A.A. (1979) Migrations of Bewick's Swans. *In* A.A. Kishchinski (ed.) *Migration of Birds in Eastern Europe and Northern Asia, Ciconiiformes-Anseriformes*, 75–79. Nauka, Moscow. [In Russian.]

Kishchinski, A.A., Zlotin, R.I. & Flint, V. Ye. (1975) Breeding of Bewick's Swans in the Soviet Union. *Zoologicheskiy Zhurnal*, 54: 1525–1528.

Klaassen, M., Lindström, A., Meltofte, H. & Piersma, T. (2001) Arctic waders are not capital breeders. *Nature*, 413: 794.

Klafs, G. & Stübs, J (1987) *Die Vogelwelt Mecklenburgs.* Aula-Verlag, Wiesbaden.

Knudsen, H.L., Laubek, B. & Ohtonen, A. (2002) Growth and survival of Whooper Swan cygnets reared in different habitats in Finland. *In* E.C. Rees, S.L. Earnst & J. Coulson (eds), Proceedings of the Fourth International Swan Symposium, 2001. *Waterbirds*, 25 (Special Publication 1): 211–220.

Koffijberg, K., Voslamber, B. & van Winden, E. (1997) *Ganzen en zwanen in Nederland: overzicht van pleisterplaatsen in de periode 1985–1994.* [*Geese and swans in the Netherlands. Overview of staging areas in 1985–1994.*] SOVON Vogelonderzoek, Beek-Ubbergen.

Kokorev, J.I. (1995) Zur Bestandssituation des Wanderfalken (*Falco perigirnus*), der Rothalsgans (*Branta ruficollis*) und des Zwergschwans (*Cygnus bewickii*) auf Taimyr. *In* P. Prokosh & H. Hotker (eds.), *Faunistik und Naturschutz auf Taimyr – Expeditionen 1989–1991*, 98–107. Corax 16, Sonderheft der Schriftenreihe Hationalpark Schleswig-Holsteinisches Wattenmeer.

Kolpashikov, L.A. (2001) Aerial surveys of the numbers and distribution of the Bewick's Swan in the typical tundra of the western and central Taimyr. *In* A.B. Popovkina (ed.) *Problems of research and conservation of Anseriformes birds of Eastern Europe and North*, pp. 66–67. Goose Swan and Duck Study Group of Northern Eurasia, Moscow. [In Russian.]

KONDRATIEV, A.YA. (1984) Migrations of eastern Bewick's Swans (*Cygnus bewickii jankowskii*) and their wintering grounds in Japan. *Zoologicheskiy Zhurnal*, 63: 1835–1848. [In Russian.]

KONDRATIEV, A.YA. (1985) Breeding biology of the Bewick's Swan *Cygnus bewickii* Yarr. in western Chukotka. *In* A.V. Andreev (ed.), *Rare and Endangered Birds of the Far East*, 29–42. Academy of Sciences USSR, Vladivostock. [In Russian.]

KONDRATIEV, A.YA. (1990) Swans at the extreme northeastern USSR: current status of populations, results and prospects of research. *In* A.I. Koshelev (ed.), *Ecology and Conservation of Swans in the USSR, Volume 2*, 75–78. Proceedings of the Second All-Union Conference on Swans of the USSR. Melitopol State Pedagogical Institute, Melitopol. [In Russian.]

KONDRATIEV, A.YA. (1991a) The distribution and status of Bewick's Swans *Cygnus bewickii*, Tundra Swans *C. columbianus* and Whooper Swans *C. cygnus* in the extreme northeast of the USSR. *In* J. Sears & P.J. Bacon (eds), Proceedings of the Third International Swan Symposium, Oxford 1989. *Wildfowl Special Supplement No. 1*: 56–61.

KONDRATIEV, A.YA. (1991b) Breeding biology of Bewick's Swans *Cygnus bewickii* in Chukota, far eastern USSR. *In* J. Sears & P.J. Bacon (eds), Proceedings of the Third International Swan Symposium, Oxford 1989. *Wildfowl Special Supplement No. 1*: 167–171.

KONDRATIEV, A.YA. (ed.) (1998) *Red Data Book of the Northern Far East Russia. Animals.* Penta, Moscow. [In Russian.]

KONDRATIEV, A.YA. & KISHCHINSKI, A.A. (1987) Marking of *Cygnus bewickii*, in western Chukotka. *In* E. V. Syroechkovski (ed.), *Swan Ecology and Migration in the USSR*, 166–167. Nauka, Moscow.

KONDRATYEVA, L.F. (1987) *Cygnus bewickii* nutrition in the breeding grounds. *Zoologicheskiy Zhurnal*, 66: 1224–1228. [In Russian.]

KOTKIN, N.A. (2000) Development of network of protected areas in the Nenets Autonomous Okrug. *In* B.S. Ebbinge *et al.* (eds), *Heritage of the Russian Arctic: Research, Conservation and International Cooperation*, 595–599. Ecopros Publishers, Moscow.

KRECHMAR, A.V. & KONDRATIEV, A. YA. (1986) Comparative and ecological analysis of Bewick's and Whooper Swan nesting biology. *In* A.V. Andreev & A.V. Krechmar (eds), *Experimental methods and results of their application in northern bird studies*, 34–58. Academy of Sciences USSR, Vladivostock. [In Russian.]

KRECHMAR, A.V. & KRECHMAR, E.A. (1997) Waterfowl of the Kava River drainage. *In* A. V. Andreev (ed.), *Species diversity and population status of waterside birds in northeast Asia. A Collection of Study Reports*, 89–124. North-Eastern Scientific Centre, Far Eastern Branch of Russian Academy of Sciences, Magadan. [In Russian.]

KRECHMAR A.V., ANDREEV A.V. & KONDRATYEV, A.YA. (1991) *Birds of the Northern Plains.* Nauka, Leningrad. [In Russian.]

KRIVENKO, V.G. (ed.) (1999) *Wetlands in Russia, Volume 1: Wetlands of international importance.* Wetlands International – AEME Publication No. 52, Wageningen, The Netherlands.

KUCHERUK, V.V., KOVALEVSKI, YU.V. & SURBANOS, A.G. (1975) Changes in population and bird fauna in the southern Yamal in the last 100 years. *IUCN Bulletin* (Bio Dept, 19#5), 80: 112–118.

Kurlavicius, P. (2003) *Field Guide to the birds of Lithuania*. Latute, Kaunas. [In Lithuanian.]
Labutin, Yu.V., Degtyarev, A.G. & Blokhin, Yu. Yu. (1984) Numbers and distribution of waterfowl in the Lena Delta and adjacent territories. *In* V. G. Krivenko (ed.), *Abstracts of the All-Union Seminar: Current Status of Waterfowl Resources*, 180–182. VNIIPrirody, Moscow. [In Russian.]
Lack, D. (1968) *Ecological Adaptations for Breeding Birds*. Chapman & Hall, London.
Latham, J.A. (1821–1824). *A General History of Birds*. Jacob & Johnson, Winchester.
Laubek, B. (1995a) Udbredelse og fæologi hos rastende og overvintrende Sang- og Pibesvaner i Danmark 1991–93. *Dansk Ornitologisk Forenings Tidsskrift*, 89: 67–82.
Laubek, B. (1995b) Habitat use by Whooper Swans *Cygnus cygnus* and Bewick's Swans *Cygnus columbianus bewickii* wintering in Denmark: increasing agricultural conflicts. *Wildfowl* 46: 8–15.
Laubek, B. (1998) The northwest European Whooper Swan (*Cygnus cygnus*) population: ecological and management aspects of an increasing waterfowl population. Unpublished Ph.D. thesis, University of Åarhus, Denmark.
Laubek, B., Nilsson, L., Wieloch, M., Koffijberg, K., Sudfelt, C. & Follestad, A. (1999) Distribution, numbers and habitat choice of the NW European Whooper Swan *Cygnus cygnus* population: results of an international census in January 1995. *Vogelwelt*, 120: 141–154.
Leito, A. & Luigujõe, L. (1991) Results of the Whooper and Bewick's Swan census. *Eesti Loodus*, 6: 388–390. [In Estonian with English summary.]
Lensink, C.H. (1968) Neckbands as an inhibitor of reproduction in Black Brant. *Journal of Wildlife Management* 32: 418–420.
Lessells, C.M. (1985) Natal and breeding dispersal of Canada Geese *Branta canadensis*. *Ibis* 127: 31–41.
Li, X. (1996) Status and conservation of wildfowl (*Anatidae*) in the Heilongjiang Province, China. *In* M. Birkan, J. van Vessem, P. Havet, J. Madsen, B. Trolliet & M. Moser (eds), Proceedings Anatidae 2000 Conference, Strasbourg, France, 5–9 December 1994. *Gibier Faune Sauvage, Game Wildlife*, 13: 1111–1118.
Livezey, B.C. (1996) A phylogenetic analysis of geese and swans (Anseriformes: Anserinae), including selected fossil species. *Systematic Biology* 45: 415–50.
Limpert, R.J., Allen, H.A. & Sladen, W.J.L. (1987) Weights and measurements of wintering Tundra Swans. *Wildfowl* 38: 108–113.
Limpert, R.J. & Earnst, S.L. (1994) Tundra Swan (*Cygnus columbianus*). *In* A. Poole & F. Gill (eds), *The Birds of North America*, No. 89. The Academy of Natural Sciences, Philadelphia and The American Ornithologists' Union, Washington D.C..
Lindell, L., Welander, B. & Dahlfors, S. (2000). Sweden. *In* M.F. Heath & M.I. Evans (eds) *Important Bird Areas in Europe: Priority Sites for Conservation. Volume 1: Northern Europe*, 673–697. BirdLife International (BirdLife Conservation Series No. 8), Cambridge, U.K.
Litvin, K.E., Gurtovaya, E.N. & Syroechkovsky, E.V. (1999) The structure of the Bewick's Swan population in the breeding period. *Casarca*, 5: 185–202.
Lofts, B. & Murton, R.K. (1968) Photoperiodic and physiological adaptations regulating avian breeding cycles and their ecological significance. *Journal of Zoology* 155: 327–94.
Lu, J. (1990) *Wetlands in China*. East China Normal University Press, Shanghai.

LU, J. (1995) The status and conservation needs of Anatidae and their habitat in China. Unpublished report for the Workshop on Action Plan for Anatidae during the Northeast Asia and North Pacific Environment Forum, Kushiro, Japan, 25–29 September 1995. Northeast Asia and North Pacific Environment Forum, Kushiro.

LUIGUJÕE, L., KURESOO, A., KESKPAIK, J., ADER, A. & LEITO, A. (1996) Migration and staging of the Bewick's Swan (*Cygnus columbianus*) in Estonia. *In* M. Birkan, J. van Vessem, P. Havet, J. Madsen, B. Trolliet & M. Moser (eds), Proceedings Anatidae 2000 Conference, Strasbourg, France, 5–9 December 1994. *Gibier Faune Sauvage, Game Wildlife* 13: 451–461.

LUTSIUK, O.B. & SYCHEV, YE.V. (1974) Materials for the studies of bird fauna on the Chukotka Peninsula. *In* N. G. Solomonov and M. V. Popov (eds), *Proceedings of Sixth Symposium on Biological Problems of the North*, 147–150. Issue I, Yakutian Filial of Siberian Branch of the USSR Academy of Sciences, Yakutsk. [In Russian.]

MA, M. & CAI, D. (2000) *Swans in China.* The Trumpeter Swan Society, Maple Plain, Minnesota. [First published in Chinese as Ma, M. & Cai, D. 1993. *Wild Swans.* China Meteorological Press, Beijing]

MA, M. & CAI, D. (2002) Threats to Whooper Swans in Xinjiang, China. *In* E.C. Rees, S.L. Earnst & J. Coulson (eds), Proceedings of the Fourth International Swan Symposium, 2001. *Waterbirds*, 25 (Special Publication 1): 331–333.

MACDONALD, J.W., CLUNAS, A.J., EATOUGH, C.J., ISHERWOOD, P. & RUTHVEN, A.D. (1987) Causes of death in Scottish swans (*Cygnus* spp). *State Veterinary Journal*, 41: 165–175.

MADDOCK, L. (1979) The "migration" and grazing succession. *In* A.R.E. Sinclair & M. Norton-Griffiths (eds), *Serengeti. Dynamics of an Ecosystem.* 104–129. University of Chicago Press, Chicago.

MADSEN, J. & NOER, H. (1996) Decreased survival of pink-footed geese *Anser brachyrhynchus* carrying shotgun pellets. *Wildlife Biology* 2: 75–82.

MATHIASSON, S. (1973) Moulting grounds of Mute Swans (*Cygnus olor*) in Sweden, their origin and relation to the population dynamics of Mute Swans in the Baltic area. *Viltrevy*, 8: 399–452.

MATSUI, S., YAMANOUCHI, N. & SUZUKI, T. (1981) On the migration route of swans in Hokkaido, Japan. *In* G.V.T. Matthews & M. Smart (eds), *Proceedings of the Second International Swan Symposium, Sapporo 1980*, 60–70. International Waterfowl Research Bureau, Slimbridge.

MATTHEWS, G.V.T. (1968) *Bird Navigation. Second Edition.* Cambridge University Press, Cambridge.

MATTHEWS, G.V.T. & CAMPBELL, C.R.G. (1969) Weights and measurements of Greylag Geese in Scotland. *Wildfowl* 20: 86–93.

MAYHEW, P.W. (1988) The daily energy intake of European Wigeon in winter. *Ornis Scandinavica*, 19: 217–223.

MAYNARD-SMITH, J. & PARKER, G.A. (1976) The logic of asymmetric contests. *Animal Behaviour* 24: 159–175.

MAYR, E. & SHORT, L.L. (1970) Species taxa of North American birds. *Bull. Nuttall Ornithol. Club*, No.9.

MCCEERY, R.H., PERRINS, C., WHEELER, D. & GROVES, S. (2002) Population structure, survival rates and productivity of Mute Swans breeding in a colony in Abbotsbury, Dorset, England. *In* E.C. Rees, S.L. Earnst & J. Coulson (eds), Proceedings of the

Fourth International Swan Symposium, 2001. *Waterbirds*, 25 (Special Publication 1): 192–201.

McWilliam, J.M. (1936) *The Birds of the Firth of Clyde.* H.F. & G. Witherby Ltd., London.

Meier, W. (1974) Über jahresperiodisches Auftraten der Wasser- und Watvögel an dem Elbabschnitt zwischen Penkefitz und Hitzacker. *Vogelkundliche Berichte aus Niedersachsen*, 6: 33–76.

Melter, J. & Schreiber, M. (2000) Wichtige Brut- und Rastvogelgebiete in Niedersachsen. Vogelkundliche Berichte aus Niedersachsen 32, Sonderheft.

Merne, O.J. (1972) Bewick's Swans feeding on waste potatoes and other agricultural crops. *British Birds* 65: 394–395.

Merne, O.J. (1977) The changing status and distribution of the Bewick's Swan in Ireland. *Irish Birds* 1: 3–15.

Merne, O.J. & Walsh, A. (1991) An orange-legged Bewick's Swan in Co. Wexford. *Irish Birds* 4: 421–422.

Mikami, S. (1989) First Japanese records of crosses between Whistling *Cygnus columbianus columbianus* and Bewick's Swans *Cygnus columbianus bewickii. Wildfowl* 40: 131–133

Mineyev, Y.N. (1986) Seasonal movements and numbers of swans in the northeast of the European part of the USSR. *Bulletin Moscow Naturalists' Society* (Biology Department), 91 (3): 19–28.

Mineyev, Y.N. (1987) *Waterfowl of the Bolshezemelskaya tundra: Fauna and ecology.* Nauka, Leningrad.

Mineyev, Y. N. (1990) Autumn migration of swans in tundras of the European northwest of the USSR. *In* A.I. Koshelev (ed.), *Ecology and Conservation of Swans in USSR, Volume 2*, 98–100. Proceedings of the Second All-Union Conference on Swans of the USSR. Melitopol State Pedagogical Institute, Melitopol. [In Russian.]

Mineyev, Y.N. (1991) Distribution and numbers of Bewick's Swans *Cygnus bewickii* in the European northeast of the USSR. *In* J. Sears & P.J. Bacon (eds), Proceedings of the Third International Swan Symposium, Oxford 1989. *Wildfowl Special Supplement No. 1*: 62–67.

Mineyev, Y.N. (1994) *Waterfowl of the Yugorsky Peninsula.* Komi Science Centre, Ural Branch of the Russian Academy of Sciences, Syktyvkar.

Mineyev Y. N. (2003) *Anseriformes Birds of the Eastern-European Tundra.* Ural Branch of the Russian Academy of Sciences, Ekateringburg.

Mineyev, Y. N. & Kondratiev, A. Ya. (2001) Bewick's Swan. *In* D. S. Pavlov *et al.* (eds), *Red Data Book of the Russian Federation. Animals*, 406–408. Astrel, Moscow

Miyabayashi, Y. & Mundkur, T. (1999) *Atlas of Key Sites for Anatidae in the Eastern Flyway.* Wetlands International - Japan, Tokyo, and Wetlands International - Asia Pacific, Kuala Lumpur.

Møller, A.P. (1990) Fluctuating asymmetry in male sexual ornaments may reliably reveal male quality. *Animal Behaviour* 40: 1185–1187.

Monda, M.J., Ratti, J.T. & McCabe, T.R. (1994) Reproductive ecology of Tundra Swans on the Arctic National Wildlife Refuge, Alaska. *Journal of Wildlife Management* 58: 757–73.

Monval, J-Y. & Pirot, J-Y. (1989) *Results of the IWRB International Waterfowl Census 1967–1986.* IWRB Special Publication No. 8. International Waterfowl Research Bureau, Slimbridge.

Moores, N. (2002) Wetlands: Korea's most threatened habitat. *Oriental Bird Club Bulletin*, 36: 54–60.

Moores, N., Kim, S-K., Park, S-B. & Tobai, S. (2001) The Yellow Sea Eco-region: Reconnaissance Report of Important Wetland and Marine Areas for Biodiversity. Volume 2: South Korea. WWF-Japan, Wetlands and Birds Korea, and Wetlands International–China Programme, Busan.

Morozov V. V. (1996) Where are Bewick's Swan wintering grounds located? *Casarca*, 2: 237–243.

Morozov, V.V. (2001) Materials to the knowledge of ornitofauna of Vaygach Island. *Ornitologia*, 29: 29–46.

Morozov, V.V. & Syroechkovki, E.E. (In press) The materials on bird fauna of Kolguev Island. *Ornithologia* 31.

Mortensen, H.C.C. (1950) *Studies in Bird Migration.* Danish Ornithological Society, Munksgaard, Copenhagen.

Mullie, W.C. & Poorter, E.P.R. (1977) Aantallan, verspreiding en terreinkeus van de Kleine Zwaan bij vijf landelijke tellingen in 1976 en 1977. *Watervogels*, 2: 85–96.

Munro, R.E., Smith, L.T. & Kupa, J.J. (1968) The genetic basis of colour differences in the Mute Swan (*Cygnus olor*). *Auk*, 85: 504–505.

Murton, R.K. & Kear, J. (1973) The nature and evolution of photoperiodic control of reproduction in wildfowl of the family Anatidae. *J. Reprod. Fert., Suppl.* 19: 67–84.

Murton, R.K. & Kear, J. (1978) Photoperiodism in waterfowl: phasing of breeding cycles and zoogeography. *Journal of Zoology* 186: 243–283.

Murton, R.K. & Westwood, N.J. (1977) *Avian breeding cycles.* Oxford University Press, Oxford.

Nature Conservation Bureau, Ministry of the Environment, Japan. (2003) *The report of the 34th annual census of waterfowl (Anatidae) in January 2003.* Ministry of the Environment, Tokyo. [In Japanese with English summary.]

Newton, I. (2004) Population limitation in migrants. *Ibis* 146: 197–226.

Nisbet, I.T.C. (1959) Bewick's Swans in the British Isles in the winters of 1954–55 and 1955–56. *British Birds* 52: 393–416.

Nisbet, I.C.T. & Drury, W.H. (1968) Short-term effects of weather on bird migration: a field study using multivariate statistics. *Animal Behaviour* 16: 496–530.

Noer, H. & Madsen, J. (1996) Shotgun pellet loads in pink-footed geese *Anser brachyrhynchus. Wildlife Biology* 2: 65–73.

Nolet, B.A. & Drent, R.H. (1998) Bewick's Swans refuelling on pondweed tubers in the Dvina Bay (White Sea) during their spring migration: first come, first served. *Journal of Avian Biology* 29: 574–581.

Nolet, B.A. & Mooij, W.M. (2002) Search paths of swans foraging on spatially autocorrelated tubers. *Journal of Animal Ecology* 71: 451–462.

Nolet, B.A., Langevoord, O., Bevan, R.M., Engelaar, K.R. Klaassen, M., Mulder, R.J.W. & Van Dijk, S. (2001a) Spatial variation in tuber depletion by swans explained by differences in net intake rates. *Ecology* 82: 1655–1667.

Nolet, B.A., Andreev, V.A., Clausen, P., Poot, M.J.M. & Wessel, E.G.J. (2001b) Significance of the White Sea as a stopover for Bewick's Swans *Cygnus columbianus bewickii* in spring. *Ibis* 143: 63–71.

NOLET, B.A., BEVAN, R.M., KLAASSEN, M., LANGEVOORD, O. & VAN DER HEIJDEN, Y.G.J.T. (2002) Habitat switching by Bewick's Swans: maximization of average long-term energy gain? *Journal of Animal Ecology* 71: 979–993.

NOLET, B.A., FULD, V.N. & VAN RIJSWIJK, M.E.C. (In press a). Foraging costs and accessibility as determinants of giving-up densities in a swan-pondweed system. *Oikos* (Special edition).

NOLET, B.A., KLAASSEN, R.H.G. & MOOIJ, W.M. (In press b) The use of a flexible patch leaving rule under exploitative competition: a field test with swans. *Oikos* (Special edition).

NOORDHUIS, R. (ed.) (2000) *Watersysteemrapportage Randmeren.* RIZA, Lelystad, the Netherlands.

NORBERG, U.M. (1996) Energetics of flight. *In* C. Carey (ed.), *Avian Genetics and Nutritional Ecology*, 199–249. Chapman & Hall, New York.

NOWAK, E., BERTHOLD, P. & QUERNER, U. (1990) Satellite tracking of migrating Bewick's Swans: a European pilot study. *Naturwissenschaften* 77: 549–550.

O'DONOGHUE, P.D. & O'HALLORAN, J. (1994) The behaviour of a wintering flock of Whooper Swans *Cygnus cygnus* at Rostellan Lake, Cork. *Proceedings of the Royal Irish Academy* 94: 109–118.

OGILVIE, M.A. (1968) The Bewick's Swans, 1967–68. *Wildfowl* 19: 162–164.

OGILVIE, M.A. (1969). Bewick's Swans in Britain and Ireland during 1956–69. *British Birds* 62: 505–522.

OGILVIE, M.A. (1972a). Large numbered leg bands for individual identification of swans. *Journal of Wildlife Management* 36: 1261–1265.

OGILVIE, M.A. (1972b) Distribution, numbers and migration. *In* P. Scott & The Wildfowl Trust (eds), *The Swans.* Michael Joseph, London.

OGILVIE, M.A. (1986) Editorial note. *British Birds* 79: 206.

O'HALLORAN, J., SMIDDY, P., QUISHI, X., O'LEARY, R. & HAYES, C. (2002) Trends in Mute Swan blood levels: evidence of grit reducing lead poisoning. *In* E.C. Rees, S.L. Earnst & J. Coulson (eds), Proceedings of the Fourth International Swan Symposium, 2001. *Waterbirds*, 25 (Special Publication 1): 363–367.

OHMORI, T. (1981) Artificial feeding of swans in Japan. *In* G.V.T. Matthews & M. Smart (eds), *Proceedings of the Second International Swan Symposium, Sapporo 1980*, 244–246. International Waterfowl Research Bureau, Slimbridge.

OSIECK, E. (2000) The Netherlands. *In* M.F. Heath & M.I. Evans (eds), *Important Bird Areas in Europe: Priority Sites for Conservation. Volume 1: Northern Europe*, 465–507. BirdLife International (BirdLife Conservation Series No. 8), Cambridge, U.K.

OSTAPENKO, V.A. (1991). Migration of Bewick's Swans *Cygnus bewickii* and Whooper Swans *C. cygnus* wintering in Japan through Sakhalin Island and adjacent territories, USSR. *In* J. Sears & P.J. Bacon (eds), Proceedings of the Third International Swan Symposium, Oxford 1989. *Wildfowl Special Supplement No. 1*: 224–226.

O'SULLIVAN, O. & SMIDDY, P. (1991) Thirty-eighth Irish Bird Report, 1990. *Irish Birds* 4: 423–462.

O'SULLIVAN, O. & SMIDDY, P. (1992) Thirty-ninth Irish Bird Report, 1991. *Irish Birds* 4: 571–620.

OTS, M. & KALAMEES, A. (2000) Estonia. *In* M.F. Heath & M.I. Evans (eds), *Important Bird Areas in Europe: Priority Sites for Conservation. Volume 1: Northern Europe*, 205–224. BirdLife International (BirdLife Conservation Series No. 8), Cambridge, U.K.

Owen, M. (1972) Some factors affecting food intake and site selection in White-fronted geese. *Journal of Animal Ecology* 41: 79–92.

Owen, M. (1977) *Wildfowl of Europe.* Macmillan, London.

Owen, M. (1980) *Wild Geese of the World.* Batsford, London.

Owen, M. (1981) Abdominal profile – a condition index for wild geese in the field. *Journal of Wildlife Management* 45: 227–30.

Owen, M. & Black, J.M. (1990) *Waterfowl Ecology.* Blackie, Glasgow.

Owen, M. & Cadbury, C.J. (1975) The ecology and mortality of swans on the Ouse Washes, England. *Wildfowl* 26: 31–42.

Owen, M. & Cook, W.A. (1977) Variation in body weight, wing length and condition of Mallard *Anas platyrhynchos platyrhyncos* and their relationship to environmental change. *Journal of Zoology* 183: 377–395.

Owen, M. & Kear, J. (1972) Food and feeding habits. *In* P. Scott & The Wildfowl Trust (eds), *The Swans.* Michael Joseph, London.

Owen, M. & Kerbes, R.H. (1971) On the autumn food of Barnacle Geese at Caerlaverock National Nature Reserve. *Wildfowl* 22: 114–119.

Owen, M. & Mitchell, C. (1988) The movements and migrations of Wigeon *Anas penelope* wintering in Britain and Ireland. *Bird Study* 35: 47–59.

Owen, M. & Norderhaug, M. (1977) Population dynamics of Barnacle Geese *Branta leucopsis* breeding in Svalbard, 1946–1976. *Ornis Scandinavica* 8: 161–174.

Owen, M. & Ogilvie, M. (1979) Wing moult and weights of Barnacle Geese in Spitzbergen. *Condor* 81: 42–52.

Owen, M., Gullestad, N. & St. Joseph, A.K..M. (1977) Barnacle Goose project 1976–77. Unpublished report to the Wildfowl Trust.

Owen, M., Atkinson-Willes, G.L. & Salmon, D.G. (1986) *Wildfowl in Great Britain, second edition.* Cambridge University Press, Cambridge.

Paakspuu, V. (1967) Observations of the transit of swans along the west coast of Estonia. *Communication of the Baltic Commision for the Study of Bird Migration* 4: 46–52.

Paakspuu, V. & Kastepold, T. (1985) The waterfowl, marsh and shorebird fauna of the Matsalu wetland. *In* E. Kumari (ed.), *Matsalu - a Wetland of International Importance.* Valgus, Tallinn.

Pak, U-I. (1995) On the distribution and ecology of Anatidae in the D.P.R. Korea. Unpublished report for the Workshop of Action Plan for Anatidae during the 1995 Northeast Asia and North Pacific Environment Forum, Kushiro, Japan, 25–29 September 1995.

Pallas, P.S. (1811) *Zoographia Rosso-Asiatica.* Petrpolis: In Officina Caes. Academiae Scientiarum impress 1811 [St. Petersburg: Imperial Academy of Sciences 1811], Volume II: 214–215.

Palmer, R.S. (ed.) (1976) *Handbook of North American birds, Volume 2.* Yale University Press, New Haven.

Pang, B. (1974) Edible birds in southern Jiangsu Province. *Chinese Journal of Zoology* 4: 26–29.

Pang, B. (1976) Birds in "Compendium of Materia Medica". *Chinese Journal of Zoology* 2: 35–37.

Park, J-Y. (2002) Current status and distribution of birds in Korea. Unpublished Ph.D. thesis, Kyung-Hee University, Korea. [In Korean.]

Parkes, K.C. (1958) Systematic notes on North American birds. 2. The waterfowl (Anatidae). *Annals of the Carnegie Museum,* 35: 117–125.

PARKER, G.A. (1974) Assessment strategy and the evolution of fighting behaviour. *Journal of Theoretical Biology* 47: 223–243.

PASKHALNY, S.P. (1996) Geese breeding in the central Yamal in 1995 (the Khalevto Lake area). *Casarca*, 2: 305–306.

PAULUS, S.L. (1984) Activity budgets of non-breeding Gadwalls in Louisiana. *Journal of Wildlife Management* 48: 371–380.

PENNYCOTT, T., YOUNG, F.M., METCALF, J.S. & CODD, G.A. (2004) Necrotic enteritis in Mute Swans associated with cynobacterial toxins. *The Veterinary Record*, May 1: 575–576.

PENNYCUICK, C.J. (1978) Identification using natural markings. *In* B. Stonehouse (ed.), *Animal Marking.* Macmillan, London.

PENNYCUICK, C.J. (1989) *Bird Flight Performance: A practical calculation manual.* Oxford University Press, Oxford.

PENNYCUICK, C.J., EINARSSON, Ó., BRADBURY, T.A.M. & OWEN, M. (1996) Migrating Whooper Swans (*Cygnus cygnus*): satellite tracks and flight performance calculations. *Journal of Avian Biology* 27: 118–134.

PENNYCUICK, C.J., BRADBURY, T.A.M., EINARSSON, Ó. & OWEN, M. (1999) Response to weather and light conditions of migrating Whooper Swans *Cygnus cygnus* and flying height profiles, observed with the Argos satellite system. *Ibis* 141: 434–443.

PERENNOU, C., ROSE, P. & POOLE, C. (1990) Asian Waterfowl Census 1990: mid winter counts, January 1990. The International Waterfowl and Wetlands Research Bureau, Slimbridge, and the Asian Wetland Bureau, Kuala Lumpur.

PERENNOU, C., MUNDKUR, T., SCOTT, D.A., FOLLESTAD, A. & KVENILD, L. (1994) The Asian Waterfowl Census 1987–91. Distribution and status of Asian Waterfowl. AWB Publ. No. 86; IWRB Publ. No. 24. Asian Wetland Bureau, Kuala Lumpur, and the International Waterfowl and Wetlands Research Bureau, Slimbridge.

PERFILIEV, V.I. (1987) Whooper Swan and Bewick's Swan in Northern Yakutia. *In* E.V. Syroechkovski (ed.), *Ecology and Migrations of Swans in the USSR.* Nauka, Moscow.

PERRINS, C.M. & REYNOLDS, C.M. (1967) A preliminary study of the Mute Swan *Cygnus olor. Wildfowl Trust Annual Report* 18: 74–84.

PERRINS, C.M. & SEARS, J. (1991) Collisions with overhead wires as a cause of mortality in Mute Swan *Cygnus olor. Wildfowl* 42: 5–11.

PETRIE, S.A. & WILCOX, K.L. (2003) Migration chronology of eastern-population Tundra Swans. *Canadian Journal of Zoology* 81: 861–870.

POLLITT, M., HALL, C., HOLLOWAY, S., HEARN, R., MARSHALL, P., MUSGROVE, A., ROBINSON, J., & CRANSWICK, P. (2003) *The Wetland Bird Survey 2000–01. Wildfowl and Wader Counts.* BTO/WWT/RSPB/JNCC, Slimbridge.

POORTER, E.P.R. (1978) De Kleine Zwanen van Arkemheen, De Haar en Zeldert. *De Lepelaar* 54: 14–17.

POORTER, E.P.R. (1981) *Cygnus columbianus bewickii* in the border lakes of the IJsselmeer polders. *In* G.V.T. Matthews & M. Smart (eds), *Proceedings of the Second International Swan Symposium, Sapporo 1980*, 49–57. International Waterfowl Research Bureau, Slimbridge.

POORTER, E.P.R. (1991) *Bewick's Swans* Cygnus columbianus bewickii*, an analysis of breeding success and changing resources.* Ministerie van Verkeer en Waterstaat, Rijkswaterstaat, Directie Flevoland.

PORTENKO, V.I. (1972) *Birds of the Chukotski Peninsula and Wrangel Island.* Part 1. Nauka, Leningrad. [In Russian.]

Poyarkov, N. D. & Johnson, S.R. (1996) The unique productivity of the Whooper Swan in the middle reaches of the Ob River (Khanty-Mansy Autonomous district). *Casarca*, 2: 230–235.

Poyarkov, N. D., Hodges, J. & Eldridge, W. (2000) *Atlas of the Distribution of Birds of the Coastal Tundra of Northeastern Asia (materials of 1993–1995 aerial surveys)* (A. B. Popovkina, ed.). Centre for Nature Conservation and Goose, Swan and Duck Study Group of Northern Eurasia, Moscow. [In Russian with English summary.]

Pozdnyakov, V.I. (2001) On the study of migration of Bewick's Swans in the Eastern Palearctic. *Casarca*, 7: 424–425.

Pozdnyakov, V.I. (2002) Status and Breeding Ecology of Bewick's Swans in the Lena Delta, Yakutia, Northern Asia. *In* E.C. Rees, S.L. Earnst & J. Coulson (eds), Proceedings of the Fourth International Swan Symposium, 2001. *Waterbirds*, 25 (Special Publication 1): 95–99.

Pozdnyakov, V.I. & Degtyarev, A.G. (2001) Bewick's Swans in Yakutia. *In* A.B. Popovkina (ed.), *Problems of research and conservation of Anseriformes birds of eastern Europe and the north*, 98–99. Goose, Swan and Duck Study Group of Northern Eurasia, Moscow. [In Russian.]

Pritchard, G., Ainsworth, H., Brown, M. & Duff, J.P. (2004) Suspected necrotic enteritis in wild swans. *The Veterinary Record*, April 10: 480.

Ptushenko, E. C. (1952) Anseriformes. *In* G.P. Dementiev & N.A. Gladkov (eds), *The Birds of the Soviet Union*, 247–344. Nauka, Moscow. [In Russian.]

Rasmussen, J.F., Neilsen, M. & Flensted, K.N. (2000) Denmark. *In* M.F. Heath & M.I. Evans (eds), *Important Bird Areas in Europe: Priority Sites for Conservation. Volume 1: Northern Europe*, 137–177. BirdLife International (BirdLife Conservation Series No. 8), Cambridge, U.K.

Raudonikis, L. (2004) *Important Bird Areas of EU Importance in Lithuania.* Latute, Kaunas. [In Lithuanian.]

Raveling, D.G. (1976) Do neckbands contribute to starvation of Lesser Snow Geese? *Journal of Wildlife Management* 40: 571–572.

Ravkin, Y.S. (1991) Numbers and distribution of Mute Swans *Cygnus olor*, Whooper Swans *C. cygnus* and Bewick's Swans *C. bewickii* in the West Siberian plain. *In* J. Sears & P.J. Bacon (eds), Proceedings of the Third International Swan Symposium, Oxford 1989. *Wildfowl Special Supplement No. 1*: 68–72.

Rees, E.C. (1981) The recording and retrieval of bill pattern variations in the Bewick's Swan. *In* G.V.T. Matthews & M. Smart (eds), *Proceedings of the Second International Swan Symposium, Sapporo 1980*, 105–119. International Waterfowl Research Bureau, Slimbridge.

Rees, E.C. (1982) The effect of photoperiod on the timing of spring migration in the Bewick's Swan. *Wildfowl* 33: 119–132.

Rees, E.C. (1987) Conflict of choice within pairs of Bewick's Swans regarding their migratory movement to and from the wintering grounds. *Animal Behaviour* 35: 1685–1693.

Rees, E.C. (1988) Aspects of the migration and movements of individual Bewick's Swans. Unpublished Ph.D. thesis, University of Bristol, U.K.

Rees, E.C. (1989) Consistency in the timing of migration for individual Bewick's Swans. *Animal Behaviour* 38: 384–393.

Rees, E.C. (1990) Bewick's Swans: their feeding ecology and coexistence with other grazing anatidae. *Journal of Applied Ecology* 27: 939–951.

Rees, E.C. (1991) Distribution within the USSR of Bewick's Swans *Cygnus columbianus bewickii* marked in Britain. *In* J. Sears & P.J. Bacon (eds), Proceedings of the Third International Swan Symposium, Oxford 1989. *Wildfowl Special Supplement No. 1*: 209–213.

Rees, E.C. & Bacon, P.J. (1996) Migratory tradition in Bewick's Swans *Cygnus columbianus bewickii*. *In* M. Birkan, J. van Vessem, P. Havet, J. Madsen, B. Trolliet & M. Moser (eds), Proceedings Anatidae 2000 Conference, Strasbourg, France, 5–9 December 1994. *Gibier Faune Sauvage, Game Wildlife* 13: 407–420.

Rees, E.C. & Bowler, J.M. (1991) Feeding ecology of Bewick's Swans *Cygnus columbianus bewickii* at a migratory site in the Estonian SSR. *In* J. Sears & P.J. Bacon (eds), Proceedings of the Third International Swan Symposium, Oxford 1989. *Wildfowl Special Supplement No. 1*: 249–255.

Rees, E.C. & Bowler, J.M. (1996) Fifty years of swan research and conservation by the Wildfowl and Wetlands Trust. *Wildfowl* 47: 248–263.

Rees, E.C. & Bowler, J.M. (2002) Bewick's Swan *Cygnus columbianus*. *In* C.V. Wernham, M.P. Toms, J.H. Marchant, J.A. Clark, G.M. Siriwardena & S.R. Baillie (eds), *The Migration Atlas: Movements of the Birds of Britain and Ireland*, 149–153. T. & A D. Poyser, London.

Rees, E.C., Owen, M., Gitay, H. & Warren, S. (1990) The fate of plastic leg rings used on geese and swans. *Wildfowl* 41: 43–52.

Rees, E. C., Black, J.M., Spray, C.J. & Thorisson, S. (1991) Comparative study of the breeding success of Whooper Swans *Cygnus cygnus* nesting in upland and lowland regions of Iceland. *Ibis* 133: 365–373.

Rees, E.C., Lievesley, P., Pettifor, R. & Perrins, C. (1996) Mate fidelity in swans: an inter-specific comparison. *In* J.M. Black (ed.), *Partnerships in Birds: The study of monogamy*, 118–137. Oxford University Press, Oxford.

Rees, E.C., Bowler, J.M. & Beekman, J.H. (1997a) *Cygnus columbianus* Bewick's Swan and Whistling Swan. *Birds of the Western Palearctic (BWP) Update* 1: 63–74.

Rees, E.C., Kirby, J.S. & Gilburn, A. (1997b) Site selection by swans wintering in Britain; the importance of habitat and geographic location. *Ibis* 139: 337–352.

Rees, E.C., Colhoun, K., Einarsson, Ó., McElwaine, G., Petersen, Æ. & Thorstensen, S. (2002) Whooper Swan *Cygnus cygnus*. *In* C.V. Wernham, M.P. Toms, J.H. Marchant, J.A. Clark, G.M. Siriwardena and S.R. Baillie (eds), *The Migration Atlas: Movements of the Birds of Britain and Ireland*, (ed.), 154–157. T & A.D. Poyser, London.

Rees, E.C., Matthews, G.V.T., C.R. Mitchell, C.R. & Owen, M. (2005a) Movements and migrations. *In* J. Kear (ed.), *Bird Families of the World: Ducks, Geese and Swans.* Oxford University Press, Oxford.

Rees, E.C., Bruce, J.H. & White, G.T. (2005b) Factors affecting the behavioural responses of Whooper Swans (*Cygnus c. cygnus*) to various human activities. *Biological Conservation*, 121: 369–382.

Ritchie, R.J. & King, J.G. (2000) Tundra Swans. *In* J.C. Truett & S.R. Johnson (eds), *The Natural History of an Arctic Oil Field*, 197–220. Academic Press, New York.

Ritchie, R. J., King, J.G., Stickney, A.A., Anderson, B.A., Rose, J.R., Wildman, A.M. & Hamilton, S. (2002) Population trends and productivity of Tundra Swans on the central Arctic coastal plain, northern Alaska, 1989–2000. *In* E.C. Rees, S.L. Earnst & J. Coulson (eds), Proceedings of the Fourth International Swan Symposium, 2001. *Waterbirds*, 25 (Special Publication 1): 22–31.

Robinson, J.A., Colhoun, K., McElwaine, J.G. & Rees, E.C. (2004) *Bewick's Swan* Cygnus columbianus bewickii *(Northwest Europe population) in Britain and Ireland 1960/61 – 1999/2000*. Waterbird Review Series, The Wildfowl & Wetlands Trust/Joint Nature Conservation Committee, Slimbridge.

Rogacheva, H. (1992) *The Birds of Central Siberia.* Husum Druck- und Verlagsgesellschaft mbH & Co. KG, Husum.

Rogeron, G. (1883) Le cygne de Bewick (*Cygnus minor*). *Bulletin de la Société d'Acclimation de France.* 3 serié, 10: 220.

Rogers, M.J. & the Rarities Committee. (1999) Report on rare birds in Great Britain in 1998. *British Birds*, 92: 554–609.

Rootsmae, L. (1990) On migration of swans in Estonia. *Communications of Baltic Commision for Study of Bird Migration* 23: 86–104.

Rose, P.M. (1995) Western Palearctic and South-West Asia Waterfowl Census 1994. IWRB Publication No. 35, International Waterfowl and Wetlands Research Bureau (IWRB), Slimbridge

Rose, P.M. & Scott, D.A. (1994) *Waterfowl population estimates.* IWRB Publication No. 29, International Waterfowl and Wetlands Research Bureau (IWRB), Slimbridge.

Rose, P.M. & Scott, D.A. (1997) *Waterfowl population estimates—Second edition.* Wetlands International Publication 44. Wetlands International, Wageningen, The Netherlands.

Rüger, A., Prentice, C. & Owen, M. (1986) *Results of the IWRB International Census 1967 – 1983*. IWRB Special Publication No. 6, International Waterfowl and Wetlands Research Bureau (IWRB), Slimbridge.

Ryley, K., & Bowler, J.M. (1994) A change of moulting site for Mute Swans *Cygnus olor* in Gloucestershire. *Wildfowl* 45: 15–21.

Sangster, G., Collinson, J.M., Helbig, A.J., Knox, A.G. & Parkin, D.T. (2004) Taxonomic recommendations for British birds: second report. *Ibis* 146: 153–157.

Santamaría, L. & Rodríguez-Gironés, M.A. (2002) Hiding from swans: optimal burial depth of sago pondweed tubers foraged by Bewick's Swans. *Journal of Ecology* 90: 303–315.

Scheller, W., Strache, R.-R., Eichstädt, W. & Schmidt, E. (2002) *Important Bird Areas (IBA) in Mecklenburg-Vorpommern – die wichtigsten Brut- und Rastvogelgebiete Mecklenburg-Vorpommerns.* cw Obotritendruck, Schwerin.

Schlede, O. (1961) Pibesvanens (*Cygnus columbianus bewickii* Yarrell) forekomst i Danmark. *Dansk Ornitologisk Forenings Tidsskrift*, 55: 185–196.

Schmidt, G.A.J. (1965) Der Zwerg- oder Tundraschwan *Cygnus bewickii* in Schleswig-Holstein und Hamburg. *Corax*, 1: 10–38.

Schubert, M. (1974) Der Zwergschwan 1961 bis 1971 in den drei Nordbezirken der DDR. *Falke*, 21: 114–119.

Scott, D.A. (1980) A preliminary inventory of wetlands of international importance for waterfowl in Western Europe and North-west Africa. IWRB Special Publication No. 2, International Waterfowl Research Bureau, Slimbridge.

Scott, D.A. & Rose, P.M. (1996) *Atlas of Anatidae Populations in Africa and Western Eurasia. Wetlands International Publication 41*, Wetlands International, The Netherlands.

Scott, D.K. (1978) Social behaviour of wintering Bewick's Swans. Unpublished Ph.D. thesis, University of Cambridge, U.K.

Scott, D.K. (1980a) The behaviour of Bewick's Swans at the Welney Wildfowl Refuge, Norfolk, and on the surrounding fens: a comparison. *Wildfowl* 31: 5–18.

SCOTT, D.K. (1980b) Functional aspects of prolonged parental care in Bewick's Swans. *Animal Behaviour* 28: 938–952.
SCOTT, D.K. (1980c) Functional aspects of the pair bond in winter in Bewick's Swans (*Cygnus columbianus bewickii*). *Behavioural Ecology and Sociobiology*, 7: 323–327.
SCOTT, D.K. (1981) Geographical variation in the bill patterns of Bewick's Swans. *Wildfowl* 32: 123–128.
SCOTT, D.K. (1988) Reproductive success in Bewick's Swans. *In* T.H. Clutton-Brock (ed.), *Reproductive Success*, 220–236. University of Chicago Press, Chicago.
SCOTT, P. (1966) The Bewick's Swans at Slimbridge. *Wildfowl Trust Annual Report*, 17: 20–26.
SCOTT, P. (1980) *Observations of Wildlife*. Phaidon, Oxford.
SCOTT, P. (1985) *Travel Diaries of a Naturalist. II.* Collins, London.
SCOTT, P. & SCOTT, PH. (1989) *The Swans Fly In (Second edition)*. Wildfowl & Wetlands Trust, Slimbridge.
SEEBOHM, H. (1901) *The Birds of Siberia: to the Petchora Valley and the Yenesei.* New edition published by Allan Sutton Publishing Ltd., 1985.
SEEGAR, W.S. (1979) Prevelance of parasitic heartworms in swans in England. *Wildfowl* 30: 147–150.
SELBY, P.J. (1831) Observations on the new species of swan, discovered by Mr Richard Wingate, of Newcastle upon Tyne. *Transactions of the Natural History Society of Northumberland, Durham and Newcastle upon Tyne.* Volume I, pp 17–25. [Issue with this paper produced 1830; whole volume dated 1831]
SERMET, E. (1963) Des cygnes de Bewick à Yverdon. *Nos Oiseaux*, 27: 181.
SHCHADILOV, Y. M. (1984) Bewick's Swan. *In* A. M. Borodin, A. G. Bannikov & V. E. Solokov (eds), *Red Data Book of the USSR: Rare and Endangered Species of Animals and Plants. Vol. 1, Animals. 2nd Edition.* Lesnaya Promyshlennost, Moscow. [In Russian.]
SHCHADILOV, Y. M. & ORLOV, V.A. (1987) Swan numbers, distribution and ecology in the breeding period in the north of the Nenetski Autonomous Okrug. *In* Y. E. Syroechkovski (ed.), *Ecology and migration of swans of the USSR*, 77–84. Nauka, Moscow.
SHCHADILOV, Y.M., BELOUSOVA, A.V., REES, E.C. & BOWLER, J.M. (1998) Long-term study of the nesting success in the Bewick's Swans in the coastal tundra in the Nenetskiy Autonomous Okrug. *Casarca*, 4: 217–228.
SHCHADILOV, Y.M., REES, E.C., BELOUSOVA, A.V. & BOWLER, J.M. (2002) Annual variation in the proportion of Bewick's Swans and Whooper Swans breeding in northern European Russia. *In* E.C. Rees, S.L. Earnst & J. Coulson (eds), Proceedings of the Fourth International Swan Symposium, 2001. *Waterbirds*, 25 (Special Publication 1): 86–94.
SINCLAIR, A.R.E. (1979) Dynamics of the Serengeti ecosystem: process and pattern. *In* A.R.E. Sinclair & M. Norton-Griffiths (eds), *Serengeti. Dynamics of an Ecosystem.* 1–30. University of Chicago Press, Chicago.
SLADEN, W.J.L. (1973) A continental study of Whistling Swans using neck collars. *Wildfowl* 24: 8–14.
SOLOVIEVA, D.V. (2000) New ringing programme in the Lena-Delta State Reserve. *Swan Specialist Group Newsletter*, 9: 11.
SOVON GANZEN- EN ZWANENWERKGROEP (1997) *Ganzen en zwanentellingen in Nederland in 1995/96.* SOVON monitoringrapport 97/05, RIZA-rapport BM96.20, IKC Natuurbeheer coproductie 16. SOVON Vogelonderzoek Nederland, Beek-Ubbergen.
SOVON GANZEN- EN ZWANENWERKGROEP (1998) *Ganzen en zwanentellingen in Nederland in 1996/97.* SOVON monitoringrapport 98/06, RIZA-rapport BM97.17,

IKC Natuurbeheer coproductie 20. SOVON Vogelonderzoek Nederland, Beek-Ubbergen.

SOVON Ganzen- en Zwanenwerkgroep (1999) *Ganzen en zwanentellingen in Nederland in 1997/98.* SOVON monitoringrapport 99/06, RIZA-rapport BM98.06, IKC Natuurbeheer coproductie 24. SOVON Vogelonderzoek Nederland, Beek-Ubbergen.

SOVON Ganzen- en Zwanenwerkgroep (2000) *Ganzen en zwanentellingen in Nederland in 1998/99.* SOVON monitoringrapport 2000/03, RIZA-rapport BM99.15, Expertisecentrum LNV coproductie 33. SOVON Vogelonderzoek Nederland, Beek-Ubbergen.

SOVON Ganzen- en Zwanenwerkgroep (2001) *Ganzen- en zwanentellingen in Nederland in 1999/00.* SOVON monitoringrapport 2001/06, RIZA-rapport BM01.17, SOVON Vogelonderzoek Nederland, Beek-Ubbergen.

Spray, C.J. & Bayes, K. (1992) The effect of neck-collars on the behaviour, weight and breeding success of Mute Swans *Cygnus olor. Wildfowl* 43: 49–57.

Spray, C.J. & Chisholm, H.K. (In press) Whooper Swan *Cygnus cygnus.* In *Birds of Scotland.*

Spray, C., Coleman, B. & Coleman, J. (2002) Mute Swan *Cygnus olor. In* C.V. Wernham, M.P. Toms, J.H. Marchant, J.A. Clark, G.M. Siriwardena & S.R. Baillie (eds), *The Migration Atlas: Movements of the Birds of Britain and Ireland,* 146–148. T & AD Poyser, London.

Stallcup, R. & Winter, J. (1975) Middle Pacific Coast region. *American Birds* 29: 735–740.

Stejneger, L. (1882) Outlines of a monograph of the Cygninae. *Proceedings of the U.S. National Museum,* 5: 147–221.

Stewart, A.G. (1978) Swans flying at 8,000 metres. *British Birds* 71: 459–460.

Stewart, R.E. & Manning, J.H. (1958) Distribution and ecology of Whistling Swans in the Chesapeake Bay region. *Auk,* 75: 203–212.

Stickney, A.A., Anderson, B.A., Ritchie, R. J. & King, J.G. (2002) Spatial distribution, habitat characteristics and nest-site selection by Tundra Swans on the Central Arctic Coastal Plain, Northern Alaska. *In* E.C. Rees, S.L. Earnst & J. Coulson (eds), Proceedings of the Fourth International Swan Symposium, 2001. *Waterbirds,* 25 (Special Publication 1): 227–235.

Stratford, J. (1997) Bewick's Swan: a new breeding species in Lithuania. *Acta Zoologica Lituanica* 7: 151–152.

Sutherland, W.J. & Allport, G.A. (1994) A spatial depletion model of the interactions between Bean Geese and Wigeon with consequences for habitat management. *Journal of Animal Ecology* 63: 51–59.

Sutherland, W.J. & Parker, G.A. (1985) Distribution of unequal competitors. *In* R.M. Sibly and R.H. Smith (eds), *Behavioural Ecology,* 255–274. Blackwell Scientific Publications, Oxford.

Švažas, S. (2001) Possible impacts of climatic conditions on changes in numbers and in distribution of certain breeding and staging wildfowl species in Lithuania. *Acta Zoologica Lituanica* 11: 163–183.

Švažas, S. & Kozulin, A. (2002) *Waterbirds of Large Fishponds of Belarus and Lithuania.* OMPO Vilnius, Vilnius.

Švažas, S., Drobelis, E., Balciauskas, L. & Raudonikis, L. (1999) *Important Wetlands in Lithuania.* OMPO Vilnius, Vilnius.

Sviridova, T. (2000). Russia. *In* Heath, M.F. & Evans, M.I. (eds), *Important Bird Areas in Europe: Priority Sites for Conservation. Volume 1: Northern Europe*, 581–652. BirdLife International (BirdLife Conservation Series No. 8), Cambridge, U.K.

Swinhoe, R. (1870) Zoological notes on a journey from Canton to Pekin and Kalgan. *Proceedings Zoological Society, London*: 427–451.

Syroechkovski, E.E. (2002) Distribution and population estimates for swans in the Siberian Arctic in the 1990s. *In* E.C. Rees, S.L. Earnst & J. Coulson (eds), Proceedings of the Fourth International Swan Symposium, 2001. *Waterbirds*, 25 (Special Publication 1): 100–113.

Syroechkovski, E.E., Rogacheva, E.V., Savchenko, A.P., Sokolov, G.A., Baranov, A.A. & Yemelyanov, V.I. (2000) *Red Data Book of the Krasnoyarsk Territory. Rare and Endangered Species of Animals.* A.P. Savchenko (ed.). Institute of Physics, Russian Academy of Sciences, Krasnoyarsk. [In Russian.]

Syroechkovsky, E. B. & Krechmar, A.V. (1981) The main factors affecting the number of Snow Geese. *In* V. G. Krivosheev (ed.), *Ecology of the Mammals and Birds of Wrangel Island*, 3–36. Far East Branch of Russian Academy of Science, Vladivostock. [In Russian.]

Syroechkovsky, E.V., Litvin, K.E. & Ebbinge, B.S. (1991) Breeding success of geese and swans on Vaygach Island (USSR) during 1986–88; interplay of weather and arctic fox predation. *Ardea*, 79: 373–382.

Syroechkovsky, E.V., Litvin, K.E. & Gurtovaya, E.N. (2002) Nesting ecology of Bewick's Swans on Vaygach Island, Russia. *In* E.C. Rees, S.L. Earnst & J. Coulson (eds), Proceedings of the Fourth International Swan Symposium, 2001. *Waterbirds*, 25 (Special Publication 1): 221–226.

Ten Kate, C.G.B. (1930) Over de talrijkheid van *Cygnus bewickii bewickii* Yarr. In den winter langs de Oost-kust van de Zuiderzee. *Org. Club Ned. Vogelk.* II: 187–198.

Timmerman, A. (1977) De Kleine Zwaan. *Vogeljaar*, 25: 113–123.

Trinder, M., Rowcliffe, M., Pettifor, R., Rees, E., Griffin, L., Ogilvie, M. & Percival, S. (2005) *Status and population viability analyses of geese in Scotland.* Scottish Natural Heritage Commissioned Report No. 107 (ROAME No. F03AC302).

Tucker, G.M. & Heath, M.F. (eds.) (1994) *Birds in Europe: their conservation status.* BirdLife International (BirdLife Conservation Series No. 3), Cambridge.

Tunnicliffe, C.F. (1979) *A Sketchbook of Birds.* Victor Gollancz Ltd., London.

Tugarinov, A.Ya. & Buturlin, S.A. (1911) Materials on birds of the Yenissey Gubernia (Territory). *Transactions of the Krasnoyarsk Subsection of East Siberian Section of the Russian Geographic Society, Krasonyarsk.* Volume 1., Parts 2–4.

Ubels, R., Vulink, J.T. & van Eerden, M.R. (2000) Diet of vertebrate herbivores. *In* M.R. van Eerden (ed.), *Pechora Delta: Structure and dynamics of the Pechora Delta ecosystems (1995–1999)*, 223–241. RIZA report nr: 2000.037 and MD report nr.: MD GAE 2000.39. Institute for Inland Water Management and Waste Water Treatment/RIZA, Lelystad, the Netherlands.

Unselt, C., Mayr, C. & Bauer, H-G. (2000) Federal Republic of Germany. *In* M.F. Heath & M.I. Evans (eds), *Important Bird Areas in Europe: Priority sites for conservation. Volume 1: Northern Europe*, 263–340. BirdLife International (BirdLife Conservation Series No. 8), Cambridge, U.K.

Uspenski, S.M. & Kishchinski, A.A. (1972) Experience of aerial counts of breeding populations of waterfowl on tundra. In: *Transactions of the Central Laboratory of*

Game Management and Zapovedniks, 210–234. Game Biology Volume 1. "Lesnaya pormyshlennost" Publication, Moscow. [In Russian.]

van den Berg, A.B. & Bosman, C.A.W. (1999) *Rare birds of the Netherlands*. Avifauna van Nederland 1. GMB Uitgeverij/KNNV Uitgeverij, Haarlem/Utrecht, the Netherlands.

van Eerden, M. R. (ed.). (2000) *Pechora Delta: Structure and dynamics of the Pechora Delta ecosystems (1995–1999)*. RIZA report number 2000.037. RIZA Institute for Inland Water Management, Lelystad, The Netherlands.

van Eerden, M. & Lavrinenko, I.A. (2000) Ecological investigations using remote sensing techniques in the Pechora Delta, Northern Russia. *In* B. Ebbinge, Yu.L. Mazourov & P.S. Tomkovich (eds), *Heritage of the Russian Arctic: Research conservation and international co-operation*, 131–135. Ecopros, Moscow.

van Eerden, M. R., Beekman, J.H., Smit, M. & Oosterbeek, K. (1997) Patch use by Bewick's Swans Cygnus columbianus bewickii feeding on Sago Pondweed *Potamogeton pectinatus* in shallow lakes in the Netherlands: variation in exploitation threshold caused by social, environmental and time dependent factors. *In* M.R. van Eerden (ed.), *Patchwork: Patch use, habitat exploitation and carrying capacity for water birds in Dutch freshwater wetlands*, 187–214. Ph.D. thesis, Rijksuniversiteit Groningen and Rijkswaterstaat Directie IJsselmeergebied, Lelystad, the Netherlands.

van Roomen, M. & Madsen, J. (eds) (1991). Waterfowl and agriculture: review an future perspective of the crop damage conflict in Europe. IWRB Special Publication No. 21, International Waterfowl and Wetlands Research Bureau, Slimbridge.

van Roomen, M. van Winden, E., Koffijberg, K., Voslamber, B., Kleefstra, R., Ottens, G. & SOVON Ganzen- en zwanenwerkgroep (2002) *Watervogels in Nederland in 2000/2001*. SOVON-Monitoringrapport 2002/04, RIZA-rapport BM02.15, SOVON Vogelonderzoek Nederland, Beek-Ubbergen.

van Roomen, M. van Winden, E., Koffijberg, K., Kleefstra, R., Ottens, G., Voslamber, B. & SOVON Ganzen- en zwanenwerkgroep (2003) *Watervogels in Nederland in 2001/2002*. SOVON-Monitoringrapport 2004/01, RIZA-rapport BM04/01, SOVON Vogelonderzoek Nederland, Beek-Ubbergen.

van Roomen, M., van Winden E., Koffijberg K., Boele A., Hustings F., Kleefstra R., Schoppers J., van Turnhout C., SOVON Ganzen- en Zwanenwerkgroep & Soldaat, L. (2004) *Watervogels in Nederland in 2002/2003*. Sovon-Monitoringrapport 2004/02. RIZA-rapport BM04/09. SOVON, Beek-Ubbergen.

van der Vliet, R.E., van der Laan, J. & CDNA. (2002) Rare birds in the Netherlands in 2001. *Dutch Birding*, 23: 315–347.

van der Wal, R., Knust, P. & Drent, R. (1998) Interactions between hare and Brent Geese in a salt marsh system: evidence for food competition? *Oceologia* 117: 227–234.

van der Wal, R., van Wijnen H., van Wieren, S., Beucher, O. & Bos, D. (2000) On facilitation between herbivores: how Brent Geese profit from Brown Hares. *Ecology* 81: 969–980.

van Wijk, R.J. (1988) Ecological studies on *Potamogeton pectinatus L.* I. General characteristics, biomass and life cycles under field conditions. *Aquatic Botany*, 31: 211–258.

Vaurie, C. (1965) *The Birds of the Palearctic Fauna, non Passeriformes.* J.F. & G. Witherby, London.

Vinokurov, A. A. (1987) Bewick's Swans in the Taimyr. *In* E. V. Syroechkovski (ed.), *Ecology and Migration of Swans in the USSR*, 138–139. Nauka, Moscow. [In Russian.]

WALLRAFF, H.G. (1972) An approach toward an analysis of the pattern recognition involved in the stellar orientation of birds. *In* S.R. Galler, K. Scmidt-Koenig, G.J. Jacobs & R. E. Belleville (eds), *Animal Orientation and Navigation*, 211–222. NASA, Washington D.C.

WALTER, P.J., BACON, P.J. & SEARS, J. (1991) An analysis of Mute Swan, *Cygnus olor*, breeding data. *In* J. Sears & P.J. Bacon (eds), Proceedings of the Third International Swan Symposium, Oxford 1989. *Wildfowl Special Supplement No. 1*: 151–156.

WATERS, R.J., CRANSWICK, P.A., MUSGROVE, A.J. & POLLITT, M.S. (1998) *The Wetland Bird Survey 1996–97 Wildfowl and Wader Counts.* BTO/WWT/RSPB/JNCC, Slimbridge.

WETMORE, A. (1951) Observations on the genera of the waterfowl. *Journal of the Washington Academy of Sciences* 41: 338–340.

WIELOCH, M. (1991) Population trends of the Mute Swan *Cygnus olor* in the Palearctic. *In* J. Sears & P.J. Bacon (eds), Proceedings of the Third International Swan Symposium, Oxford 1989. *Wildfowl Special Supplement No. 1*: 22–32.

WINGATE, R.R. (1831) Notice of a new species of swan. *Transactions of the Natural History Society of Northumberland, Durham and Newcastle upon Tyne.* Volume I, pp 1–2. [Issue with this paper produced 1830; whole volume dated 1831.]

WINTER, J. & MANOLIS, T. (1978) Middle Pacific Coast region. *American Birds*, 32: 394–397.

WITHERBY, H.F. (1939) The influx of Bewick's and Whooper Swans, winter 1938–39. *British Birds*, 32: 378–381.

WON, P-O. (1981) Present status of the swans wintering in Korea. *In* G.V.T. Matthews & M. Smart (eds), *Proceedings of the Second International Swan Symposium, Sapporo 1980*, 15–18. International Waterfowl Research Bureau, Slimbridge.

YARRELL, W. (1830) On a new species of wild swan taken in England (*C. bewickii*), and hitherto confounded with the Hooper. *Transactions of the Linnean Society*, 16: 445–453.

YU, K. (1991) Wintering waterfowl at eastern Chongming Island and their protection. *Chinese Wildlife*, 2: 15–18.

YU, K. (1992) The Bewick's Swans in Chongming Island, Shanghai. *China Nature*, 1: 24–25

YU, X. & LIU, Y. (1987) Sentence of Wu Yuying *et al.* for their unlawful hunting of swans and Gray Cranes. *Chinese Wildlife* 1: 37.

ZHENG, G. & WANG, Q. (comp.) (1998) *China Red Data Book of Endangered Animals: Aves.* National Environmental Protection Agency, Endangered Species Scientific Commission, P.R.C. Science Press, Beijing. [In Chinese and in English.]

ZALOGIN, B.S. & RODIONOV, N.A. (1969) *Estuaries of the USSR Rivers.* Mysl, Moscow. [In Russian.]

Index

Page references in *italics* indicate figures.